Ernst Schering Research Foundation Workshop 55
Chronic Viral and Inflammatory Cardiomyopathy

Ernst Schering Research Foundation
Workshop 55

Chronic Viral and Inflammatory Cardiomyopathy

H.-P. Schultheiss, J.-F. Kapp, G. Grötzbach
Editors

With 68 Figures

Series Editors: G. Stock and M. Lessl

ISSN 0947-6075

ISBN-10 3-642-42143-1 Springer Berlin Heidelberg New York
ISBN-13 978-3-642-42143-3 Springer Berlin Heidelberg New York

Springer is a part of Springer Science+Business Media
springeronline.com

Softcover re-print of the Hardcover 1st edition 2006

Editor: Dr. Ute Heilmann, Heidelberg
Desk Editor: Wilma McHugh, Heidelberg
Production Editor: Michael Hübert, Leipzig
Cover design: design & production, Heidelberg
Typesetting and production: LE-TEX Jelonek, Schmidt & Vöckler GbR, Leipzig
21/3152/YL – 5 4 3 2 1 0 Printed on acid-free paper

Preface

Following its tradition of promoting novel areas of scientific discourse, the Ernst Schering Research Foundation (ESRF) hosted this workshop on chronic viral and inflammatory cardiomyopathy. In late October 2004, scientists from Canada, Germany, the Georgian Republic, Great Britain, Italy, Japan, the Netherlands, Israel, Sweden, and the United States gathered in Berlin to discuss their concepts, hypotheses, and latest findings on myocarditis and cardiomyopathy.

This expert meeting was held in cooperation with the German Research Foundation, which in the same year had supported transregional collaborative research activities entitled "Inflammatory Cardiomyopathy – Molecular Pathogenesis and Therapy."

Organizing the workshop, our efforts strove to render tighter the network between the distinct disciplines involved in cardiomyopathy research, building bridges between its molecular, pathogenetic, diagnostic, and therapeutic determinants.

It all began as a story of a neighborhood, and we would like to express our hope that this long-term project, which will require much and intensive cooperative work among the participants, will evolve to become a story of a good neighborhood.

Schering and the Charité, with their four campuses – the Mitte, Benjamin Franklin, Berlin-Buch, and Virchow clinics – are good neighbors and cooperate in many research and development tasks. Equally, on a personal level, two of us (H.P. Schultheiss and J-F Kapp) are neigh-

bors ourselves, and the proximity of our houses makes it possible that we sometimes meet when we are out for a walk. On one of these occasions, Schering's expert meeting on interferon-beta in multiple sclerosis

1. Karin Klingel, Tübingen, Germany
2. Kirk U. Knowlton, San Diego, USA
3. Jeffrey Towbin, Houston, USA
4. Akira Matsumori, Kyoto, Japan
5. Thomas-C. Bock, Tübingen, Germany
6. Stephane Heymans, Maastricht, Netherlands
7. Hugo A. Katus, Heidelberg, Germany
8. Finn Waagstein, Gothenburg, Sweden
9. Mathias Pauschinger, Berlin, Germany
10. Vincenz Hombach, Ulm, Germany
11. Nodar N. Kipshidze, Tbilissi, Georgia
12. Ruprecht Zierz, Schering AG, Berlin, Germany
13. Susanne Modrow, Regensburg, Germany
14. Udo Sechtem, Stuttgart, Germany
15. Andre Keren, Jerusalem, Israel
16. Wolfgang Poller, Berlin, Germany
17. Uwe Kühl, Berlin, Germany
18. Peter Liu, Toronto, Canada
19. Bernhard Maisch, Marburg, Germany
20. Andrea Frustaci, Milan, Italy
21. William McKenna, London, UK
22. Noel Rose, Baltimore, USA
23. Heinz-Peter Schultheiss, Berlin, Germany
24. Stefan Felix, Greifswald, Germany
25. Hans-Dieter Volk, Berlin, Germany
26. Claus-Steffen Stürzebecher, Schering AG, Berlin, Germany
27. Georg Grötzbach, Schering AG, Berlin, Germany
28. Christian Kraus, Schering AG, Berlin, Germany

came up, and on the next walk, we pondered the concept of treating virus-induced cardiomyopathy. Becoming quite enthusiastic at the end of the walk, the budget for a pilot study using interferon-beta to verify this concept had been set. The outcome of the pilot study encouraged us to proceed to the currently ongoing formal phase II protocol.

Cardiac inflammation is difficult to diagnose, and even if diagnosed, can we treat it effectively? This question was raised in 1772 by Jean-Baptiste de Sénac, physician to Louis XV. At that time, he could have chosen almost any disease and proper diagnosis would not have made any difference as far as effective treatment was concerned – independent of whether or not the diagnosis itself was rather straightforward or difficult.

Now, 230 years after de Sénac's doubts, there is some hope – maybe even some confidence – that we might have indeed finally identified a potential treatment for cardiac inflammation, and we have also managed to make its diagnosis an almost routine procedure.

When talking about dilated cardiomyopathy, precise diagnosis is important, because differentiation between autoimmune inflammatory cardiomyopathy on the one hand, and cardiomyopathy with viral persistence on the other, allows therapeutic benefit from immunosuppressive treatment for one disease and from antiviral interferon-beta treatment for the other.

Constituting an invasive tool, endomyocardial biopsy provides valuable diagnostic information including virus identification, and morphological and immunohistochemical patterns. For the long-term, it would be desirable to develop serological markers for gaining prognostic and etiologic data. Experience from predicting transplant rejection was presented, hoping to be adopted for the identification of biomarkers. In view of the limitations of the current classification, a new histomorphological staging of myocarditis and inflammatory cardiomyopathy was proposed. As a new non-invasive approach, magnetic resonance imaging may gain importance in the diagnosis of cardiomyopathy, yet for the time being, this technique is still in too early a stage.

The interconnectedness of genetics and susceptibility to disease, of viral and non-viral inflammation, and of the role of immunity and the development of autoimmunity is a fascinating and much-discussed labyrinth. Cardiotropism needs to be verified for parvovirus B19, HHV6, and HVC.

Pilot experience in treating virus-positive patients with interferon-beta was presented. The audience appreciated these results hinting at a first causal treatment of viral cardiomyopathy. Other approaches dealt with receptor binding and protease inhibitors. Immunosuppression was found to be beneficial to virus-negative myocarditis patients.

In a special session it was controversially and spiritedly discussed whether or not (1) virus persistence was a determinant for progression, (2) autoimmunity was of significance, (3) parvovirus was relevant, and (4) there was a specific matrix destruction in the course of cardiomyopathy. The pertaining statements have also been included in this volume.

We are convinced that the unique composition of and the broad spectrum covered by this two-day workshop will be of exceptional value to the readers of volume 55.

Heinz-Peter Schultheiss
Joachim-Friedrich Kapp
Georg Grötzbach

Contents

V Diagnosis and Treatment

List of Editors and Contributors

Editors

Schultheiss, H.-P.
Department of Cardiology and Pneumology, Campus Benjamin Franklin,
Charité University Medicine Berlin,
Hindenburgdamm 30, Germany
(e-mail: heinz-peter.Schultheiss@charite.de)

Kapp, J.-F.
Berlex Inc. Specialized Therapeutics,
M1/3-9, Montville NJ 07045, USA
(e-mail: joachim-friedrich.kapp@schering.de)

Grötzbach, G.
Schering AG, Medical Development, Specialized Therapeutics,
Cardiovascular Europe,
13342 Berlin, Germany
(e-mail: georg.groetzbach@schering.de)

Contributors

Angelini, A.
University of Padua Medical School,
Via A. Gabelli, 61, 35121 Padova, Italy

Anker, S.D.
Department of Clinical Cardiology, Imperial College School of Medicine,
National Heart & Lung Institute,
Dovehouse Street, London, SW3 6LY, UK

and
Division of Applied Cachexia Research, Department of Cardiology,
Charité Medical School, Campus Virchow-Klinikum, Berlin, Germany

Bock, C.-T.
Department of Molecular pathology, Institute of Pathology,
University Hospital of Tübingen,
Liebermeisterstr. 8, 72076 Tübingen, Germany
(e-mail: thomas.bock@med.uni-tuebingen.de)

Caforio, A.L.P.
Division of Cardiology, Department of Cardiological,
Thoracic and Vascular Sciences
Centro "V. Gallucci", Policlinico University of Padua,
Via Giustiniani, 2, 35128 Padova, Italy
(e-mail: alida.caforio@unipd.it)

Calabrese, F.
University of Padua Medical School,
Via A. Gabelli, 61, 35121 Padova, Italy

Carturan, E.
University of Padua Medical School,
Via A. Gabelli, 61, 35121 Padova, Italy

Chimenti, C.
Cardiology Department, Catholic University Largo A. Gemelli 8,
00168 Rome, Italy

Chu, G.
Heart and Stroke, Richard Lewar Centre of Excellence,
NCSB 11-1266, Toronto General Hospital,
200 Elizabeth Street, Toronto, Ontario, M5G 2C4, Canada

Doehner, W.
Division of Applied Cachexia Research, Department of Cardiology,
Charité Medical School, Campus Virchow-Klinikum, Berlin, Germany

Fechner, H.
Department of Cardiology and Pneumology, Campus Benjamin Franklin,
Charité University Medicine Berlin,
Hindenburgdamm 30, Germany

Felix, S.B.
Department of Internal Medicine B,
Ernst-Moritz-Arndt University Greifswald,
Friedrich-Loeffler Str. 23a, 17487 Greifswald, Germany
(e-mail: felix@uni-greifswald.de)

Frustaci, A.
Cardiology Department, Catholic University Largo A. Gemelli 8,
00168 Rome, Italy
(e-mail: biocard@rm.unicatt.it)

Fuse, K.
Heart and Stroke, Richard Lewar Centre of Excellence,
NCSB 11-1266, Toronto General Hospital,
200 Elizabeth Street, Toronto, Ontario, M5G 2C4, Canada

von Haehling, S.
Department of Clinical Cardiology, Imperial College School of Medicine,
National Heart & Lung Institute,
Dovehouse Street, London, SW3 6LY, UK
(e-mail: stephan.von.haehling@web.de)

Hager, S.
Division of Cardiology
Robert-Bosch-Krankenhaus
Auerbachstraße 110, 70876 Stuttgart, Germany
(e-mail: stefan.hager@rbk.de)

Heymans, S.
Experimental and Molecular Cardiology Laboratory/CARIM,
Department of Cardiology, University Hospital Maastricht,
PO Box 5800, 6202 AZ Maastricht, The Netherlands
(e-mail: s.heymans@cardio.unimaas.nl)

Kallwellis-Opara, A.
Department of Cardiology and Pneumology, Campus-Benjamin Franklin,
Charité University Medicine Berlin,
Hindenburgdamm 30, 12200 Berlin, Germany
(e-mail: angela.kallwellis-opara@charite.de)

Keren, A.
Heiden Department of Cardiology, Bikur Cholim Hospital,
Jerusalem, Israel
(e-mail: andrek@cc.huji.ac.il)

Knowlton, K.U.
Department of Medicine, University of California,
San Diego School of Medicine,
9500 Gilman Drive, 0613K, La Jolla, CA 92093, USA
(e-mail: kknowlton@ucsd.edu)

Kühl, U.
Department of Cardiology and Pneumology, Campus Benjamin Franklin,
Charité University Medicine Berlin,
Hindenburgdamm 30, Germany
(e-mail: uwe.kuehl@charite.de)

Liu, P.
Heart and Stroke, Richard Lewar Centre of Excellence,
NCSB 11-1266, Toronto General Hospital,
200 Elizabeth Street, Toronto, Ontario, M5G 2C4, Canada
(e-mail: peter.liu@utoronto.ca)

Liu, Y.
Heart and Stroke, Richard Lewar Centre of Excellence,
NCSB 11-1266, Toronto General Hospital,
200 Elizabeth Street, Toronto, Ontario, M5G 2C4, Canada

Mahon, N.G.
Cardiovascular Department, Mater Misericordiae Hospital,
Dublin, Ireland

Mahrholdt, H.
Division of Cardiology
Robert-Bosch-Krankenhaus
Auerbachstraße 110, 70876 Stuttgart, Germany
(e-mail: heike.mahrholdt@rbk.de)

Matsumori, A.
Department of Cardiovascular Medicine,
Kyoto University Graduate School of Medicine,
54 Kawahara-cho Shogoing, Sakyo-ku, Kyoto 606-8507, Japan
(e-mail: amat@kuhp.kyoto-u.ac.jp)

McKenna, W.J.
The Heart Hospital,
University College London Hospitals NHS Foundation Trust,
London, United Kingdom

Modrow, S.
Institut für Medizinische Mikrobiologie und Hygiene,
Universität Regensburg,
Franz-Josef-Strauß Allee 11, 93053 Regensburg, Germany
(e-mail: susanne.modrow@klinik.uni-regensburg.de)

Opavsky, A.
Heart and Stroke, Richard Lewar Centre of Excellence,
NCSB 11-1266, Toronto General Hospital,
200 Elizabeth Street, Toronto, Ontario, M5G 2C4, Canada

Pauschinger, M.
Department of Cardiology and Pneumology, Campus-Benjamin Franklin,
Charité University Medicine Berlin,
Hindenburgdamm 30, 12200 Berlin, Germany
(e-mail: matthias.pauschinger@charite.de)

Pieroni, M.
Cardiology Department, Catholic University Largo A. Gemelli 8,
00168 Rome, Italy

Poller, W.
Department of Cardiology and Pneumology, Campus-Benjamin Franklin,
Charité University Medicine Berlin,
Hindenburgdamm 30, 12200 Berlin, Germany
(e-mail: wolfgang.poller@charite.de)

Rose, N.R.
Departments of Pathology and of Molecular Microbiology and Immunology,
Bloomberg School of Public Health – E5014,
615 North Wolfe Street, Baltimore, MD 21205, USA
(e-mail: nrrose@jhsph.edu)

Sechtem, U.
Cardiology and Pulmology, Robert-Bosch-Krankenhaus,
Auerbachstrasse 110, 70376 Stuttgart, Germany,
(e-mail: udo.sechtem@rbk.de)

Thiene, G.
University of Padua Medical School,
Via A. Gabelli, 61, 35121 Padova, Italy
(e-mail: gaetano.thiene@unipd.it)

Towbin, J.A.
Baylor College of Medicine, Texas Children's Hospital,
6621 Fannin Street, MC 19345-C, Houston, TX 77030, USA
(e-mail: jtowbin@bcm.tmc.edu)

Vogelsberg, H.
Division of Cardiology
Robert-Bosch-Krankenhaus
Auerbachstraße 110, 70876 Stuttgart, Germany
(e-mail: holger.vogelsberg@rbk.de)

Waagstein, F.
Wallenberg Laboratory, Sahlgrenska Academy,
Göteborg University,
Göteborg, Sweden
(e-mail: finn.waagstein@wlab.gu.se)

I Chronic Viral and Inflammatory Cardiomyopathy

Overview and Outlook

1 Overview on Chronic Viral Cardiomyopathy/Chronic Myocarditis

H.-P. Schultheiss, U. Kühl

Abstract. Myocarditis is most often induced by cardiotropic viruses and often resolves with minimal cardiac remodelling and without discernable prognostic impact. Acute myocarditis has a highly diverse clinical presentation (asymptomatic, infarct-like presentation, atrioventricular (AV)-block, atrial fibrillation, sudden death due to ventricular tachycardia, fulminant myocarditis with severely depressed contractility). Progression of myocarditis to its sequela, dilated cardiomyopathy (DCM), has been documented in 20% of cases and is pathogenically linked to chronic inflammation and viral persistence. Persistence of cardiotropic viruses (enterovirus, adenovirus) constitutes one of the predominant aetiological factors in DCM. Additionally, circulating autoantibodies to distinct cardiac autoantigens have been described in patients with DCM, providing evidence for autoimmune involvement. Since clinical complaints of myocarditis and DCM are unspecific, a positive effect of any specific therapy depends on an accurate biopsy-based diagnosis and characterization of the patients with histological, immunohistological and molecular biological methods (PCR), which have developed into sensitive tools for the detection of different viruses, active viral replication, and myocardial inflammation. The immunohistochemical characterization of infiltrates has supported a new era in the diagnosis of myocardial inflammation compared with the Dallas criteria, which has led to a new entity of secondary cardiomyopathies acknowledged by the WHO, the inflammatory cardiomyopathies (DCMi). Immunohistochemically quantified lymphocytes significantly better reflect troponin levels and correlate with findings by anti-myosin scintigraphy compared with the histological analysis. Furthermore, the orchestrated induction of endothelial cell adhesion molecules (CAMs) in 65% of DCM patients has confirmed that CAM induction is a prerequisite for lymphocytic infiltration in DCMi. The combination of these immunohistological with molecular biological diagnostic techniques of virus analysis allows a further classification of dilated cardiomyopathy by differentiating the disease entity in subgroups of virus-positive and virus-negative patients with or without cardiac inflammation. Further analysis of the predominant Th1-/Th2-immune response may provide additional prognostic information on the natural course of the disease. This differential analysis improves the clinical management of patients and is an indispensable prerequisite for the development of specific antiviral or immunomodulatory treatment strategies.

1.1 Introduction

Cardiomyopathies are diseases of the heart characterized by ventricular dysfunction that is not caused by primary heart diseases, e.g. hypertension or congenital, valvular, coronary, arterial or pericardial abnormalities. They are classified as primary cardiomyopathies if the origin of contractile dysfunction is unknown (dilated cardiomyopathy, hypertrophic cardiomyopathy) and as secondary or specific cardiomyopathies if the heart is affected in association with specific infectious, immunological, metabolic, neuromuscular or toxic diseases.

Dilated cardiomyopathy (DCM) is one of the most common cardiomyopathy entities. Its yearly incidence is 5–8 cases per 100,000, the age-corrected prevalence is 36 cases with 17 hospitalizations, and 3.8 deaths per 100,000 per year are due to DCM. DCM affects males most frequently, with a sex ratio of 3:1, manifesting predominantly between the third and fifth decades (Olbrich 2001). In a homogeneous population of young military servicemen, the incidence of myocarditis was reported 0.17 per 1,000 man-years (Karjalainen and Heikkila 1999), but the real numbers are expected to be substantially higher due to the often subclinical presentation of acute myocarditis and misinterpretation of unspecific symptoms (Kühl et al. 1997a).

During the past decade, the basis of left ventricle (LV) dysfunction has begun to unravel. In approximately 30%–40% of cases, the disorder is inherited; autosomal- dominant inheritance is most common (although X-linked, autosomal recessive and mitochondrial inheritance occurs) (Towbin and Bowles 2000). In the remaining patients, the disorder is presumed to be acquired, with inflammatory heart disease playing an important role (Towbin and Bowles 2002; Towbin and Bowles 2001). Evolution of acute myocarditis to dilated cardiomyopathy (DCM), which occurs in 21% of the patients within a mean follow-up of 33 months (D'Ambrosio et al. 2001), gave rise to the hypothesis that certain DCM cases might be due to sequela of a "chronic myocarditis" (Fig. 1). Acknowledging the unequivocal evidence on the chronic inflammatory process involved in DCM, the 1995 report of the WHO/ISFC Task Force

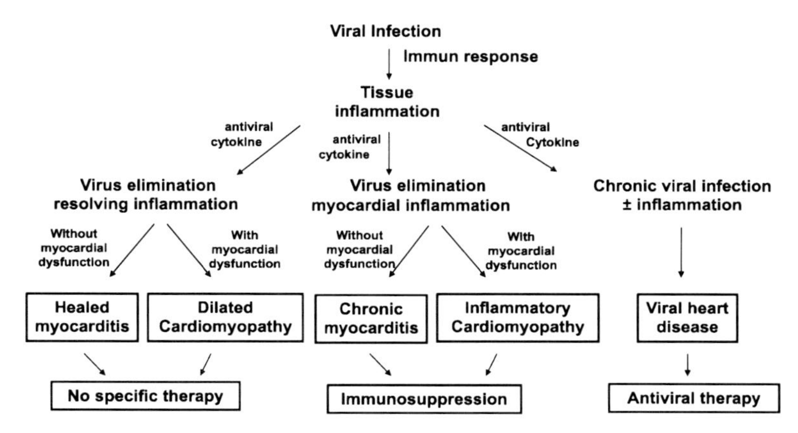

Fig. 1. Natural course of virus induced myocarditis developing DCM/ inflammatory cardiomyopathy

on the Definition and Classification of Cardiomyopathies introduced the new entity "inflammatory cardiomyopathy" among other specific cardiomyopathies (DCMi) (Richardson et al. 1996), which is characterized by idiopathic heart failure with evidence for intramyocardial inflammation.

Myocarditis and DCMi in the Western world are most often caused by cardiotropic viruses (Mason 2002). A number of similar viruses have been identified in biopsies of patients with myocarditis and DCM (Kühl et al. 2003a, 2005). Despite the increasing role of the systemic inflammatory response involved secondarily in the progression of congestive heart failure irrespective of the underlying pathogenesis (Anker and von Haehling 2004) DCMi refers exclusively to DCM with primary anti-cardiac inflammatory and/or viral aetiology. This differentiation between systemic and organ-specific inflammation is pivotal for the consideration of diagnostic procedures on endomyocardial biopsies (EMBs) and treatment modalities targeting intramyocardial inflammation and/or cardiotropic viral infection. The use of EMBs is indispensable for the diagnosis of DCMi, which mainly comprises the detection and characterization of intramyocardial inflammation by immunohistological evaluation, and the proof of viral infection by molecular biological techniques (Table 1). This interdisciplinary approach has expanded our understanding of the pathogenesis of DCMi, and these insights were ultimately decisive for the design of immunomodulatory trials with beneficial outcomes in DCMi patients (Noutsias et al. 2004; Mason et al. 1995; Frustaci et al. 2003; Dec and Fuster 1994a; Kühl et al. 2003b; Wojnicz et al. 2001).

1.2 Diagnosis of Myocarditis and DCMi

The prognosis, as well as the initial clinical presentation of patients with myocarditis and DCM, is highly variable and ranges from spontaneous improvement to rapidly progressive heart failure and sudden cardiac death (D'Ambrosio et al. 2001). To date, no clinical parameter has proved to be a relevant discriminator of prognosis or pathogenesis. In contrast, both of the major pathogenic hallmarks of DCMi, namely viral persistence and intramyocardial inflammation diagnosed in EMBs, have

Table 1. Analysis of endomyocardial biopsies

	Histology	Immunohistology	PCR
Morphological evaluation of myocardial tissue	+	–	–
Acute myocarditis/necrosis	(+)*	–	–
Detection and quantification of inflammation	–	+	–
Detection of adhesion molecules (ICAM, VCAM)	–	+	–
Subtyping of infiltrating cells (cytotoxic cells)	–	+	–
Differentiation of immune response (Th1/2)	–	+	–
Detection of different viruses (subtypes)	–	–	+
Detection of viral replication/load	–	–	+

*Low sensitivity, high sampling error

been associated with adverse prognosis in DCM (Why et al. 1994a; Fujioka et al. 2000; 21 Angelini et al. 2002; Kanzaki et al. 2001).

- The clinical symptoms of myocarditis and DCMi are unspecific (Table 2).
- A diagnosis of inflammation based only on clinical history/presentation and non-invasive examinations is not possible.
- The unequivocal diagnosis of chronic myocarditis/inflammatory cardiomyopathy can be only achieved by analysis of endomyocardial biopsies including :
  1. Histology (in the acute phase)
  2. Immunohistology (in the chronic phase)
  3. Molecular biology/virology

Table 2. Clinical symptoms of virus-negative and virus-positive patients

	Negative	EV	PVB19	HHV6	PVB/HHV
Total (n)	260	54	210	89	56
Infection	48%	45%	49%	35%	42%
Tiredness	69%	82%	72%	59%	82%
Angina pectoris	40%	10%	47%	36%	40%
Dyspnea on exertion	55%	50%	60%	64%	77%
Pericardial effusion	6%	2%	7%	5%	13%
Impaired contractility (global)	68%	71%	68%	69%	69%
Impaired contractility (region.)	43%	24%	31%	44%	25%
Rhythm disturbances	46%	50%	51%	38%	62%
SVES	7%	9%	11%	4%	6%
VES	23%	9%	24%	19%	18%
Atrial fibrillation	19%	25%	16%	18%	23%
Ventricular tachycardia	9%	0%	13%	4%	7%

SVES, supraventricular extra beats; VES, ventricular extra beats

1.3 Histology

Histological analysis has been considered the "gold standard" for diagnosing cardiomyopathies. This is true for acute myocarditis; but the histological diagnosis of chronic myocarditis and DCMi causes difficulties and the results are contradictory. According to the Dallas criteria, myocarditis in its acute stage is histologically defined by lymphocytic infiltrates in association with myocyte necrosis. The histological definition of chronic myocarditis according to the Dallas classification (histologically "borderline and ongoing myocarditis") demands the presence of infiltrating lymphocytes without further histomorphological signs of myocyte injury or immunohistological features of a persisting, activated inflammatory process in the first and the control biopsy. These cellular infiltrates in chronic heart failure, however, are often sparse or focal and therefore might be missed by sampling error (Shanes et al. 1987; Hauck et al. 1989). Moreover, it is difficult to differentiate non-inflammatory interstitial cells from infiltrating lymphocytes by light microscopy. This leads to a misinterpretation of interstitial cells as inflammatory lymphocytes and thus to an over- or underestimation of the degree of inflammation. Ultimately, one has to keep in mind that infiltrating lymphocytes, especially if not activated, are not necessarily representative for an ongoing immune process affecting the entire myocardium. Those two mechanisms (misinterpretation of interstitial cells and sampling error) are mainly responsible for the low diagnostic yield in the histological analysis and explain the often-reported high "interobserver variability".

1.4 Immunohistology

Meanwhile, immunohistological methods have been successfully introduced into the diagnosis of an inflammatory myocardial process. In contrast to routine histology with the above-explained difficulties in detecting lymphocytes, cardiac inflammation is, in addition to the quantification of lymphocytes, immunohistologically characterized by different markers of cell activation and the enhanced expression of histocompatibility antigens and adhesion molecules (Noutsias et al. 2002a,b). The sensitivity and specificity of monoclonal antibodies, directed against specific epitopes of immunocompetent cells, allow the unambiguous identification, characterization and quantification of inflammatory cells infiltrating myocardial tissues (Noutsias et al. 2003a). If the number of immunoreactive T lymphocytes is determined by the use of anti-CD2, -CD3, -CD4 or -CD8-antibodies, tissues with a mean number of lymphocytes exceeding 2.0 per high-power field (7.0 cells/mm^2) can be considered to have pathologically increased lymphocytic infiltrations (Fig. 2a, b) because additional markers of immune activation such as an expression of adhesion molecules are found to be enhanced in more than 90% of these cardiac tissues (Fig. 2d). Control tissues with mean lymphocyte counts of 0.7 ± 0.4 cells per high-power field (0.0–5.0/mm^2) express these markers in less than 30% of the tissues (Fig. 2c; see comments of Kühl et al. 1997b; also Kühl et al. 1997a; Noutsias et al. 1999; Bowles et al. 1986).

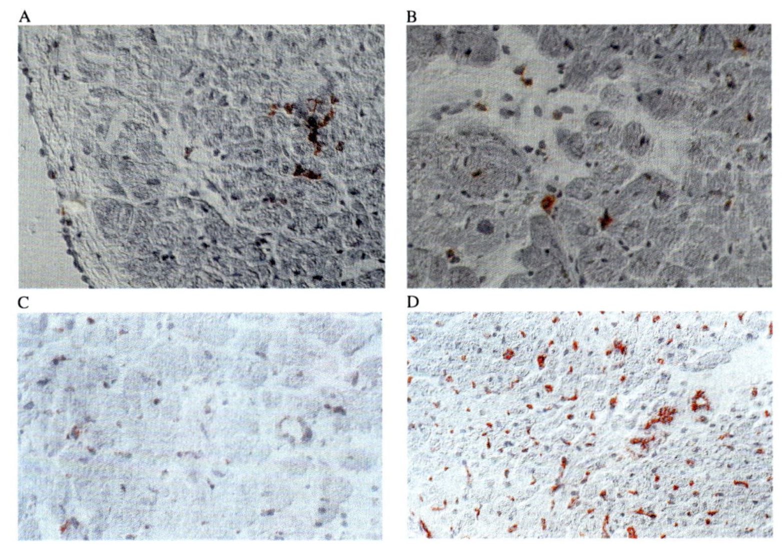

Fig. 2a–d. Immunohistological analysis of lymphocytes and adhesion molecules (ICAM-1) in endomyocardial biopsies. **a** Focal $CD3^+$-lymphocytic infiltration. **b** Diffuse $CD3^+$-lymphocytic infiltration. **c** Normal expression of ICAM-1 in a CD3-negative tissue. **d** Enhanced expression of ICAM-1 in a CD3-positive tissue

1.5 Molecular Biology/Virology

Positive serological tests cannot prove virus infection of the myocardium. The proof of a viral affliction of the myocardium warrants analysis of endomyocardial biopsies with more sensitive molecular biological methods, such as in situ hybridization or polymerase chain reaction (PCR) (Bowles et al. 1986; Why et al. 1994b). These methods enable the detection of viral RNA and DNA in tissues with even low numbers of viral copies (Fig. 3). Enteroviral genome was confirmed by PCR amplification in ca. 15% of EMBs from myocarditis and DCM patients (Baboonian and Treasure 1997). The adverse prognostic impact of enteroviral persistence in DCM was identified early (Dec and Fuster 1994a), and recent results indicate a special importance of the replicative infection

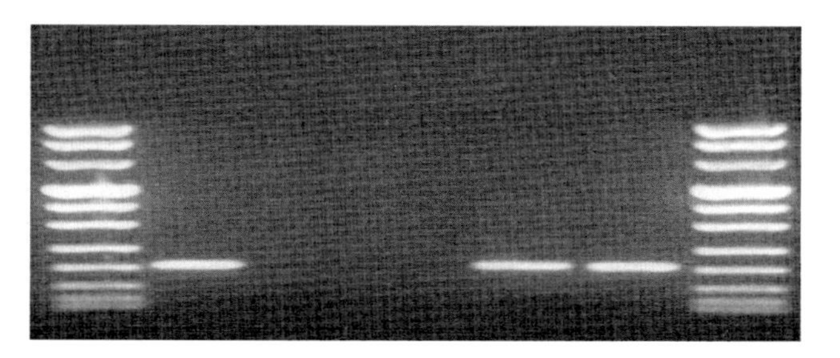

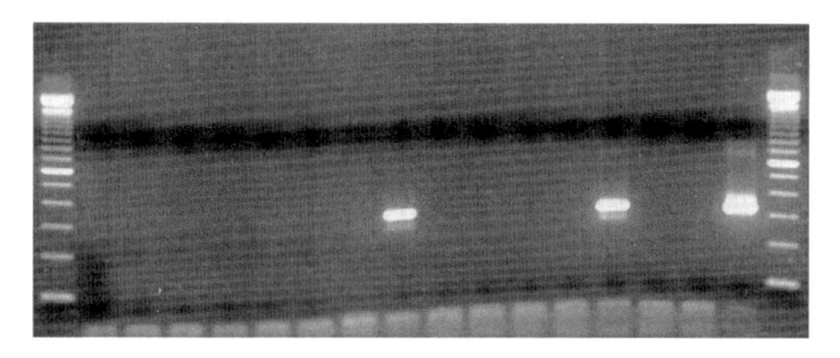

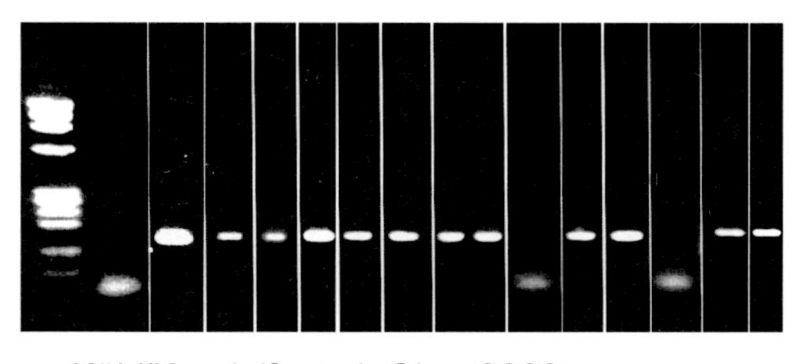

Fig. 3. Detection of different viral genomes by nested polymerase chain reaction (nPCR). *EV*, enterovirus (from Pauschinger et al. 1999a); *ADV*, adenovirus (from Pauschinger et al. 1999b); *PVB19*, Parvivirus B19 (from Kühl et al. 2003b)

mode in a significant proportion of DCM patients (Fujioka et al. 2000; Pauschinger et al. 1999a). Adenovirus has been elucidated as a further cardiotropic virus in DCM (Pauschinger et al. 1999b). There is an evolving relevance of further cardiotropic viruses, especially parvovirus B19, human herpes-virus 6, hepatitis C and Epstein-Barr virus (Kühl et al. 2003a; Frustaci et al. 2003; Matsumori 2001).

Of note, investigations on the tissue distribution of these viruses revealed profound differences: Whereas the enteroviral genomes have been detected within cardiomyocytes by in situ hybridization, endothelial cells are the primary targets of parvovirus B19 infection (Kandolf et al. 1987; Li et al. 2000; Bultmann et al. 2003). In this context, the pattern of viral receptors might be a decisively important issue (Noutsias et al. 2001; Poller et al. 2002). At present, we are lacking detailed data on the prognostic impact, pathways of viral entry and persistence of these recently identified cardiotropic viruses. However, the retrospective analysis by Frustaci et al. confirmed that patients with persistence of these viruses (except for hepatitis C) do not improve, or even deteriorate under immunosuppression, which infers that myocardial persistence of these viruses may have a similar pathogenic and prognostic relevance as coxsackievirus (Frustaci et al. 2003).

Employing group-specific primers, homology screening for various enteroviral and adenoviral strains can be conducted in addition to the differentiation of actively replicating virus from non- replicating, resting viral genomes (Pauschinger et al. 1999a; Calabrese and Thiene 2003). Quantification of virus loads by real-time PCR and sequence analysis of possibly pathogenic virus subtypes complete both the diagnostic accuracy and pathophysiological understanding of cardiotropic viral infections.

1.6 Biopsy-Based Classification of DCMi

The combined use of quantitative and qualitative histological and immunohistological analysis of myocardial inflammation with quantitative and qualitative molecular biological analytical methods allows a more detailed differential diagnosis in patients with clinically suspected myocarditis or DCMi (Noutsias et al. 2004; Pauschinger et al. 2004). As far as acute viral myocarditis is concerned, the Dallas classification has enabled a more standardized assessment of the histological changes in the myocardium. The clinical benefit, however, is limited by the fact that histological recording of myocytolysis is only possible for a short period of time during the acute disease (10–14 days) and usually no longer present at time points when later biopsies are taken in patients with chronic DCMi, which is considered a late sequela of virus-induced myocarditis. In chronic disease, this use of a combined set of analytical methods allows the identification of viral persistence and chronic-immunological processes as a possible cause for heart muscle injury in a considerable number of these patients (Kühl et al. 1996; Noutsias et al. 2003b). Three different entities of viral-induced dilated cardiomyopathy have been recognized (Fig. 1; Kühl et al. 1997b).

1. Postmyocarditic viral heart disease. These patients clinically present as dilated cardiomyopathy. Immunohistologically, they are characterized by an inconspicuous endomyocardial biopsy without chronic inflammation. Molecular biological methods do not indicate viral persistence.
2. Chronic (auto)immune-mediated myocarditis/inflammatory cardiomyopathy. Immunohistochemical analysis of endomyocardial biopsy reveals an active inflammatory process within the myocardium in the absence of viral persistence.
3. Chronic viral heart disease. Histologically, these patients cannot be differentiated from postmyocarditic viral heart disease (healed myocarditis/DCM). Viral persistence can be detected by in situ hybridization or PCR. Virus-induced myocardial inflammation may be present.

The rationale for this aetiological differentiation of "dilated cardiomyopathy" into different phases of the disease is to better understand the pathomechanism that might give rise to a more specific treatment strategy. If the developmental process of postmyocarditic heart failure is accelerated by chronic inflammation and/or viral persistence, prognosis of patients might be dependent on whether virus and inflammation persist in the myocardium (Fig. 2).

Recent data obtained by this combined mode of diagnostic procedures indicate a high percentage of single and multiple viral infections of the myocardium in patients with clinically suspected myocarditis of recent onset, myocarditis in the past and DCMi. Follow- up analysis of the patients reveals that spontaneous virus clearance is often associated with a spontaneous recovery of myocardial function. On the other hand, progression of myocardial dysfunction occurs in those patients who develop virus persistence, independent of the virus subtype involved. The patients with chronic virus persistence, however, often get benefit from a specific anti-viral treatment if this has been initiated early, before an irreversible virus-induced myocardial dysfunction has developed (Kühl et al. 2003b; Deonarain et al. 2004).

1.7 Unresolved Issues

With respect to the prevalence of virus-induced myocarditis/cardiomyopathies, the natural course of DCMi and their treatment responses, a number of open questions still exist.

Unknown are:

- The prevalence and clinical significance of various cardiotropic viruses
- The significance of endogenous and exogenous factors for viral infection, virus persistence or virus clearance
- The relevance of multiple viral infections, human pathogenic virus subtypes, genomic mutations predisposing for the disease, virus replication rate, or virus load, for the clinical spontaneous course of the disease
- The cause of highly variable virus receptor expression and its significance for the primary virus infection and the further course of the disease and
- The influence of the local and systemic antiviral immune response for the natural course of the disease?

1.8 Future Goals

- Development of risk response prognosis parameters
- Standardization of diagnostic methods
- Establishment of a causal differential therapy
- Development of diagnostic and therapeutic guidelines

Acknowledgements. This publication has been supported by the Deutsche Forschungsgemeinschaft through Sonderforschungsbereich/Transregio 19.

References

Angelini A, Crosato M, Boffa GM, Calabrese F, Calzolari V, Chioin R, Daliento L, Thiene G (2002) Active versus borderline myocarditis: clinicopathological correlates and prognostic implications. Heart 87:210–215

Anker SD, von Haehling S (2004) Inflammatory mediators in chronic heart failure: an overview. Heart 90:464–470

Baboonian C, Treasure T (1997) Meta-analysis of the association of enteroviruses with human heart disease. Heart 78:539–543

Bowles NE, Richardson PJ, Olsen EG, Archard LC (1986) Detection of coxsackie-B-virus-specific RNA sequences in myocardial biopsy samples from patients with myocarditis and dilated cardiomyopathy. Lancet 1:1120–1123

Bultmann BD, Klingel K, Sotlar K, Bock CT, Baba HA, Sauter M, Kandolf R (2003) Fatal parvovirus B19-associated myocarditis clinically mimicking ischemic heart disease: an endothelial cell-mediated disease. Hum Pathol 34:92–95

Calabrese F, Thiene G (2003) Myocarditis and inflammatory cardiomyopathy: microbiological and molecular biological aspects. Cardiovasc Res 60:11–25

D'Ambrosio A, Patti G, Manzoli A, Sinagra G, Di Lenarda A, Silvestri F, Di Sciascio G (2001) The fate of acute myocarditis between spontaneous improvement and evolution to dilated cardiomyopathy: a review. Heart 85:499–504

Dec GW, Fuster V (1994a) Idiopathic dilated cardiomyopathy [see comments]. N Engl J Med 331:1564–1575

Deonarain R, Cerullo D, Fuse K, Liu PP, Fish EN (2004) Protective role for interferon-b in coxsackievirus B3 infection. Circulation 110:3540–3543

Frustaci A, Chimenti C, Calabrese F, Pieroni M, Thiene G, Maseri A (2003) Immunosuppressive therapy for active lymphocytic myocarditis: virological and immunologic profile of responders versus nonresponders. Circulation 107:857–863

Fujioka S, Kitaura Y, Ukimura A, Deguchi H, Kawamura K, Isomura T, Suma H, Shimizu A (2000) Evaluation of viral infection in the myocardium of patients with idiopathic dilated cardiomyopathy. J Am Coll Cardiol 36:1920–1926

Hauck AJ, Kearney DL, Edwards WD (1989) Evaluation of postmortem endomyocardial biopsy specimens from 38 patients with lymphocytic myocarditis: implications for role of sampling error. Mayo Clin Proc 64:1235–1245

Kandolf R, Ameis D, Kirschner P, Canu A, Hofschneider PH (1987) In situ detection of enteroviral genomes in myocardial cells by nucleic acid hybridization: an approach to the diagnosis of viral heart disease. Proc Natl Acad Sci U S A 84:6272–6276

Kanzaki Y, Terasaki F, Okabe M, Hayashi T, Toko H, Shimomura H, Fujioka S, Kitaura Y, Kawamura K, Horii Y, Isomura T, Suma H (2001) Myocardial inflammatory cell infiltrates in cases of dilated cardiomyopathy as a determinant of outcome following partial left ventriculectomy. Jpn Circ J 65:797–802

Karjalainen J, Heikkila J (1999) Incidence of three presentations of acute myocarditis in young men in military service. A 20-year experience. Eur Heart J 20:1120–1125

Kühl U, Noutsias M, Seeberg B, Schultheiss HP (1996) Immunohistological evidence for a chronic intramyocardial inflammatory process in dilated cardiomyopathy. Heart 75:295–300

Kühl U, Pauschinger M, Schultheiss HP (1997a) [Etiopathogenetic differentiation of inflammatory cardiomyopathy. Immunosuppression and immunomodulation]. Internist (Berl) 38:590–601

Kühl U, Pauschinger M, Schultheiss HP (1997b) Neue Konzepte zur Diagnostik der entzündlichen Herzmuskelerkrankung [New concepts in the diagnosis of inflammatory myocardial disease (see comments)]. Dtsch Med Wochenschr 122:690–698

Kühl U, Pauschinger M, Bock T, Klingel K, Schwimmbeck PL, Seeberg B, Krautwurm L, Schultheiß HP, Kandolf R (2003a) Parvovirus B19 infection mimicking acute myocardial infarction. Circulation 108:945–950

Kühl U, Pauschinger M, Schwimmbeck PL, Seeberg B, Lober C, Noutsias M, Poller W, Schultheiss HP (2003b) Interferon-beta treatment eliminates cardiotropic viruses and improves left ventricular function in patients with myocardial persistence of viral genomes and left ventricular dysfunction. Circulation 107:2793–2798

Kühl U, Pauschinger M, Noutsias M, Seeberg B, Bock T, Lassner D, Poller W, Kandolf R, Schultheiss HP (2005) High prevalence of viral genomes and multiple viral infections in the myocardium of adults with "idiopathic" left ventricular dysfunction. Circulation 111:887–893

Li E, Brown SL, Von Seggern DJ, Brown GB, Nemerow GR (2000) Signaling antibodies complexed with adenovirus circumvent CAR and integrin interactions and improve gene delivery. Gene Ther 7:1593–1599

Mason JW (2002) Viral latency: a link between myocarditis and dilated cardiomyopathy? J Mol Cell Cardiol 34:695–698

Mason JW, JB OC, Herskowitz A, Rose NR, McManus BM, Billingham ME, Moon TE (1995) A clinical trial of immunosuppressive therapy for myocarditis. The Myocarditis Treatment Trial Investigators [see comments]. N Engl J Med 333:269–275

Matsumori A (2001) Hepatitis C virus and cardiomyopathy. Intern Med 40:78–79

Noutsias M, Seeberg B, Schultheiss HP, Kühl U (1999) Expression of cell adhesion molecules in dilated cardiomyopathy: evidence for endothelial activation in inflammatory cardiomyopathy. Circulation 99:2124–2131

Noutsias M, Fechner H, de Jonge H, Wang X, Dekkers D, Houtsmuller AB, Pauschinger M, Bergelson J, Warraich R, Yacoub M, Hetzer R, Lamers J, Schultheiss HP, Poller W (2001) Human coxsackie-adenovirus receptor is colocalized with integrins alpha(v)beta(3) and alpha(v)beta(5) on the cardiomyocyte sarcolemma and upregulated in dilated cardiomyopathy: implications for cardiotropic viral infections. Circulation 104:275–280

Noutsias M, Pauschinger M, Ostermann K, Escher F, Blohm JH, Schultheiss H, Kühl U (2002a) Digital image analysis system for the quantification of infiltrates and cell adhesion molecules in inflammatory cardiomyopathy. Med Sci Monit 8:MT59–71

Noutsias M, Pauschinger M, Schultheiss H, Kühl U (2002b) Phenotypic characterization of infiltrates in dilated cardiomyopathy – diagnostic significance of T-lymphocytes and macrophages in inflammatory cardiomyopathy. Med Sci Monit 8:CR478–CR487

Noutsias M, Pauschinger M, Schultheiss HP, Kühl U (2003a) Advances in the immunohistological diagnosis of inflammatory cardiomyopathy. Eur Heart J 4:154–162

Noutsias M, Pauschinger M, Poller WC, Schultheiss HP, Kühl U (2003b) Current insights into the pathogenesis, diagnosis and therapy of inflammatory cardiomyopathy. Heart Fail Monit 3:127–135

Noutsias M, Pauschinger M, Poller WC, Schultheiss HP, Kühl U (2004) Immunomodulatory treatment strategies in inflammatory cardiomyopathy: current status and future perspectives. Expert Rev Cardiovasc Ther 2:37–51

Olbrich HG (2001) [Epidemiology-etiology of dilated cardiomyopathy]. Z Kardiol 90 Suppl 1:2–9

Pauschinger M, Doerner A, Kuehl U, Schwimmbeck PL, Poller W, Kandolf R, Schultheiss HP (1999a) Enteroviral RNA replication in the myocardium of patients with left ventricular dysfunction and clinically suspected myocarditis. Circulation 99:889–895

Pauschinger M, Bowles NE, Fuentes-Garcia FJ, Pham V, Kühl U, Schwimmbeck PL, Schultheiss HP, Towbin JA (1999b) Detection of adenoviral genome in the myocardium of adult patients with idiopathic left ventricular dysfunction. Circulation 99:1348–1354

Pauschinger M, Chandrasekharan K, Noutsias M, Kühl U, Schwimmbeck LP, Schultheiss HP (2004) Viral heart disease: molecular diagnosis, clinical prognosis, and treatment strategies. Med Microbiol Immunol (Berl) 193:65–69

Poller W, Fechner H, Noutsias M, Tschoepe C, Schultheiss HP (2002) Highly variable expression of virus receptors in the human cardiovascular system. Implications for cardiotropic viral infections and gene therapy. Z Kardiol 91:978–991

Richardson P, McKenna W, Bristow M, Maisch B, Mautner B, J OC, Olsen E, Thiene G, Goodwin J, Gyarfas I, Martin I, Nordet P (1996) Report of the 1995 World Health Organization/International Society and Federation of Cardiology Task Force on the Definition and Classification of cardiomyopathies [news]. Circulation 93:841–842

Shanes JG, Ghali J, Billingham ME, Ferrans VJ, Fenoglio JJ, Edwards WD, Tsai CC, Saffitz JE, Isner J, Furner S, et al. (1987) Interobserver variability in the pathologic interpretation of endomyocardial biopsy results. Circulation 75:401–405

Towbin JA, Bowles NE (2000) Genetic abnormalities responsible for dilated cardiomyopathy. Curr Cardiol Rep 2:475–480

Towbin JA, Bowles NE (2001) Molecular genetics of left ventricular dysfunction. Curr Mol Med 1:81–90

Towbin JA, Bowles NE (2002) Molecular diagnosis of myocardial disease. Expert Rev Mol Diagn 2:587–602

Why HJ, Archard LC, Richardson PJ (1994a) Dilated cardiomyopathy–new insights into the pathogenesis. Postgrad Med J 70:S2–7

Why HJ, Meany BT, Richardson PJ, Olsen EG, Bowles NE, Cunningham L, Freeke CA, Archard LC (1994b) Clinical and prognostic significance of detection of enteroviral RNA in the myocardium of patients with myocarditis or dilated cardiomyopathy. Circulation 89:2582–2589

Wojnicz R, Nowalany-Kozielska E, Wojciechowska C, Glanowska G, Wilczewski P, Niklewski T, Zembala M, Polonski L, Rozek MM, Wodniecki J (2001) Randomized, placebo-controlled study for immunosuppressive treatment of inflammatory dilated cardiomyopathy: two-year follow-up results. Circulation 104:39–45

2 Unsolved Medical Issues and New Targets for Further Research in Viral Myocarditis and Dilated Cardiomyopathy

K.U. Knowlton

Abstract. Meaningful advances have been made in understanding the mechanisms that contribute to dilated cardiomyopathy and myocarditis. Our data confirmed the hypothesis that there is an interaction of genetic predisposition and acquired factors, in that both can affect the dystrophin–glycoprotein complex. We could show that dystrophin deficiency increases susceptibility to viral infection. Our experiments addressed the role of coxsackievirus in the pathogenesis of cardiomyopathy, while other viruses may be involved, such as adenovirus, parvovirus, influenza virus, etc. Furthermore, we could demonstrate that cardiac myocyte-specific transgenic expression of SOCS1 inhibited coxsackievirus-induced signaling of Janus kinase (JAK) and signal transducer and activator of transcription (STAT), with accompanying increases in viral replication, cardiomyopathy, and mortality in infected mice. Future treatment strategies may include the development of coxsackie–adenovirus receptor (CAR) inhibitors and enteroviral protease 2A inhibitors. Additional studies are ongoing to determine the effectiveness of these inhibitors on viral infection in culture and in the intact heart.

This symposium reflected on the meaningful advances that have been made in understanding the mechanisms that contribute to dilated cardiomyopathy and myocarditis. However, there remain important issues that will benefit from continued investigation. If we look at myocarditis and dilated cardiomyopathy from the perspective of a patient recently diagnosed with this condition, there are four questions that the patient is likely to ask:

- What caused this?
- Why me?
- What can be done to treat me?
- Is there something that could have been done to prevent this?

Attention to each of these questions can help us focus our thoughts on areas of research that would benefit from our attention in the future. For example, the question "What caused this?" should stimulate us to find improved strategies to identify the etiology of the disease. The question of "Why me?" directs our attention to the issue of susceptibility factors for myocarditis and dilated cardiomyopathy. The question of "What can be done to treat me?" should focus our efforts on identifying etiologically specific therapies. And finally, the question of prevention takes us in the direction of whether there is a role for vaccination or other preventive measures. Each of these questions is interrelated, and the answers will be more easily addressed as we continue to understand the basic mechanisms that are important in myocarditis and dilated cardiomyopathy.

2.1 Causes of Cardiomyopathy

While there is debate regarding the exact percentage of patients that have a hereditary or acquired form of cardiomyopathy, it is clear that both are important. Genetic abnormalities in the dystrophin-glycoprotein complex have been well-established as causing cardiomyopathy in Duchenne and Becker muscular dystrophies, limb-girdle muscular dystrophies, and some forms of X-linked dilated cardiomyopathy. These have been discussed by others during this symposium.

Over the last years, it has been made clear that virus infection can cause a subset of cardiomyopathies in the form of acute myocarditis or chronic dilated cardiomyopathies. The exact incidence is of considerable debate. However, among the cardiotropic viruses, coxsackievirus is the most thoroughly studied virus that can cause acquired cardiomyopathies and has been reported to account for 10%–25% of "idiopathic" dilated cardiomyopathies. Coxsackievirus is a member of the picornavirus family, closely related to poliovirus and rhinovirus, and bears a small positive strand of RNA. Other viruses are also likely to have an important role. Many presentations addressed at this symposium the relatively high incidence of parvoviruses in patients with cardiomyopathy.

Given the importance of both genetic and acquired forms of cardiomyopathy, in our laboratory we set out to determine whether there is any relationship between the genetic and acquired forms of dilated

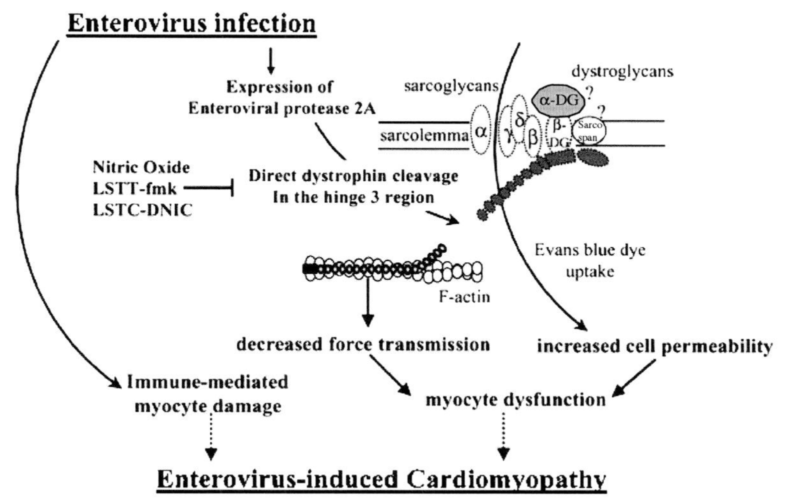

Fig. 1. Scheme summarizing the putative role of the dystrophin cleavage by the enteroviral protease 2A in the pathogenesis of the enterovirus-induced cardiomyopathy. (Badorff and Knowlton 2004)

cardiomyopathy. Guided by computer neural network prediction of protease 2A cleavage sites (Duckert et al. 2004), we found that the coxsackievirus protease 2A is able to specifically and directly cleave the dystrophin molecule in the hinge 3 region (Badorff et al.). Mutation of the hinge 3 region at the cleavage site prevents cleavage. Cleavage results in disruption of the cell membrane, allowing molecules both to enter and exit the myocardial cell, thus contributing to the pathogenesis of the cardiomyopathy. Figure 1 summarizes this pathogenetic pathway of enterovirus-induced cardiomyopathy.

The physiological localization of dystrophin at the cell membrane is disrupted after cleavage by the enteroviral protease 2A as demonstrated by immunostaining for the carboxyl terminus of dystrophin. Figure 2 shows clearly that in a mouse model of coxsackievirus infection, the same cells that are infected also have disruption of the dystrophin–

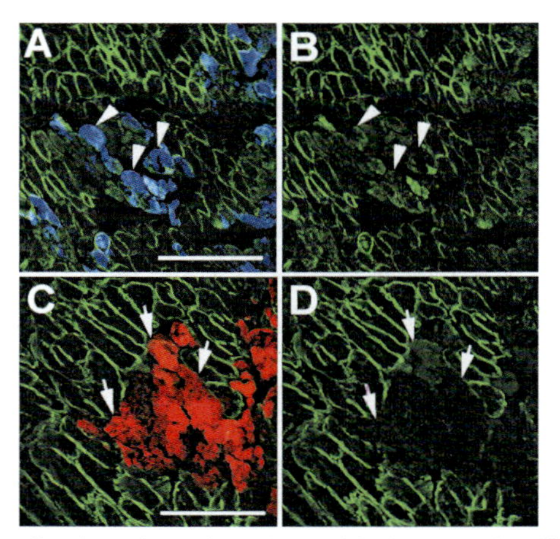

Fig. 2a–d. Loss of carboxyl-terminal dystrophin fragment localization to membrane and sarcolemmal integrity in coxsackievirus group B type 3 (CVB3)-infected myocytes in vivo. Sarcolemmal localization of the carboxyl terminus of dystrophin (*green*) is lost in CVB3-infected myocytes (*blue*, *arrowheads*). Disruption of dystrophin in the sarcolemma is associated with a loss of sarcolemmal membrane integrity as assessed with Evans blue dye staining (*red*, *arrows*). (Lee et al. 2000)

glycoprotein complex. In fact, these same cells are able to take up large molecules such as Evans blue dye showing disruption of the cell membrane.

While this disruption of the dystrophin–glycoprotein complex is clear in mice, a similar phenomenon has also been demonstrated in humans with acute myocarditis caused by coxsackievirus (Badorff and Knowlton 2004). Additional analysis is underway. Taken together, these data indicate that the combination of genetic and acquired causes of cardiomyopathy could be related in that both can affect the dystrophin–glycoprotein complex. This raises the question as to whether there are important interactions between acquired and hereditary causes of disease and whether these interactions could affect the severity of the disease. In addition, it suggests that analysis of cardiomyopathies as either hereditary or acquired rather than both may be overly simplistic.

To asses the interaction between acquired and genetic causes of cardiomyopathy, we evaluated whether dystrophin deficiency increased susceptibility to viral infection. To do this, we infected both dystrophin-competent mice and dystrophin-deficient mice with coxsackievirus B3 (CVB3). In normal, infected dystrophin-competent mice, there was very little Evans blue dye positive uptake, indicating little viral infection. However, in CVB3-infected, dystrophin-deficient mice there

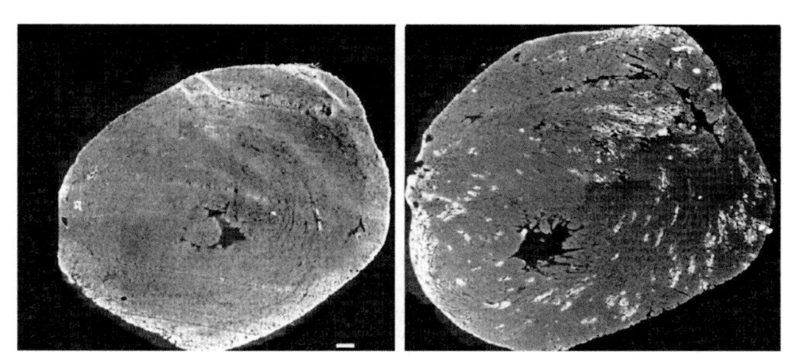

Fig. 3. Dystrophin deficiency markedly increases enterovirus-mediated disruption of the sarcolemma. Staining with Evans blue dye (*white on dark background*), showing sarcolemmal disruption in the heart of CVB3-infected $C57^{wt}$/C3H mice (*left*) and $C57^{mdx}$/C3H (*right*) on day 6 after infection. (Xiong et al. 2002)

was considerably more Evans blue dye uptake when compared to the infected, dystrophin- competent mice. The dystrophin-deficient mice without coxsackievirus infection did not show Evans blue dye positive staining at the 4- to 6-week time point studied (see Fig. 3).

In addition, there was an increase in serum cardiac troponin I levels released from the infected, dystrophin-deficient cells. There were very low levels of troponin I in the serum of infected, dystrophin-competent mice (see Fig. 4). This is consistent with a process that would allow an increase in release of virus from the cells.

There is also an increase of virus titer in the infected, dystrophin-deficient mice compared to the infected, dystrophin-competent mice (see Fig. 5).

In summary, while dystrophin-competent cells have a stable membrane, dystrophin-deficient membranes are more fragile. With coxsackievirus infection in the presence or absence of dystrophin, there is viral replication. Cleavage of dystrophin leads to increased membrane permeability in the dystrophin-competent cells, and some virus is able to exit the cells and infect adjacent cells, contributing to the observed cardiomyopathy. In dystrophin-deficient cells, however, there is more efficient disruption of the membrane following viral replication, without need for dystrophin cleavage. Essentially, all the virus leaves the myocyte to infect adjacent cells, which increases viral replication (Fig. 6).

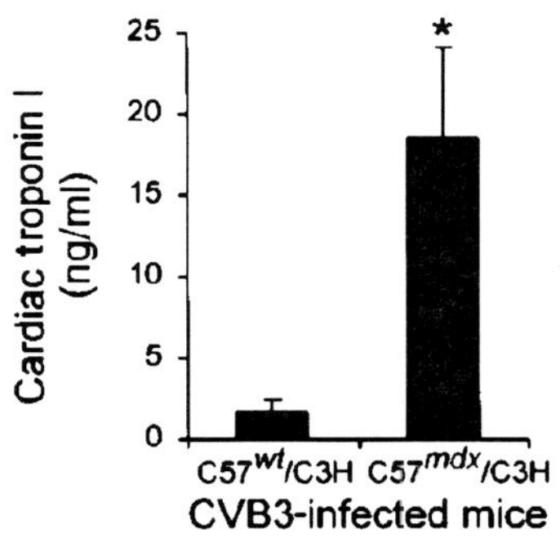

Fig. 4. Serum concentration of cardiac troponin I (cTnI) after infection with CVB3. Concentrations of cTnI were significantly higher in $C57^{mdx}$/C3H than $C57^{wt}$/C3H mice (*, $p < 0.05$, $n = 4$). (Xiong et al. 2002)

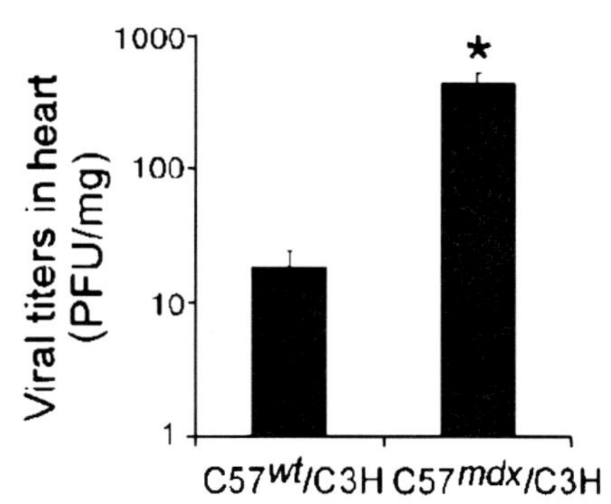

Fig. 5. Viral titers in the hearts of CVB3-infected mice 6 days after infection. The viral titers in the hearts of $C57^{mdx}$/C3H were significantly higher than in $C57^{wt}$/C3H mice (*, $p < 0.05$, $n = 3$). *PFU*, plaque-forming units. (Xiong et al. 2002)

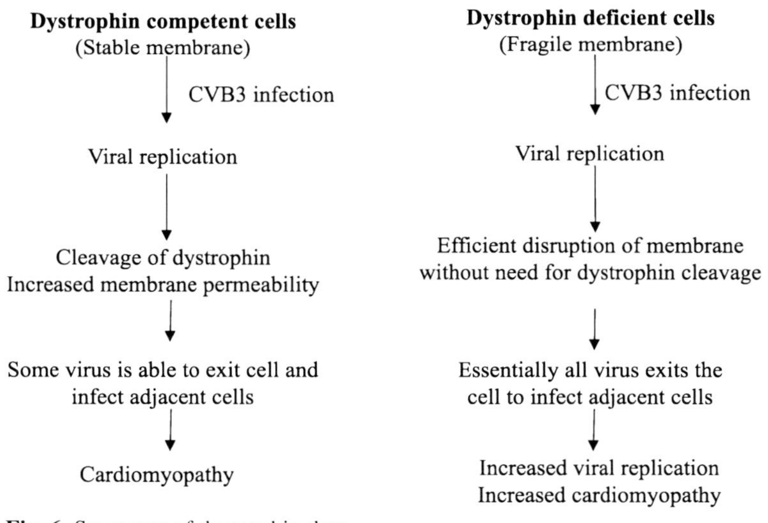

Fig. 6. Summary of dystrophin data

The hypothesis of the interaction of a genetic disease and an acquired disease has thus been verified, at least in mice. Our experiments addressed the role of coxsackievirus in the pathogenesis of cardiomyopathy, while other viruses may be involved, such as adenovirus, parvovirus, influenza virus, etc.

2.2 Susceptibility Factors for Cardiomyopathy

In trying to better understand the pathways and molecules that are involved in the susceptibility to viral infection and why it occurs in some patients but not others, the following are worth noting:

- Nutritional factors are likely to be important as shown in animal models where selenium deficiency increases the virulence of coxsackievirus infection (Beck et al. 1995). In addition, it has been reported in the Keshan province of China that selenium deficiency in the soil may have a role in a high incidence of viral cardiomyopathy (Peng et al. 2000).
- Susceptibility has been linked to the major histocompatibility locus (Wolfgram et al. 1986).
- The function and effect of mitogen-activated protein (MAP) kinases on susceptibility to virus infection have been thoroughly discussed in P. Liu and colleagues' contribution to this volume.
- The role of dystrophin deficiency as a factor in viral susceptibility has been illustrated above.

To further address susceptibility factors to viral infection, we sought to determine whether there is an innate immune response within the cardiac myocyte that can affect its susceptibility to viral infection. The specific question we focused on was whether JAK-STAT (Janus kinase and signal transducer and activator of transcription) signaling within the cardiac myocyte can affect viral infection.

Figure 7 briefly summarizes the JAK–STAT and SOCS cascade. Cytokines that initiate JAK–STAT signaling include cardiotrophin (CT-1), leukemia inhibitory factor (LIF) and interferons (IFNs), all of which have slightly different downstream pathways mediated by a different receptors, such as gp130 for CT-1 and LIF and IFN type I and type II receptors that bind to α- and β-interferons or γ-interferons. Cytokine binding re-

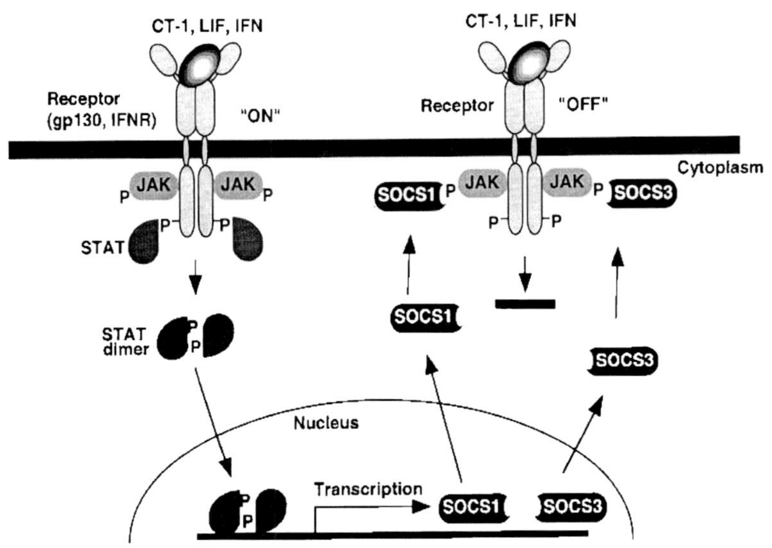

Fig. 7. The JAK-STAT-SOCS signaling cascade. *JAK*, Janus kinase; *STAT*, signal transducers and activators of transcription; *SOCS*, suppressor of cytokine signaling; *CT-1*, cardiotrophin; *LIF*, leukemia inhibitory factor; *IFN*, interferon; *IFNR*, interferon receptor; *gp130*, glycoprotein 130

sults in phosphorylation of the JAK molecules, STAT phosphorylation, and dimerization of STAT, which then translocates to the nucleus leading to transcription of STAT-dependent genes. One of the STAT-dependent families of genes comprises the SOCS molecules, which serve as negative feedback regulators for the JAK–STAT signaling cascade. SOCS can inhibit the phosphorylation of JAK and thus prevent the JAK–STAT signaling. Induction of SOCS expression is an essential, auto-regulatory process.

We set out to test the hypothesis that activation of JAK–STAT signaling within the cardiac myocyte is important for antiviral defense. To begin, we infected mice with CVB3 and found that JAK–STAT signaling and SOCS signaling were induced in hearts of infected wild-type Balb/c mice with upregulated STAT1 phosphorylation and STAT3 phosphorylation, as well as upregulated SOCS1 and 3 molecules (Fig. 8).

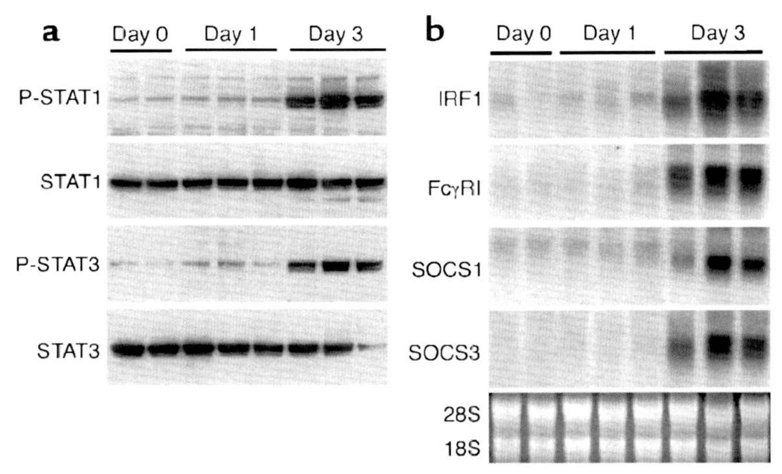

Fig. 8a,b. Correlation of CVB3-induced cardiac injury and JAK-STAT activation. **a** Mice were infected with CVB3. Protein lysate from the heart was blotted at the indicated days after CVB3 inoculation and probed with the antibodies indicated. **b** Northern blot of total RNA from the heart after CVB3 inoculation was probed for IRF1, FcγRI, SOCS1, and SOCS3 expression. 28S and 18S RNA are shown as controls. (Yasukawa et al. 2003)

Subsequently, we inhibited JAK–STAT signaling in the cardiac myocyte by expressing a myc-tagged SOCS1 using the α-MHC (myosin heavy chain) promoter.

The transgenic mice without infection were normal. We have not been able to detect any histologic abnormalities or echocardiographic abnormalities, and there were no readily identifiable abnormalities found with pressure overload.

But when the SOCS1 transgenic mice were infected with CVB3, we found a dramatic increase in susceptibility to viral infection. Approximately 12% of the area of the heart on histologic section stained positive for Evans blue dye (shown as the bright red in Fig. 9), in the wild-type mice. However, with expression of SOCS1 in the cardiac myocyte, the area of Evans blue dye staining increased to almost 60%.

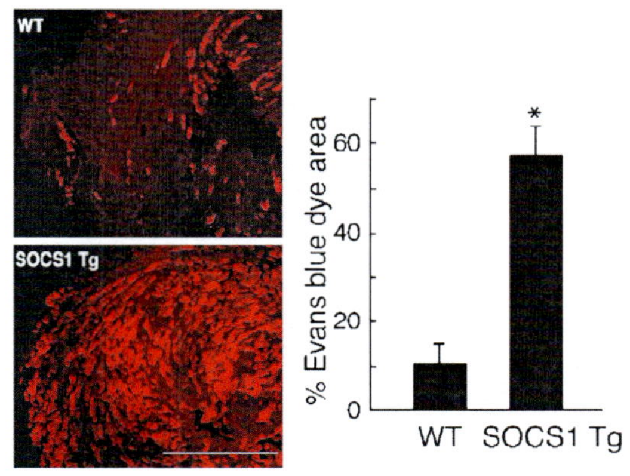

Fig. 9. Increased myocardial injury in SOCS1 transgenic mice. Evans blue dye uptake in the heart was markedly increased in surviving SOCS1 transgenic mice on day 4 after infection (*left*; *red stain*). The area of Evans blue staining in the hearts is shown (*right*; mean $\pm$ SE, $n = 3$, $*p < 0.0001$) as a percentage. (Yasukawa et al. 2003)

In regards to the mechanisms by which this occurred, we did not see an increase in inflammatory infiltrates. However, the virus titer in the heart in SOCS1 transgenic mice on days 4 and 5 after CVB3 infection was significantly higher when compared with that of wild-type mice. As expected, the viral titer in the liver was not elevated (Fig. 10).

Additional work will be required to determine the exact pathways that are inhibited by SOCS1 that lead to such a marked increase in viral infection.

Although there has been considerable work on the subject of susceptibility, there is still a need to identify families who have increased susceptibility to viral heart disease and to identify genetic susceptibility factors. In anecdotal reports, there is evidence of familial aggregation of cardiomyopathy, but precise quantitation is still required. If we are able to address more stringently the increased susceptibility to viral myocarditis in families, we may be better able to answer the individual patient's question of "Why me".

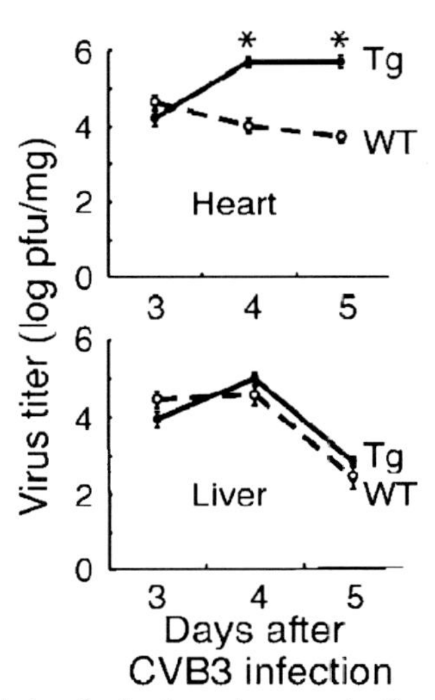

Fig. 10. Increased viral titer in the heart but not the liver of SOCS1 transgenic mice. Virus titers in SOCS1 transgenic and wild-type hearts and livers from 3 to 5 days after infection (mean ± SE, $n = 3$, $*p < 0.01$ comparing SOCS1 transgenic mice with wild-type littermates). (Yasukawa et al. 2003)

Another task lies in identifying more specifically molecules that contribute to susceptibility and studying them as candidate genes in populations with myocarditis or dilated cardiomyopathy. Some of the susceptibility factors that might be important could include interleukin (IL)-10, IL-1, and IL-6 receptors.

2.3 Identifying the Cause of the Disease

The answer to the question "What caused this?" is closely related to how we can improve our diagnostic strategy in the future. Endomyocardial biopsies have been helpful, as we have heard about in this symposium. While myocardial biopsy can be performed safely, it is invasive, has a low but significant risk, and histologic analysis of the specimen gives results with relatively low sensitivity, requiring multiple samples to be taken.

The full diagnostic procedure is expensive and calls for a specialized molecular laboratory to perform the sophisticated diagnostics.

So, are there better ways to make a diagnosis? In theory, the ideal test would be a biomarker that could be assayed in a blood sample with high specificity and sensitivity for infection or inflammation in the heart. During the symposium, we have heard about mRNA microarrays that address the pattern of lymphocyte activation, etc. There may also be sensitive or specific footprints of protein detection in the serum that could lead to the development of protein chips of viral heart disease. Perhaps a protein serum marker could be developed from an understanding of the pathogenesis of viral infection of the heart and its interactions with myocardial proteins. Ideally, a biomarker should combine prognostic and etiologic aspects of the disease. In this context, a recently published paper reported that IL-10 levels were predictive for subsequent cardiogenic shock, requiring a ventricular assist device and mortality of fulminant myocarditis patients (Nishii et al. 2004; Knowlton and Yajima 2004).

A non-invasive imaging strategy like MRI or nuclear scintigraphy could address the unmet diagnostic need. Some exciting data on the gadolinium-enhanced cardiovascular magnetic resonance were presented at this workshop. Nuclear scintigraphy may also be promising if better or more specific antibodies were used to label the infected heart.

2.4 Etiologically Specific Therapies for Myocarditis

With regard to therapy, one treatment option under evaluation is immunomodulation, which has been discussed by A. Frustaci. The myocarditis treatment trial showed us how difficult non-specific immunosuppression can be in a multicenter, blinded, randomized fashion. Nevertheless, if we are able to select the patients in whom the immune response is important in the pathogenesis of the disease, the effect of immune modulation that is directed at individual immune mechanisms could be tested.

Antiviral therapy is another treatment strategy.

- Interferon therapy has been the focus of several presentations during this workshop, and we are looking forward to the outcome of the currently ongoing clinical trial.

- Receptor binding inhibitors like pleconaril may offer new treatment opportunities for viral myocarditis. Inhibition of coxsackie–adenovirus receptor (CAR) expression may also constitute a promising therapeutic goal. CAR expression has been shown to be increased in dilated cardiomyopathy. Given that soluble CAR isoforms can bind to the extracellular domain of CAR and the CVB3 capsid, they may have significant inhibitory or stimulatory effects on CAR signaling and may have an important role in the host defense against viral infection (Dörner et al. 2004).
- The enteroviral protease 2A, in analogy to the human immunodeficiency virus (HIV) protease, is also an attractive target for antiviral drug development.

As discussed above, dystrophin is proteolytically cleaved by the coxsackievirus protease 2A, leading to functional impairment and morphological disruption of the cell membrane. Our experiments indicated protease 2A-mediated cleavage occurs in the hinge 3 region. The cleavage site is shown in Fig. 11.

Having identified the site of dystrophin cleavage, we sought to develop a specific inhibitor. Taking advantage of the experience of scientists using protein recognition peptides for apoptosis research, we developed peptides z-LSTT-fmk and z-LSTL-fmk and demonstrated that they could inhibit proteolytic degradation of dystrophin as shown in Fig. 12. Additional studies are ongoing to determine the effectiveness of these inhibitors on viral infection in culture and in the intact heart.

In summary, during this symposium, there has been a large amount of data presented that will help us understand the key questions that face patients, physicians, and scientists that seek to understand the key questions surrounding myocarditis and dilated cardiomyopathy. We leave with a sense of progress in understanding what causes the disease, why it affects a subset of individuals while sparing the majority, how we can develop better therapies, and ultimately whether the disease can be prevented.

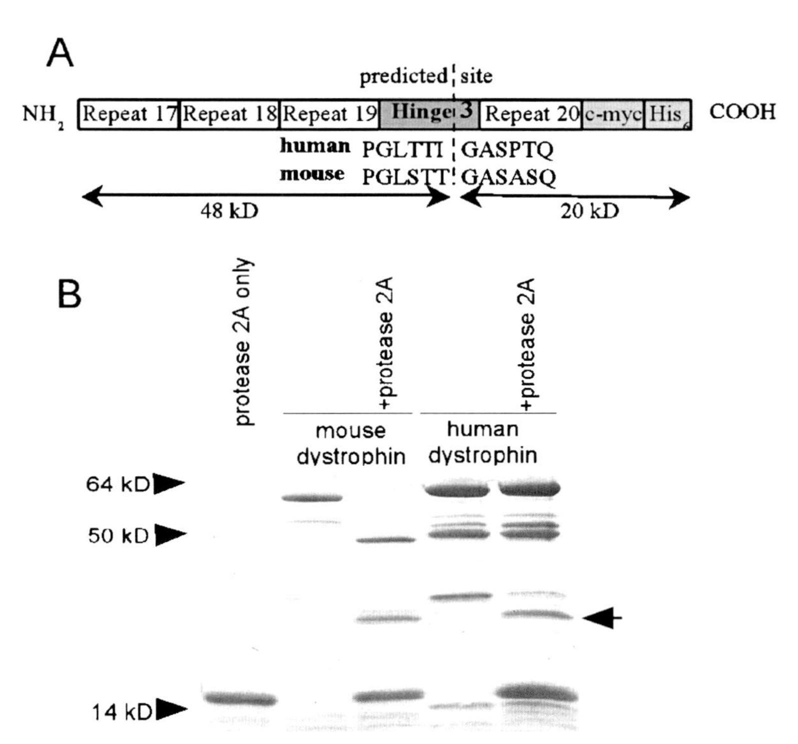

Fig. 11a,b. Direct cleavage of the hinge 3 region of dystrophin by coxsackieviral protease 2A. **a** Schematic diagram of the bacterial expression vectors for the dystrophin miniproteins containing the hinge 3 region between complete spectrin-like repeats 17–20. The sequence of the putative cleavage site in human and mouse dystrophin is shown *below* along with the molecular masses of the fragments that would result from cleavage at this site. The vectors include c-myc tag (for immunoblotting) and a His6 tag (for purification). **b** Bacterially expressed, purified dystrophin fusion proteins were incubated in the absence or presence of protease 2A for 60 min followed by sodium dodecyl sulfate polyacrylamide gel electrophoresis (SDS-PAGE) and Coomassie staining of the gel. The *arrow* indicates the C-terminal fragment that was used for N-terminal peptide sequencing. (Badorff et al. 2000)

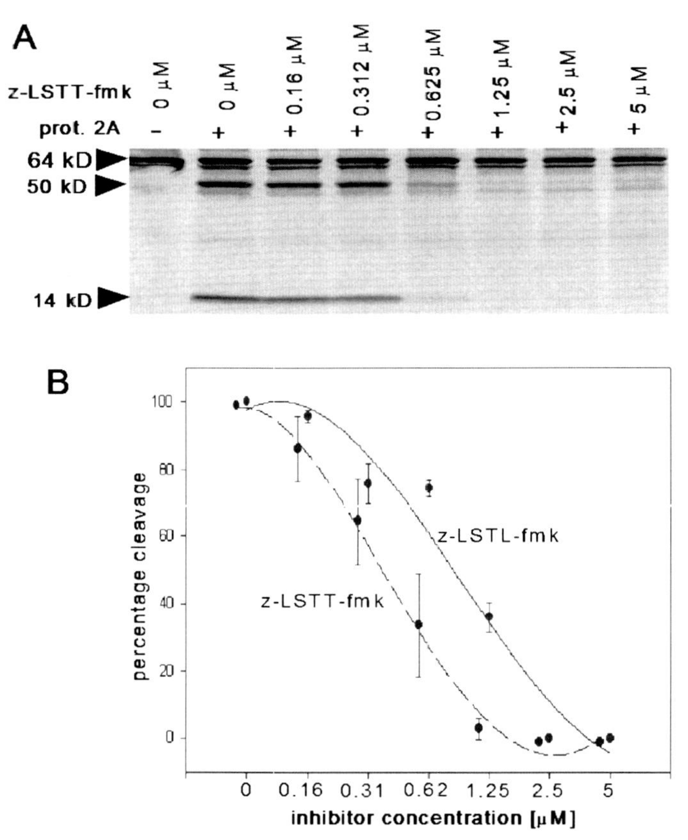

Fig. 12a,b. A dystrophin-based coxsackieviral protease 2A inhibitor. **a** The mouse dystrophin miniprotein containing the hinge 3 region was incubated with 1 μM protease 2A and the indicated concentrations of z-LSTT-fmk for 1 h in cleavage buffer without dithiothreitol followed by SDS-PAGE and autoradiography. z-LSTT-fmk dose-dependently inhibits protease 2A. **b** The cleavage inhibition by z-LSTT-fmk and z-LSTL-fmk was densitometrically calculated. Shown are the means and standard deviations from three independent experiments. The percentage of cleavage was calculated relative to the value obtained with no inhibitor. Half-maximal inhibition of 1 μM protease 2A occurs with 550 nM z-LSTT-fmk and 1050 nM z-LSTL-fmk, respectively. (Badorff et al. 2000)

References

Badorff C, Lee GH, Lamphear BJ, Martone ME, Campbell KP, Rhoads RE, Knowlton KU (1999) Enteroviral Protease 2A Cleaves Dystrophin: Evidence of Cytoskeletal Disruption in an Acquired Cardiomyopathy. Nature Medicine, 5(3):320–326

Badorff C, Knowlton KU (2004) Dystrophin disruption in enterovirus-induced myocarditis and dilated cardiomyopathy: from bench to bedside. Med Microbiol Immunol (Berl) 193:121–126

Badorff C, Berkely N, Mehrotra S, et al. (2000) Enteroviral protease 2A directly cleaves dystrophin and is inhibited by a dystrophin-based substrate analogue. J Biol Chem 275:11191–11197

Beck MA, Shi Q, Morris VC, Levander OA (1995) Rapid genomic evolution of a non-virulent coxsackievirus B3 in selenium-deficient mice results in selection of identical virulent isolates. Nat Med 1:433–436

Dörner A, Xiong D, Couch K, et al. (2004) Alternatively spliced soluble coxsackie-adenovirus receptors inhibit coxsackievirus infection. J Biol Chem 279:18497–18503

Duckert P, Brunak S, Blom N (2004) Prediction of proprotein convertase cleavage sites. Protein Eng Des Sel 17:107–112

Knowlton KU, Yajima T (2004) Interleukin-10: biomarker or pathologic cytokine in fulminant myocarditis? J Am Coll Cardiol 44:1298–1300

Lee GH, Badorff C, Knowlton KU (2000) Dissociation of sarcoglycans and the dystrophin carboxyl terminus from the sarcolemma in enteroviral cardiomyopathy. Circ Res 87:489–495

Nishii M, Inomata T, Takehana H, et al. (2004) Serum levels of interleukin-10 on admission as a prognostic predictor of human fulminant myocarditis. J Am Coll Cardiol 44:1292–1297

Peng T, Li Y, Yang Y, Niu C, Morgan-Capner P, Archard LC, Zhang H (2000) Characterization of enterovirus isolates from patients with heart muscle disease in a selenium-deficient area of China. J Clin Microbiol 38:3538–3543

Wolfgram LJ, Beisel KW, Herskowitz A, Rose NR (1986) Variations in the susceptibility to coxsackievirus B3-induced myocarditis among different strains of mice. J Immunol 136:1846–1852

Xiong D, Lee GH, Badorff C, et al. (2002) Dystrophin deficiency markedly increases enterovirus-induced cardiomyopathy: a genetic predisposition to viral heart disease. Nat Med 8:872–877

Yasukawa H, Yajima T, Duplain H, et al. (2003) The suppressor of cytokine signaling-1 (SOCS1) is a novel therapeutic target for enterovirus-induced cardiac injury. J Clin Invest 111:469–478

II Viruses

3 Frontiers in Viral Diagnostics

M. Pauschinger, A. Kallwellis-Opara

Abstract. Dilated cardiomyopathy (DCM) is a fatal myocardial disease with an incidence of 40:100,000. In recent years, viral infection as a causative agent for myocarditis followed by DCM has become a main topic of research. On the one hand, the virus violates the myocardial integrity itself; on the other hand, the virus induces inadequate local humoral and cellular defense reaction resulting in cardiomyocyte death, fibrosis, and overall cardiac dysfunction. Classical virological approaches are no longer sufficient to detect and identify the virus in the heart. The possibility of endomyocardial biopsies, as well as the further development of new high-specific and sensitive molecular approaches including real-time PCR or sequencing, allows us to detect and to identify the patient-specific causal virus and to predict the progression of disease and hopefully, in the future, to develop virus-specific treatment strategies.

3.1 Meaning of Viral Infections by Heart Failure

Acute viral myocarditis is an inflammatory heart disease that is associated with cardiac dysfunction, and it goes through different stages, which are acute viral infection, autoimmunity, and dilated cardiomyopathy (DCM) (Martino et al. 1994; Richardson et al. 1996; Sole and Liu 1993). The relevance of viral agents in this disease entity was debated for a long time because classical diagnostic tools like serology or direct virus isolation from the myocardium failed in most cases to establish the viral etiology of acute myocarditis. Advances in molecular biological techniques, which resulted in the introduction of polymerase chain reaction (PCR) and in situ hybridization as diagnostic tools, enabled the detection of viral genome in the myocardium of patients with clinically suspected myocarditis and also in cases of DCM (Baboonian and Treasure 1997; Bowles et al. 1986, 2003; Grasso et al. 1992; Jin et al. 1990; Kandolf et al. 1987; Kandolf and Hofschneider 1989; Kasper et al. 1994; Martin et al. 1994; Pauschinger et al. 1998a,b, 1999a,b; Tracy et al. 1990; Why et al. 1994; Woodruff 1980). The most prevalent cardiotropic viruses are enterovirus (coxsackievirus), adenovirus, parvovirus B19 (PVB19), and human herpes virus type 6 (HHV-6) (Kühl et al. 2005). Whereas coxsackievirus and adenovirus were the more prevalent viruses in the past decade, recent data indicate an epidemiological shift in favor of PVB19 (Kühl et al. 2003; J. Towbin, personal communication). The most detailed knowledge concerning virus-induced myocardial damage has been unraveled with coxsackievirus (Fig. 1).

The acute viral infection of a cardiotropic virus leads to the replication of the virus in the myocardium, resulting in myocardial injury and induction of a host immune response (acute phase). This biological activity has been demonstrated in murine models and in human myocardial tissue (Hohenadl et al. 1994; Kandolf et al. 1993; Klingel et al. 1998; Pauschinger et al. 1999b).

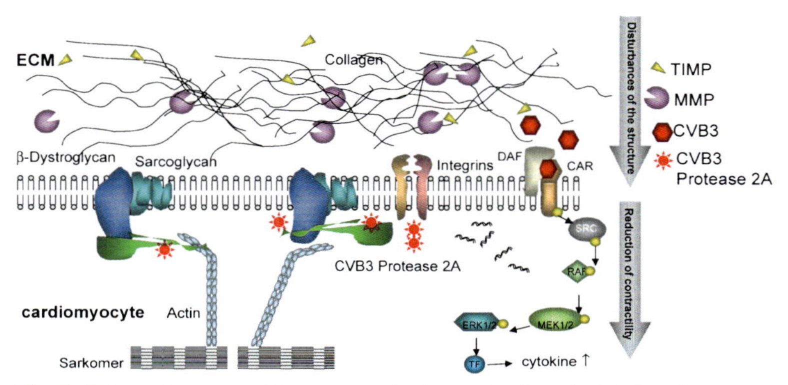

Fig. 1. Schematic graph of known mechanisms of enteroviral-induced myocardial injury causing myocardial dysfunction

3.1.1 Direct Injury of the Heart by the Virus

In the recent years several viruses were established to be cardiotropic, including PVB19, HHV6, adenovirus, and enterovirus (Bowles et al. 1986; Kühl et al. 2003, 2004; Martin et al. 1994; Pauschinger et al. 1998b; Schonian et al. 1995). The pathophysiological mechanisms of direct myocardial injury by coxsackievirus B3 (CVB3) associated with acute viral myocarditis and chronic dilated cardiomyopathy have been evaluated in detail (Andreoletti et al. 2000; Li et al. 2000; Noutsias et al. 2004). However, the detailed pathomechanisms of the other highly prevalent viruses such as PVB19 and HHV6 are yet not well described.

The majority of patients exposed to CVB3 do not develop a myocardial disease, because the viruses are not primarily cardiotropic. The coxsackievirus–adenovirus receptor (CAR) is a key-determinant for the cellular uptake and pathogenesis of these viruses. Its increase is reported in DCM patients. Furthermore, recombinant expression of CAR increased the virus uptake in rat cardiomyocytes (Noutsias et al. 2001; Fig. 1).

Subsequent to the viral attachment, viral RNA is released into the cell and is used as a template for viral protein production and replication. Wessely et al. demonstrated in a murine model that the transgenic expression of non-replicative enteroviral cDNA (CVB3rVPO) causes DCM through cytopathic effects of this non-replicative enteroviral cDNA intermediate (Wessely et al. 1998b; Fig. 1). In addition, viral-specific proteases such as coxsackieviral protease 2A cleave dystrophin in vitro and in vivo, leading to the disruption of the cytoskeleton (Badorff et al. 1999; Fig. 1).

Concerning signal transduction, the uptake of CVB3 in neonatal rat cardiomyocytes induces phosphorylation of extracellular signal regulated kinase (ERK)1/2, leading to a change of transcription patterns, including cytokines. The phosphorylation increased parallel to the virus titers and severity of myocarditis in mice models, whereas the final level of phosphorylated ERK1/2 was dependent on the strain- dependent accessibility for myocarditis (Opavsky et al. 2002). In addition, the activation of the CAR receptor caused an activation of the tyrosine kinase p56, which seems to be a key factor in signal transduction causing virus-induced myocardial dysfunction (Liu et al. 2000; Fig. 1).

Finally, a complete spontaneous elimination of viruses in these murine models results in an improvement of left-ventricular ejection fraction, which underlines the meaning of viral infection in cardiac dysfunction.

3.1.2 Injury of the Heart by Inadequate Immunological Response

Besides the direct CVB3-mediated myocardial injury, immunological responses of the infected myocardium composed of cellular, humoral immunity and cytokine expression are important in the development of myocarditis. However, this virus-induced intramyocardial inflammation aims to eliminate the virus and is therefore not primarily detrimental to the heart (Liu and Mason 2001; McCarthy et al. 2000). Concerning the effectiveness in viral elimination, a combination of lymphocytic T cell profile and an adequate lymphocytic B cell activation is relevant. Therefore, gene-targeted knockout-mice of $CD4^+$, $CD8^+$ or T cell receptor (TCR-β) demonstrated a minimal heart injury after CVB3 infection (Opavsky et al. 1999). In addition, immune-mediated myocytolysis liberates cryptic cardiac antigens, thereby evoking an anti-cardiac autoimmunity by molecular mimicry (Gauntt et al. 1995), which may ultimately persist even after a complete viral elimination ensued. In addition, genetic predisposition [i.e., certain HLA haplotypes (Carlquist et al. 1991)] might be an important contributing factor for a detrimental immune response causing a persistent type of myocardial inflammation. Activated B lymphocytes produce numerous anti-cardiac autoantibodies, which can cross-react with myocardial antigens and may thus also contribute to impairment of cardiac contractility (Liu and Mason 2001; Mason et al. 1995; Wessely et al. 1998a). Moreover, autoimmunity meditated by autoantibodies against a variety of host cell proteins, including antibodies against the cardiac β1-adrenergic receptor, contractile proteins, mitochondrial proteins, the muscarinic receptor, the calcium channel, and the sarcoplasmic reticulum may be part of the pathogenesis of DCM (Caforio et al. 1992; Magnusson et al. 1994; Nishimura et al. 2001; Okazaki et al. 2003; Schulze et al. 1999; Staudt et al. 2001; Staudt et al. 2002; Staudt et al. 2003). Another inadequate immune response is the increased expression of various cytokines including tumor necrosis factor (TNF)-α, interleukin (IL)-1β or IL-6. These enhance the immune response and activate additional enzyme expression, which take part in the cardiac remodeling process including matrix metalloproteinases (MMP) and their inhibitors [e.g., tissue inhibitors of metalloproteinases (TIMP)] (Li et al. 2002; Pauschinger et al. 2004).

3.2 Classical Virological Approaches

According to the current WHO classification of cardiomyopathies, myocarditis is an inflammatory heart disease that is associated with cardiac dysfunction. Due to an inadequate immune response with the consequence of viral persistence and/or chronic myocardial inflammation, a progressive myocardial dysfunction can be observed with the clinical phenotype of dilated cardiomyopathy (Richardson et al. 1996). The very variable clinical symptoms include subclinical asymptomatic forms, arrhythmias, chest pain, heart failure, and sudden death, all of which are dependent on such factors as age, gender, immunocompetence of the host, and infective argents.

The classical diagnostic tools for the detection of viral persistence in the myocardium can be grouped in three categories: direct virus detection, virus isolation, and serology. Direct detection of intact viral particles is only possible in highly selected patients with an extreme virus load in the acute stage of myocarditis. Therefore, this approach is not feasible in diagnosing viral persistence after the acute phase of the disease due to the low virus load. Virus isolation in cell culture and its further identification by using its properties in cell culture is also not possible in the chronic phase of the disease due to the low virus load during this phase. The classical approach using serology fails because this approach gives only an indirect method for virus detection. In addition, the infected target tissue cannot be delimited by this approach. Therefore, the classical virological tools for diagnosing viral infection give insufficient information of the cardiotrophy and identity of the virus. The introduction of molecular biological detection assays for diagnosing persistence of viral genome in the myocardium substantially broadened the diagnosis of viral persistence in the myocardium.

3.3 New High-Specific and Sensitive Molecular Approaches

3.3.1 Qualitative Virus Detection

Qualitative detection assays of viral genome in the myocardium of patients with myocarditis and dilated cardiomyopathy enabled the diagnosis of viral persistence in the myocardium of these patients. Bowles et al. used slot blot hybridization and showed, for the first time, the presence of enteroviral genome in the myocardium of a significant number of patients with myocarditis and DCM (Bowles et al. 1986; Table 1). However, slot blot hybridization caused several ambiguities because of a high grade of cross reactivity of the probe used. Therefore, an arbitrary threshold was defined for positive confirmation of the presence of viral etiology. Due to inherent methodological problems, slot blot hybridization was no longer considered a diagnostic tool in establishing viral heart disease. Soon, in situ hybridization and PCR were introduced in the diagnosis of viral heart disease and were found to be more reliable.

In situ hybridization combines morphological analysis and detection of viral genome at the molecular level, and hence attains an advantage over PCR as a diagnostic tool. In situ hybridization has proved beyond further debates that there is indeed enteroviral infection of myocytes in a significant proportion of patients with myocarditis and DCM (Ho-

Table 1. Brief summary of different molecular biological methods used for viral detection in the myocardium, with a comparison of advantages and disadvantages of the different detection assays

	Method	Sensitivity/ advantages	Disadvantages	Publication
Qualitative virus	In situ hybridization	Local detection	Not usable in routine diagnostics	Hohenadl et al. 1991; Kandolf and Hofschneider 1989; Bowles et al. 1986
	Nested polymerase chain reaction (PCR)	Detection of 1 copy usable in routine diagnostics	No local detection Semiquantitative	Baboonian and Treasure 1997
	Immunohistological detection of virus			
Quantitative virus detection	Quantitative real time PCR	5–10 copies		
		Highly specific Quantitative Usable in routine diagnostics		
Virological activity	E.g., strain-specific enteroviral RNA detection			Pauschinger et al. 1998b; Fujioka et al. 2000
Identification of subtypes	Sequencing			Pauschinger et al. 1999a

henadl et al. 1991; Kandolf and Hofschneider 1989). Nevertheless, this virological method is very time consuming and difficult to standardize. In addition, the focal pattern of myocardial infection calls for an absolute need for analysis of all cross sections of endomyocardial biopsies. Furthermore, exposition time after hybridization may be quite long owing to a very low infection.

PCR is less time consuming, gives more precise results, and also allows for the analysis of whole biopsy specimens. However, due to a probable non-specific hybridization and amplification, it is mandatory to have a distinct PCR protocol to avoid contamination, which raises the incidence of false-positive results.

In spite of technical difficulties, in situ hybridization and PCR remain the most reliable diagnostic tools in confirming viral heart disease, and PCR is ideal as a primary screening technique. Still, the specificity of PCR in diagnosing viral heart disease was under discussion for a long time. However, the meta-analysis by Baboonian et al. of different studies that employed PCR to confirm viral etiology in cardiac disease showed very clearly a significant increase in the presence of enteroviral genome in endomyocardial biopsies from patients with myocarditis and DCM when compared to controls (23% versus 6% and 23% versus 7%, respectively). The authors concluded that RT-PCR is an adequately reliable diagnostic tool to confirm viral persistence in the myocardium of patients with myocarditis or DCM (Baboonian and Treasure 1997).

These first steps in diagnosing the relevance of viral persistence in patients with dilated cardiomyopathy were confirmed by a recent study by Kühl et al. (2004). This study screened 244 patients with the clinically suspected diagnosis of dilated cardiomyopathy [median left ventricular ejection fraction (LVEF) 35.0%; range: 9% to 59%] for persistence of different cardiotropic viruses by nested PCR. Viral genomes could be amplified from 67.4% of endomyocardial biopsies. Thereby enterovirus was detected in 9.4%, adenovirus in 1.6%, PVB19 in 51.4%, HHV6 in 21.6%, Epstein-Barr-virus in 2.0%, and HCV in 0.8% of the biopsies, including 45 cases (27.3%) with multiple infections (Kühl et al. 2004). Due to these recent data, the nested RT-PCR is a powerful, fast, and sensitive tool to analyze whole biopsy specimens and to even detect a single copy of viral genome (Table 1).

3.3.2 Quantitative Virus Detection

Nested PCR as a qualitative detection assay for diagnosing persistence of viral genome is a highly sensitive and specific diagnostic tool. However, using this approach, no quantitative assessments of viral copy numbers can be made. Therefore, real-time PCR, as a well-established tool for quantitative measurement of the target DNA, was introduced to the diagnostics of viral heart disease. In a recent study, PVB19 was quantified by real-time PCR in patients with diastolic left ventricular dysfunction. This study demonstrated that different amounts of virus copy numbers could be detected in different myocardial tissue samples. Using this highly sensitive and specific detection assay, a lowest number of 5–10 viral copies can be detected in one myocardial tissue sample. However, the clinical relevance of virus load in the diagnostic approach of viral heart disease is not clear. A recent study by Barbaro et al. showed an inverse correlation between human immunodeficiency virus (HIV)-1 virus load and left ventricular function parameters (Barbaro et al. 2000). In addition, in HIV-1 mediated disease the exact quantification of the virus load is a basic marker for the rate and severity of the disease and its therapy (Gibellini et al. 2004). In heart-independent diseases, like hepatitis-C, the virus load is a reliable marker for inflammation and indicator for therapy success (Poynard et al. 2003; Sacks et al. 2004). In addition, the measurement of the virus load might be a useful tool to monitor the therapy success of antiviral therapy strategies (e.g., β-interferon) in viral heart disease. However, it is still an open question if the virus load determined in myocardial tissue has any meaning for the course of the disease (Table 1).

3.3.3 Detection of Virological Activity

PCR and in situ hybridization of endomyocardial samples obtained from patients with clinically suspected viral heart disease do not evaluate the biological activity of viral genome. To characterize the activity of enteroviral genome, we have developed an assay based on RT-PCR and Southern blot hybridization to detect plus-strand-specific and minus-strand-specific enteroviral detection. The method relies on the first step of enteroviral replication, i.e., transcription of minus-strand RNA from the plus- strand enteroviral genomic template. Selective RT-PCR detection of minus-strand enteroviral RNA is therefore an indicator of active enteroviral replication. Using this method, we analyzed endomyocardial biopsies from 45 patients with clinically suspected myocarditis. Of the patients, 18 of 45 (40%) were positive for plus-strand enteroviral RNA in the myocardium, and 56% of these 18 patients also had minus-strand enteroviral RNA, indicating that a significant fraction of patients with clinically suspected myocarditis had active enteroviral replication in the myocardium (Pauschinger et al. 1998b). Similar differences in the viral activity (either latent viral persistence or active viral replication) in the myocardium of patients with myocarditis were reported by Fujioka et al. using the same diagnostic approach (Fujioka et al. 2000). More recently, Calabrese et al. found 49% of their patient population to be positive for 23 different viral genomes (Calabrese et al. 2002). Out of the ten enteroviral plus-strand-positive patients, nine were demonstrated to have minus-strand enteroviral RNA suggestive of active viral replication (Calabrese et al. 2002). However, there are no data concerning biological activity for the other known cardiotropic viruses, especially PVB19 (Table 1).

3.3.4 Identification of Subtypes

PCR is a powerful tool for detecting virus infection but it gives no information about serotypes. Interestingly, Martin et al. reported that in a number of adenovirus-associated pediatric and adult cases of myocarditis, a reduced level of inflammation in enterovirus-positive cases was found, suggesting a virus-specific immune response and possibly a virus-specific course of disease (Martin et al. 1994). By using the powerful method of genetic sequencing, we were able to demonstrate that 1.0% of 12 adenovirus-positive cases of left ventricular dysfunction suffered from an adenovirus group C type 2 or 5 infection (Pauschinger et al. 1999a). Several studies correlated various virus serotypes, including HHV6 type A and B, CVB1, -B3, and -B5, or PVB19 types 1–3, and virus variants of PVB19 to heart diseases. To date, there is no proof for the relevance of the virus variants for the diseases, but ongoing studies are trying to link them to the clinical course of the disease. The relevance of different serotypes has already been demonstrated in hepatitis C. In this disease, different therapeutic strategies have been applied to different subtypes due to the different response rates.

However, the meaning of virus variants for the pathogenesis has been demonstrated in animal models. Only CVB3 variant M (CVB3M), but not variant CVB3o, induces myocarditis (Estrin et al. 1986). Additional studies have to prove the theory that different CVB3 variants are less pathogenic and can interact with the CAR receptor with strain-specific affinity in mice, decreasing the number of infected myocytes (Selinka et al. 2002). Furthermore comparison of CVB3-H3 and the antibody escape mutant (H3-10A1), possessing a mutation in VP2, indicate a different TNF-α expression from infected Balb/c monocytes and an absence of autoimmune response during a H3-10A1-infection (Knowlton et al. 1996; Loudon et al. 1991; Table 1).

3.4 Open Questions

Even though the spectrum of methods has become more various and the techniques have become more specific and sensitive, the clinical impact of these findings are still a major field of research activities. In addition, further studies have to clarify the meaning of virus load, viral activity, and virus variants for the course of the disease.

References

Andreoletti L, Bourlet T, Moukassa D, Rey L, Hot D, Li Y, Lambert V, Gosselin B, Mosnier JF, Stankowiak C, Wattre P (2000) Enteroviruses can persist with or without active viral replication in cardiac tissue of patients with end-stage ischemic or dilated cardiomyopathy. J Infect Dis 182:1222–1227

Baboonian C, Treasure T (1997) Meta-analysis of the association of enteroviruses with human heart disease. Heart 78:539–543

Badorff C, Lee GH, Lamphear BJ, Martone ME, Campbell KP, Rhoads RE, Knowlton KU (1999) Enteroviral protease 2A cleaves dystrophin: evidence of cytoskeletal disruption in an acquired cardiomyopathy. Nat Med 5:320–326

Barbaro G, Di LG, Soldini M, Giancaspro G, Grisorio B, Pellicelli AM, D'Amati G, Barbarini G (2000) Clinical course of cardiomyopathy in HIV-infected patients with or without encephalopathy related to the myocardial expression of tumour necrosis factor-alpha and nitric oxide synthase. AIDS 14:827–838

Bowles NE, Richardson PJ, Olsen EG, Archard LC (1986) Detection of coxsackie-B-virus-specific RNA sequences in myocardial biopsy samples from patients with myocarditis and dilated cardiomyopathy. Lancet 1:1120–1123

Bowles NE, Ni J, Kearney DL, Pauschinger M. Schultheiss HP, McCarthy R, Hare J, Bricker JT, Bowles KR, Towbin JA (2003) Detection of viruses in myocardial tissues by polymerase chain reaction. evidence of adenovirus as a common cause of myocarditis in children and adults. J Am Coll Cardiol 42:466–472

Caforio AL, Grazzini M, Mann JM, Keeling PJ, Bottazzo GF, McKenna WJ, Schiaffino S (1992) Identification of alpha- and beta-cardiac myosin heavy chain isoforms as major autoantigens in dilated cardiomyopathy. Circulation 85:1734–1742

Calabrese F, Rigo E, Milanesi O, Boffa GM, Angelini A, Valente M, Thiene G (2002) Molecular diagnosis of myocarditis and dilated cardiomyopathy in children: clinicopathologic features and prognostic implications. Diagn Mol Pathol 11:212–221

Carlquist JF, Menlove RL, Murray MB, O'Connell JB, Anderson JL (1991) HLA class II (DR and DQ) antigen associations in idiopathic dilated cardiomyopathy. Validation study and meta-analysis of published HLA association studies. Circulation 83:515–522

Estrin M, Smith C, Huber S (1986) Antigen-specific suppressor T cells prevent cardiac injury in Balb/c mice infected with a nonmyocarditic variant of coxsackievirus group B, type 3. Am J Pathol 125:578–584

Fujioka S, Kitaura Y, Ukimura A, Deguchi H, Kawamura K, Isomura T, Suma H, Shimizu A (2000) Evaluation of viral infection in the myocardium of patients with idiopathic dilated cardiomyopathy. J Am Coll Cardiol 36:1920–1926

Gauntt CJ, Arizpe HM, Higdon AL, Wood HJ, Bowers DF, Rozek MM, Crawley R (1995) Molecular mimicry, anti-coxsackievirus B3 neutralizing monoclonal antibodies, and myocarditis. J Immunol 154:2983–2995

Gibellini D, Vitone F, Gori E, La PM, Re MC (2004) Quantitative detection of human immunodeficiency virus type 1 (HIV-1) viral load by SYBR green real-time RT-PCR technique in HIV-1 seropositive patients. J Virol Methods 115:183–189

Grasso M, Arbustini E, Silini E, Diegoli M, Percivalle E, Ratti G, Bramerio M, Gavazzi A, Vigano M, Milanesi G (1992) Search for coxsackievirus B3 RNA in idiopathic dilated cardiomyopathy using gene amplification by polymerase chain reaction. Am J Cardiol 69:658–664

Hohenadl C, Klingel K, Mertsching J, Hofschneider PH, Kandolf R (1991) Strand-specific detection of enteroviral RNA in myocardial tissue by in situ hybridization. Mol Cell Probes 5:11–20

Hohenadl C, Klingel K, Rieger P, Hofschneider PH, Kandolf R (1994) Investigation of the coxsackievirus B3 nonstructural proteins 2B, 2C, and 3AB: generation of specific polyclonal antisera and detection of replicating virus in infected tissue. J Virol Methods 47:279–295

Jin O, Sole MJ, Butany JW, Chia WK, McLaughlin PR, Liu P, Liew CC (1990) Detection of enterovirus RNA in myocardial biopsies from patients with myocarditis and cardiomyopathy using gene amplification by polymerase chain reaction. Circulation 82:8–16

Kandolf R, Hofschneider PH (1989) Viral heart disease. Springer Semin Immunopathol 11:1–13

Kandolf R, Ameis D, Kirschner P, Canu A, Hofschneider PH (1987) In situ detection of enteroviral genomes in myocardial cells by nucleic acid hybridization: an approach to the diagnosis of viral heart disease. Proc Natl Acad Sci U S A 84:6272–6276

Kandolf R, Klingel K, Zell R, Canu A, Fortmuller U, Hohenadl C, Albrecht M, Reimann BY, Franz WM, Heim A (1993) Molecular mechanisms in the pathogenesis of enteroviral heart disease: acute and persistent infections. Clin Immunol Immunopathol 68:153–158

Kasper EK, Agema WR, Hutchins GM, Deckers JW, Hare JM, Baughman KL (1994) The causes of dilated cardiomyopathy: a clinicopathologic review of 673 consecutive patients. J Am Coll Cardiol 23:586–590

Klingel K, Rieger P, Mall G, Selinka HC, Huber M, Kandolf R (1998) Visualization of enteroviral replication in myocardial tissue by ultrastructural in situ hybridization: identification of target cells and cytopathic effects. Lab Invest 78:1227–1237

Knowlton KU, Jeon ES, Berkley N, Wessely R, Huber S (1996) A mutation in the puff region of VP2 attenuates the myocarditic phenotype of an infectious cDNA of the Woodruff variant of coxsackievirus B3. J Virol 70:7811–7818

Kühl U, Pauschinger M, Bock T, Klingel K, Schwimmbeck CP, Seeberg B, Krautwurm L, Poller W, Schultheiss HP, Kandolf R (2003) Parvovirus B19 infection mimicking acute myocardial infarction. Circulation 108:945–950

Kühl U, Pauschinger M, Noutsias M, Seeberg B, Bock T, Lassner D, Poller W, Kandolf R, Schultheiss H (2005) High prevalence of viral genomes and multiple viral infections in the myocardium of adults with "idiopathic" left ventricular dysfunction. Circulation 111:887–893

Li J, Schwimmbeck PL, Tschope C, Leschka S, Husmann L, Rutschow S, Reichenbach F, Noutsias M, Kobalz U, Poller W, Spillmann F, Zeichhardt H, Schultheiss HP, Pauschinger M (2002) Collagen degradation in a murine myocarditis model: relevance of matrix metalloproteinase in association with inflammatory induction. Cardiovasc Res 56:235–247

Li Y, Bourlet T, Andreoletti L, Mosnier JF, Peng T, Yang Y, Archard LC, Pozzetto B, Zhang H (2000) Enteroviral capsid protein VP1 is present in myocardial tissues from some patients with myocarditis or dilated cardiomyopathy. Circulation 101:231–234

Liu P, Aitken K, Kong YY, Opavsky MA, Martino T, Dawood F, Wen WH, Kozieradzki I, Bachmaier K, Straus D, Mak TW, Penninger JM (2000) The tyrosine kinase p56lck is essential in coxsackievirus B3-mediated heart disease. Nat Med 6:429–434

Liu PP, Mason JW (2001) Advances in the understanding of myocarditis. Circulation 104:1076–1082

Loudon RP, Moraska AF, Huber SA, Schwimmbeck P, Schultheiss P (1991) An attenuated variant of coxsackievirus B3 preferentially induces immunoregulatory T cells in vivo. J Virol 65:5813–5819

Magnusson Y, Wallukat G, Waagstein F, Hjalmarson A, Hoebeke J (1994) Autoimmunity in idiopathic dilated cardiomyopathy. Characterization of antibodies against the beta 1-adrenoceptor with positive chronotropic effect. Circulation 89:2760–2767

Martin AB, Webber S, Fricker FJ, Jaffe R, Demmler G, Kearney D, Zhang YH, Bodurtha J, Gelb B, Ni J, et al. (1994) Acute myocarditis. Rapid diagnosis by PCR in children. Circulation 90:330–339

Martino TA, Liu P, Sole MJ (1994) Viral infection and the pathogenesis of dilated cardiomyopathy. Circ Res 74:182–188

Mason JW, O'Connell JB, Herskowitz A, Rose NR, McManus BM, Billingham ME, Moon TE (1995) A clinical trial of immunosuppressive therapy for myocarditis. The Myocarditis Treatment Trial Investigators. N Engl J Med 333:269–275

McCarthy RE 3rd, Boehmer JP, Hruban RH, Hutchins GM, Kasper EK, Hare JM, Baughman KL (2000) Long-term outcome of fulminant myocarditis as compared with acute (nonfulminant) myocarditis. N Engl J Med 342:690–695

Nishimura H, Okazaki T, Tanaka Y, Nakatani K, Hara M, Matsumori A, Sasayama S, Mizoguchi A, Hiai H, Minato N, Honjo T (2001) Autoimmune dilated cardiomyopathy in PD-1 receptor-deficient mice. Science 291:319–322

Noutsias M, Fechner H, de JH, Wang X, Dekkers D, Houtsmuller AB, Pauschinger M, Bergelson J, Warraich R, Yacoub M, Hetzer R, Lamers J, Schultheiss HP, Poller W (2001) Human coxsackie-adenovirus receptor is colocalized with integrins alpha(v)beta(3) and alpha(v)beta(5) on the cardiomyocyte sarcolemma and upregulated in dilated cardiomyopathy: implications for cardiotropic viral infections. Circulation 104:275–280

Noutsias M, Pauschinger M, Poller WC, Schultheiss HP, Kühl U (2004) Immunomodulatory treatment strategies in inflammatory cardiomyopathy: current status and future perspectives. Expert Rev Cardiovasc Ther 2:37–51

Okazaki T, Tanaka Y, Nishio R, Mitsuiye T, Mizoguchi A, Wang J, Ishida M, Hiai H, Matsumori A, Minato N, Honjo T (2003) Autoantibodies against cardiac troponin I are responsible for dilated cardiomyopathy in PD-1-deficient mice. Nat Med 9:1477–1483

Opavsky MA, Penninger J, Aitken K, Wen WH, Dawood F, Mak T, Liu P (1999) Susceptibility to myocarditis is dependent on the response of alphabeta T lymphocytes to coxsackieviral infection. Circ Res 85:551–558

Opavsky MA, Martino T, Rabinovitch M, Penninger J, Richardson C, Petric M, Trinidad C, Butcher L, Chan J, Liu PP (2002) Enhanced ERK-1/2 activation in mice susceptible to coxsackievirus-induced myocarditis. J Clin Invest 109:1561–1569

Pauschinger M, Doerner A, Remppis A, Tannhauser R, Kühl U, Schultheiss HP (1998a) Differential myocardial abundance of collagen type I and type III mRNA in dilated cardiomyopathy: effects of myocardial inflammation. Cardiovasc Res 37:123–129

Pauschinger M, Kühl U, Dorner A, Schieferecke K, Petschauer S, Rauch U, Schwimmbeck PL, Kandolf R, Schultheiss HP (1998b) [Detection of enteroviral RNA in endomyocardial biopsies in inflammatory cardiomyopathy and idiopathic dilated cardiomyopathy]. Z Kardiol 87:443–452

Pauschinger M, Bowles NE, Fuentes-Garcia FJ, Pham V, Kühl U, Schwimmbeck PL, Schultheiss HP, Towbin JA (1999a) Detection of adenoviral genome in the myocardium of adult patients with idiopathic left ventricular dysfunction. Circulation 99:1348–1354

Pauschinger M, Doerner A, Kühl U, Schwimmbeck PL, Poller W, Kandolf R, Schultheiss HP (1999b) Enteroviral RNA replication in the myocardium of patients with left ventricular dysfunction and clinically suspected myocarditis. Circulation 99:889–895

Pauschinger M, Chandrasekharan K, Schultheiss HP (2004) Myocardial remodeling in viral heart disease: possible interactions between inflammatory mediators and MMP-TIMP system. Heart Fail Rev 9:21–31

Poynard T, Yuen MF, Ratziu V, Lai CL (2003) Viral hepatitis C. Lancet 362:2095–2100

Richardson P, McKenna W, Bristow M, Maisch B, Mautner B, O'Connell J, Olsen E, Thiene G, Goodwin J, Gyarfas I, Martin I, Nordet P (1996) Report of the 1995 World Health Organization/International Society and Federation of Cardiology Task Force on the Definition and Classification of cardiomyopathies. Circulation 93:841–842

Sacks SL, Griffiths PD, Corey L, Cohen C, Cunningham A, Dusheiko GM, Self S, Spruance S, Stanberry LR, Wald A, Whitley RJ (2004) Lessons from HIV and hepatitis viruses. Antiviral Res 63 Suppl 1:S11-S18

Schonian U, Crombach M, Maser S, Maisch B (1995) Cytomegalovirus-associated heart muscle disease. Eur Heart J 16 Suppl O:46–49
Schulze K, Witzenbichler B, Christmann C, Schultheiss HP (1999) Disturbance of myocardial energy metabolism in experimental virus myocarditis by antibodies against the adenine nucleotide translocator. Cardiovasc Res 44:91–100
Selinka HC, Wolde A, Pasch A, Klingel K, Schnorr JJ, Kupper JH, Lindberg AM, Kandolf R (2002) Comparative analysis of two coxsackievirus B3 strains: putative influence of virus-receptor interactions on pathogenesis. J Med Virol 67:224–233
Sole MJ, Liu P (1993) Viral myocarditis: a paradigm for understanding the pathogenesis and treatment of dilated cardiomyopathy. J Am Coll Cardiol 22:99A-105A
Staudt A, Mobini R, Fu M, Grosse Y, Stangl V, Stangl K, Thiele A, Baumann G, Felix SB (2001) beta(1)-Adrenoceptor antibodies induce positive inotropic response in isolated cardiomyocytes. Eur J Pharmacol 423:115–119
Staudt A, Bohm M, Knebel F, Grosse Y, Bischoff C, Hummel A, Dahm JB, Borges A, Jochmann N, Wernecke KD, Wallukat G, Baumann G, Felix SB (2002) Potential role of autoantibodies belonging to the immunoglobulin G-3 subclass in cardiac dysfunction among patients with dilated cardiomyopathy. Circulation 106:2448–2453
Staudt Y, Mobini R, Fu M, Felix SB, Kuhn JP, Staudt A (2003) Beta1-adrenoceptor antibodies induce apoptosis in adult isolated cardiomyocytes. Eur J Pharmacol 466:1–6
Tracy S, Wiegand V, McManus B, Gauntt C, Pallansch M, Beck M, Chapman N (1990) Molecular approaches to enteroviral diagnosis in idiopathic cardiomyopathy and myocarditis. J Am Coll Cardiol 15:1688–1694
Wessely R, Henke A, Zell R, Kandolf R, Knowlton KU (1998a) Low-level expression of a mutant coxsackieviral cDNA induces a myocytopathic effect in culture: an approach to the study of enteroviral persistence in cardiac myocytes. Circulation 98:450–457
Wessely R, Klingel K, Santana LF, Dalton N, Hongo M, Jonathan LW, Kandolf R, Knowlton KU (1998b) Transgenic expression of replication-restricted enteroviral genomes in heart muscle induces defective excitation-contraction coupling and dilated cardiomyopathy. J Clin Invest 102:1444–1453
Why HJ, Meany BT, Richardson PJ, Olsen EG, Bowles NE, Cunningham L, Freeke CA, Archard LC (1994) Clinical and prognostic significance of detection of enteroviral RNA in the myocardium of patients with myocarditis or dilated cardiomyopathy. Circulation 89:2582–2589
Woodruff JF (1980) Viral myocarditis. A review. Am J Pathol 101:425–484

4 Invited for Debate: Is Virus Persistence a Determinant for Disease Progression?

A. Keren

Abstract. Evidence-based medicine is not based on assumptions or extrapolations from experimental data in animal models but on solid evidence obtained from randomized clinical trials. There is today experimental and clinical proof of the capability of viruses to invade, persist, and replicate in the myocardium, where they can induce chronic damage. However, the pathogenetic role of viruses detected on endomyocardial biopsies is not clear, and clinical observational data published in the literature on the prognostic implications of this finding are nonuniform and need further elucidation in randomized trials, such as the currently ongoing Betaferon in Chronic Viral Cardiomyopathy (BICC) study. This is the first large randomized trial that evaluates the importance of antiviral therapy in patients with inflammatory cardiomyopathy with presence of viral genome in the myocardium. The results of the BICC trial might supply the needed scientific evidence for the clinical use of a targeted antiviral therapy for prevention of the progression of myocardial dysfunction and reversal of myocardial damage.

4.1 Background

Evidence-based medicine means integrating individual clinical expertise with the best available external clinical evidence from systematic research. It is not based on assumptions or extrapolations from experimental data in animal models but on solid evidence obtained from randomized clinical trials (Sacket et al. 1996).

In myocarditis and inflammatory cardiomyopathy, the mouse represented an important experimental model for understanding the disease process and its course (Kawai 1999). The pathologic process starts with the viremic phase, followed within a few days by initiation of the viral clearing process from the affected myocardium by activation of defense immune mechanisms. About 2 weeks after the initial infection viral clearance is usually achieved. Chronic myocardial damage is induced either by continued immune activity (autoimmunity) despite viral clearance, or with viral persistence despite activation of cellular and humoral defense immune mechanisms (Feldman and McNamara 2000; Kawai 1999; Liu and Mason 2001).

4.2 Viral Presence in the Myocardium

There is today solid experimental and clinical proof of the capability of viruses to invade, persist, and replicate in the myocardium, where they can induce chronic damage (Bowles et al. 1986; Grasso et al. 1992; Kandolf et al. 1987; Pauschinger et al. 1999). Most published studies found a better clinical outcome following myocarditis in patients without, vs. patients with, enteroviral RNA in the myocardium (Fujioka et al. 2000; Why et al. 1994). In addition, it was found that genetic defense mechanisms, like abundance of coxsackie–adenovirus receptors (CAR), are important in susceptibility to viral infection (Noutsias et al. 2001).

In view of these data, the three major questions to be addressed are whether there is currently convincing evidence for:

1. Clinical relevance of the viruses found in endomyocardial biopsies in patients with myocarditis and/or with inflammatory cardiomyopathy
2. Viral presence and adverse outcome in these patients
3. Improved outcome following viral clearance

4.2.1 Clinical Relevance of the Viruses Found

The most commonly described viral genomes in endomyocardial biopsies were enteroviral RNA (mostly coxsackievirus B in adults) and adenoviral DNA in children (Bowles et al. 1986, 2003; Grasso et al. 1992; Kandolf et al. 1987; Pauschinger et al. 1999). Parvovirus (another RNA virus) was not detected in adolescents and adults with myocarditis published by Bowles et al. (2003), but was present in 50% of a recently published series of patients with myocarditis presenting with the clinical picture of acute myocardial infarction (Kühl et al. 2003a). It is not clear whether this discrepancy is due to geographic factors, to epidemiological differences, or a changing profile over time of disease-associated viruses. In addition, parvoviruses are detected with increasing frequency in biopsies of patients with "idiopathic" left ventricular dysfunction. Still, in a recent review, myocarditis is not mentioned among the major manifestations of parvovirus infection (Young and Brown 2004).

Therefore, more research is required to clarify the clinical implications of the different types of viruses found in endomyocardial biopsies and the reasons for the changing type of viruses detected over time (including the intriguing finding of parvovirus B19).

4.2.2 Viral Presence in the Myocardium and Clinical Outcome

Several authors reported a worse outcome in patients with left ventricular dysfunction and the enteroviral genome found in endomyocardial biopsies (Fujioka et al. 2000; Why et al. 1994). However, Figulla et al. (1995) reported different findings (see Fig. 1). Among 20 enterovirus-positive patients compared with 57 enterovirus-negative patients on endomyocardial biopsy, they found significantly better preserved myofibrils, improved left ventricular contractility during follow-up, and better heart transplant-free survival in enterovirus-positive patients compared with those without enteroviral finding in biopsies. Four enterovirus-positive patients whose condition deteriorated were treated with a 6-month course of subcutaneous interferon-α. The treatment induced hemodynamic improvement in all four patients, despite the fact that the therapy eliminated the enteroviral infection only in two.

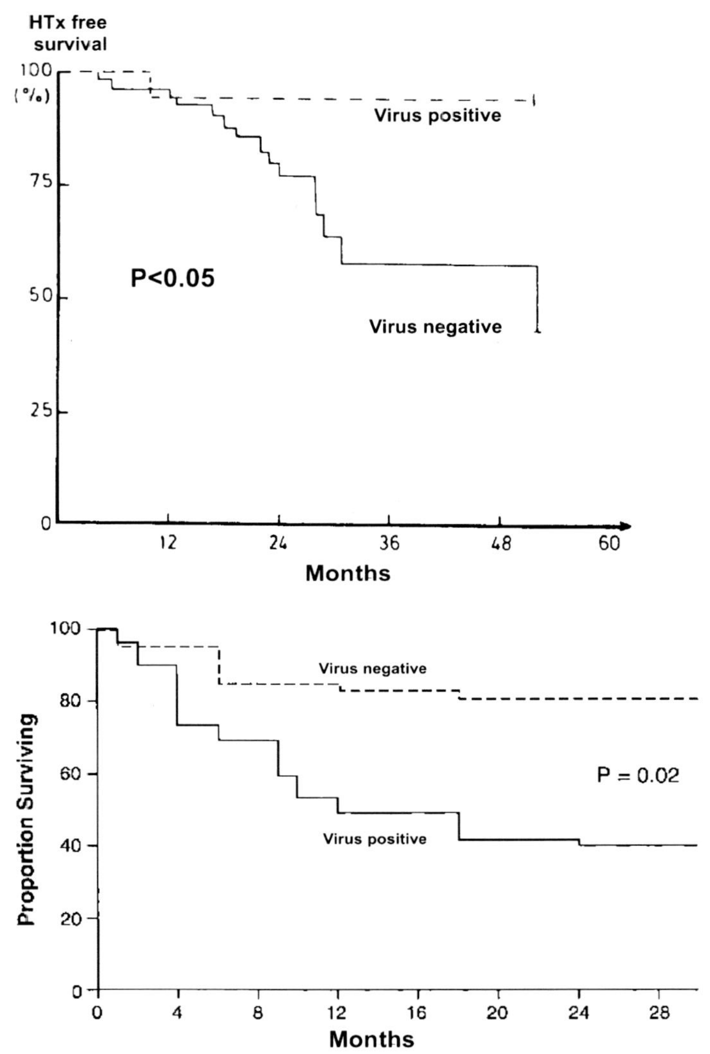

Fig. 1. Differences in the reported prognosis of enterovirus-positive patients on endomyocardial biopsies. *Upper panel*, Figulla et al. (1995); *Lower panel*, Why et al. (1994)

Therefore, despite experimental evidence of myocardial damage induced by viral persistence, clinical observational data published in the literature on the prognostic implication of this finding are non-uniform and need further elucidation in randomized trials.

4.2.3 Is There Proof of Improved Outcome Following Clearance of Viral Genome from the Myocardium?

Kühl et al. (2003b) recently reported the results of a phase II study in which 22 patients with chronic heart failure and PCR-proven myocardial adeno and enteroviral persistence, received interferon-β therapy for 6 months. In all cases, clearance of viral genome was achieved, without major complications. Left ventricular size and contractility significantly improved following therapy.

The positive results of this pilot study are concordant with the results of a retrospective analysis published by Frustaci et al. (2003) in patients with chronic heart failure who received immunosuppressive therapy. Generally, responders to therapy (patients who improved) had cardiac autoantibodies but no viral genomes in their endomyocardial biopsies. Non-responders, on the other hand, had viral persistence and no autoantibodies.

The results of these initial studies suggest that in patients with chronic heart failure, abnormal immune responses and viral persistence should be targeted by specific therapies which may result in reversal of myocardial damage, prevention of progression of myocardial dysfunction, and improvement in prognosis.

The promising methodology of interferon-β therapy in patients with chronic heart failure and viral persistence is currently being evaluated in the first large randomized multicenter study, the BICC trial – Betaferon in Chronic Viral Cardiomyopathy. A positive outcome of that trial would represent a real scientific breakthrough and may provide evidence of the need to treat myocardial viral presence and persistence in patients with inflammatory cardiomyopathy.

References

Bowles NE, Ridchardson PJ, Olsen EG, et al. (1986) Detection of coxsackie-B-virus-specific RNA sequences in myocardial biopsy samples from patients with myocarditis and dilated cardiomyopathy. Lancet 1:1120–1123

Bowles NE, Ni J, Kearnet D, et al. (2003) Detection of viruses in myocardial tissues by polymerase chain reaction. J Am Coll Cardiol 42:466–472

Feldman AM, McNamara D (2000) Myocarditis. N Engl J Med 343:1388–1398

Figulla HR, Stille-Siegener M, Mall G, et al. (1995) Myocardial enterovirus infection with left ventricular dysfunction: a benign disease compared with idiopathic dilated cardiomyopathy. J Am Coll Cardiol 25:1170–1175

Frustaci A, Chimenti C, Calabrese F, et al. (2003) Immunosuppressive therapy for active lymphocytic myocarditis: virologic and immunologic profile of responders versus nonresponders. Circulation 107:857–863

Fujioka S, Kitaura Y, Ukimura A, et al. (2000) Evaluation of viral infection in the myocardium of patients with idiopathic dilated cardiomyopathy. J Am Coll Cardiol 36:1920–1926

Grasso M, Arbustini E, Silini E, et al. (1992) Search for coxsackie virus B3 RNA in idiopathic dilated cardiomyopathy using gene amplification by polymerase chain reaction. Am J Cardiol 69:658–664

Kandolf R, Ameis D, Kirschner P, et al. (1987) In situ detection of enteroviral genomes in myocardial cells: an approach to the diagnosis of viral heart disease. Proc Natl Acad Sci U S A 84:672–676

Kawai C (1999) From myocarditis to cardiomyopathy: mechanisms of inflammation and cell death. Learning from the past for the future. Circulation 99:1091–1100

Kühl U, Pauschinger M, Bock T, et al. (2003a) Parvovirus B19 infection mimicking acute myocardial infarction. Circulation 108:945–950

Kühl U, Pauschinger M, Schwimmbeck PL, et al. (2003b) Interferon-beta treatment eliminates cardiotropic viruses and improves left ventricular function in patients with myocardial persistence of viral genomes and left ventricular dysfunction. Circulation 107:2793–2798

Liu PP, Mason JW (2001) Advances in the understanding of myocarditis. Circulation 104:1076–1082

Noutsias M, Fechner H, de Jongue H, et al. (2001) Human coxsackie-adenovirus receptor is colocalized with integrins alpha-v beta-3 and alpha-v beta-5 on the cardiomyocyte sarcolemma and upregulated in dilated cardiomyopathy. Implications for cardiotropic viral infections. Circulation 104:275–280

Pauschinger M, Doerner A, Kuehl U, et al. (1999) Enteroviral RNA replication in the myocardium of patients with left ventricular dysfunction and clinically suspected myocarditis. Circulation 99:889–895

Sacket DL, Rosenberg WMC, Muir JM, et al. (1996) Evidence based medicine: what is and what it isn't. BMJ 312:71–72
Why HJ, Meany BT, Ridchardson PJ, et al. (1994) Clinical and prognostic significance of detection of enteroviral RNA in the myocardium of patients with myocarditis or dilated cardiomyopathy. Circulation 89:2582–2589
Young NS, Brown KE (2004) Parvovirus B19. N Engl J Med 350:586–597

5 Parvovirus B19: The Causative Agent of Dilated Cardiomyopathy or a Harmless Passenger of the Human Myocard?

S. Modrow

Abstract. Parvovirus B19 infections may cause a widespread benign and self-limiting disease in children and adults known as erythema infectiosum (fifth disease). Several further manifestations are associated with B19 infections, such as arthralgias, arthritis, leucopenia and thrombocytopenia, anaemia and vasculitis and spontaneous abortion and hydrops fetalis in pregnant women. Persistent infections with continuous virus production may occur in immunocompetent as well as in immunosuppressed individuals. Parvovirus B19 infections have been frequently implicated as a cause or trigger of various forms of autoimmune diseases affecting joints, connective tissue and large and small vessels. Autoimmune neutropenia, thrombocytopenia and haemolytic anaemia are known as sequelae of B19 infections. The molecular basis of the autoimmune phenomena is unclear. Many patients with these long-lasting symptoms are not capable of eliminating the virus or controlling its propagation. Furthermore, latent viral genomes have been detected in cells of various organs and tissues by PCR. At present, it is not clear if these cells produce viral proteins and/or infectious B19 particles, if the virus genome can be reactivated to productive replication and if the presence of viral DNA indicates a causative role of parvovirus B19 with distinct diseases.

5.1 Classification, Structure and Organisation

Parvovirus B19 is a member of the genus Erythrovirus within the family Parvoviridae which comprises small (lat. parvus = small), non-enveloped particles with a diameter of 20–28 nm containing a single-stranded linear DNA genome 5.596 nucleotides in length. Erythroviruses display a preference for infection of erythroid cells, and parvovirus B19 is the only member of the Parvoviridae pathogenic for humans. The icosahedral capsid does not display prominent spikes at the three-fold axes (Kaufmann et al. 2004), and consists of two structural proteins, VP1 (83 kDa) and VP2 (58 kDa), which are identical except for 227 amino acids at the amino-terminal of the VP1-protein, the so-called VP1-unique region. Each capsid consists of a total of 60 capsomers: VP2 is the major capsid protein, and constitutes approximately 95% of the total virus particle. Based on antibody-binding analysis and first X-ray structures the VP1-unique region is assumed to be exposed at the surface of the virus particle (Kajigaya et al. 1992; Kawase et al. 1995; Rosenfeld et al. 1992; B. Kaufmann et al., in preparation). The long-lasting neutralising antibody response is mainly directed against the VP1-unique region. A phospholipase A_2-like activity has been linked to the carboxyterminal half (amino acids 130–195) of this capsid protein domain (Zadori et al. 2001; Dorsch et al. 2002). This enzyme activity may contribute to inflammatory processes induced by the production of arachidonic acid, leucotrienes and prostaglandins.

In addition to the structural proteins, the viral genome encodes a number of non-structural proteins. The major non-structural protein, NS1 (77 kDa), is a multifunctional protein. It has been shown to possess site-specific DNA-binding and helicase activities, is functionally active as a transactivator of the viral p6 and various cellular promoters (Gareus et al. 1998; Raab et al. 2001, 2002; Vassias et al. 1998), e.g. the cellular promoters for the expression of tumour necrosis factor (TNF)-α and interleukin (IL)-6 (Fu et al. 2002; Moffatt et al. 1996; Mitchell 2002). Additionally, the NS1 protein contains a well-conserved nucleoside triphosphate-binding motif, which is essential for a variety of biological functions, such as the ATPase activity and the pronounced cytotoxicity of the protein. The cytotoxic NS1-activity was demonstrated to be closely related to cell-cycle arrest (Morita et al. 2003) and apoptosis by a pathway involving caspase 3, whose activation may be a key event during NS1-induced cell death (Moffatt et al. 1996, 1998; Sol et al. 1999). The function of two small non-structural polypeptides of 11 kDa and 7.5 kDa is unknown.

The original nucleotide sequence of the parvovirus B19 genome was obtained from a virus isolate present in the serum of a child suffering from homozygous sickle cell anaemia (Shade et al. 1986). A large number of isolates have been entirely or partly sequenced resulting in the identification of three distinct genotypes of parvovirus B19 (Nguyen et al. 1999; Hokynar et al. 2002; Servant et al. 2002). Parvovirus B19, genotype 1 is the most common genotype in Europe, Asia and the USA, whereas genotype 2 has been reported preferentially in skin biopsies in Scandinavia (Hokynar et al. 2002), occasionally in immunosuppressed patients in Denmark and Germany (Nguyen et al. 2002; Liefeldt et al. 2005) and in coagulation factor concentrates and in few liver explants in Germany (Schneider et al. 2004). Parvovirus B19 genotype 3 has been occasionally observed in patients with transient aplastic anaemia in France (Nguyen et al. 1999; Heegaard et al. 2002a,b). Between the different genotypes, nucleotide variations of up to 11% are observed that are preferentially localised in genomic regions encoding the carboxyterminal part of the NS1-protein and the VP1-unique region (Servant et al. 2002). Despite the fact that nucleotide variations of up to 2% or 3% may be observed when comparing members of the same genotype, the viral genomes show a rather high degree of amino acid sequence conservation (Dorsch et al. 2001; Hemauer et al. 1996; Erdman et al. 1996). The limited number of point mutations resulting in amino acid variations in the VP1-unique region of genotype 1 are not associated with differences in immunological recognition (Dorsch et al. 2001). Furthermore, a high degree of serological cross-reactivity was observed between the VP2-proteins of genotypes

1 and 3 (Heegaard and Brown 2002) and the VP1-unique region of genotypes 1, 2 and 3 (R. Rabl and S. Modrow, unpublished observation).

5.2 Pathogenesis

Parvovirus B19 occurs worldwide but is restricted exclusively to human hosts. The majority of infections occur during childhood and adolescence. Seroprevalence (specific IgG against the capsid proteins) is approximately 2%–15% in children at an age of 1 to 5 years, 30%–40% in adolescents (15 years of age) and 40%–60% in young adults (20 years of age) (Cohen and Buckley 1988; Kelly et al. 2000; Tsujimura et al. 1995). The prevalence of IgG-antibodies reaches maximal levels in the elderly, in which more than 90% are positive.

The pathogenesis of a parvovirus B19 infection is dependent on both viral and host factors, the interplay of which will determine the particular outcome resulting from a particular virus–host interaction. These aspects and the clinical symptoms have been reviewed in detail (Heegaard and Brown 2002; Lehmann and Modrow 2003; Kerr and Modrow 2005). Various studies document a direct pathogenic or 'lytic' effect of parvovirus B19 infection on erythroid cell precursors (Takahashi et al. 1990; Koduri 1998), which are destroyed as a consequence of the infection. The giant pronormoblast (lantern cell) is virtually pathognomic of, but not invariable in, B19 infection (Morey et al. 1992a; Caper and Kurtzman 1996, Infectious parvovirus B19 virus has been demonstrated during the acute phase at various sites in the human body, including peripheral blood, urine, nasopharynx, bone marrow, liver, myocardium, skin and synovium. Cell types infected by B19 virus include erythroblasts, megakaryoblasts, granulocytes, macrophages, follicular dendritic cells, T and B lymphocytes, hepatocytes and endothelial cells (Takahashi et al. 1998; Moffatt et al. 1996; Saal et al. 1992). The viral DNA has been shown to persist in various sites including bone marrow (Kerr et al. 1997), synovium (Kerr et al. 1995; Söderlund et al. 1997), testis (Diss et al. 1999), myocardium (Bültmann et al. 2003b; Pankuweit et al. 2003), liver (Eis-Hübinger et al. 2001) and skin (Nikkari et al. 1999; Hokynar et al. 2002).

The pathogenesis of thrombocytopenia associated with acute B19 infection is thought to be explained by expression of the viral gene encoding the NS1-protein in the absence of replication (Pallier et al. 1997). The cytotoxicity of the NS1 protein was first demonstrated by Ozawa and colleagues (1987) using cells transfected with an NS1 expression plasmid and later by using stable cell lines and inducible expression of the NS1-gene (Moffatt et al. 1998; Sol et al. 1999). The gradual cytocidal effect mediated by the NS1-protein (Ozawa et al. 1986) that is observed during parvovirus B19 infection of erythroid lineage cells is combined with features of apoptosis including marginated chromatin, cytoplasmic vacuolisation and nuclear blebbing (Morey et al. 1993). Activation of caspases 3, 6 and 8 was induced by NS1-protein synthesis in UT7/Epo-cells (Moffatt et al. 1998). NS1 expression results in an increase in sensitivity to apoptosis induced by TNF-α (Sol et al. 1999). Furthermore the viral NS1-protein is a transactivator of both the viral p6-promoter as well as a variety of cellular promoters. These include the promoter region controlling the expression of TNF-α (Fu et al. 2002) and IL-6 genes (Moffatt et al. 1996). Elevated levels of TNF-α have been shown to be present in patients during the acute and convalescent phases of B19 infection (Kerr et al. 2001). In combination with the phospholipase A $(PLA)_2$-like activity of the VP1-unique region, the prolonged or continuous presence of these proinflammatory cytokines during acute-convalescent and persistent B19-infection may contribute to the induction of long-lasting clinical symptoms and autoimmune reactions.

5.3 Clinical Aspects

Parvovirus B19 infection may be asymptomatic in up to 50% of children and 25% of adults (Woolf et al. 1989). However, reticulocytopenia occurs with both symptomatic and asymptomatic acute parvovirus B19 infection. During this phase, reticulocyte numbers drop to undetectable levels for about 7 days, followed by a mild and transient depression of haemoglobin lasting for 3 to 7 days in normal persons (Young 1988).

In common parvovirus B19, infection may result in a flu-like disease. Particularly children may suffer from erythema infectiosum (fifth disease), a self-limiting rash illness (Cherry 1994; Heegaard and Brown 2002). Parvovirus B19 has been further associated with a wide spectrum of diseases (Table 1). Transient anaemia, leucocytopenia or thrombocytopenia may occur without requiring any therapy. However, in some patients, severe thrombocytopenia, pure red cell aplasia or pancytopenia were observed. Besides this, haematological sequelae of acute infection, hepatitis, myocarditis, myositis, acute lung injury and neurological disease may occur occasionally (Bültmann et al. 2003b; Langnas et al. 1995; Bousvaros et al. 1998; Wardeh and Marik 1998; Yoto et al. 2001). In patients with undiagnosed acute meningoencephalitis, parvovirus B19 DNA has been shown to be detectable in 4.3% of the cerebrospinal fluid samples (Barah et al. 2001; Haseyama et al. 1997). In pregnant women, spontaneous abortion and non-immune hydrops fetalis have been reported as clinical manifestations (Rogers et al. 1993; Nyman et al. 2002). In non-immune pregnant women with acute infection, a fetal death rate of about 9% has been found (Yaegashi et al. 1998; Miller et al. 1998). Depending on the haematological status of the host, e.g. patients with sickle cell anaemia or thalassaemia, B19 infection results in aplastic crisis. Persistent parvovirus B19 infection has been reported in patients both with and without underlying immunodeficiencies (Kurtzman et al. 1987; Pont et al. 1992).

Furthermore, parvovirus B19 infection may induce a broad spectrum of autoimmune phenomena. The clinical spectrum ranges from mild arthralgias to severe necrotising vasculitis (Table 2). Parvovirus B19 infection may have an impact on the development of arthritis via several different pathogenic mechanisms. In persons with a genetically determined susceptibility for the development of rheumatic diseases, an acute B19 infection is directly followed by long-lasting arthritis. In these cases the infection is the initial trigger for the establishment of the rheumatic disease. In B19-negative patients with pre-existing arthritis, the clinical status worsens co-incidentally with onset of B19 infection. This phenomenon has not only been observed in patients with idiopathic rheumatic disease but also in arthritis caused by other infectious agents, e.g. *Borrelia* or *Streptococci*. The presence of parvovirus B19 in the peripheral blood or synovial fluid combined with IgM- and/or IgG-antibodies against structural proteins VP1 and/or VP2 has been shown in patients with long-lasting polyarthralgia/polyarthritis and indicates

Table 1. Disease manifestations associated with parvovirus B19 infection

Frequency of association	System	Disease
Common		Non-specific illness
		Erythema infectiosum (EI)
		Transient aplastic crisis (TAC)
		Transient arthropathy
		Fetal death
		Chronic PRCA in immunocompromised persons
Less common	Cutaneous	Henoch-Schonlein purpura
		Papular-purpuric gloves and socks syndrome (PPGSS)
		Gianotti-Crosti syndrome
		Desquamation
		Erythema multiforme
		Livedo reticularis
		Erythema nodosum
	Haematological	Aplastic anaemia
		Thrombocytopenia (including ITP)
		Neutropenia
		Transient erythroblastopenia of childhood (TEC)
		Virus-associated haemophagocytic syndrome (VAHS)
		Acute leukaemia and myelodysplasia (MDS)
		Lymphadenopathy
		Kikuchi's disease (with SLE)
		Hypersplenism
		Congenital red cell aplasia
	Hepatobiliary	Hepatitis
		Acute liver failure
	Cardiovascular	Myocarditis
		Pericarditis
		Vasculitis
		Acute heart failure

Table 1. (continued)

Frequency of association	System	Disease
Less common	Neurological	Meningitis
		Encephalitis
		Guillain-Barré
		Cerebellar ataxia
		Transverse myelitis
		Peripheral neuropathy
		Carpal tunnel syndrome
		Congenital neurological disease
	Rheumatological	Arthritides (including adult, juvenile, RA, JIA)
		Vasculitides (including SLE)
		Chronic fatigue syndrome (CFS)
		Systemic sclerosis (SS)
		Myositis
		Uveitis
	Renal	Glomerulonephritis

an ongoing continuous virus production (Lehmann et al. 2003). The persistent B19 infection is frequently associated with the production of anti-phospholipid antibodies (Loizou et al. 1997; von Landenberg et al. 2003). Some of these cases meet criteria for the diagnosis of rheumatoid arthritis (Naides et al. 1990; Murai et al. 1999; Stahl et al. 2000). However, development of rheumatoid arthritis after acute parvovirus infection appears to be rare (Nikkari et al. 1994; Moore 2000).

Parvovirus B19 infection has been associated with various forms of collagenosis and may mimic systemic lupus erythematosus (SLE) in children and adults (Moore 2000; Nikkari et al. 1995; Negro et al. 2001; Narvaez Garcia et al. 2001; Tovari et al. 2002). Similar to the situation in arthritis patients and also in SLE patients, parvovirus B19 infection has been described both as the agent causing or triggering the rheumatic disease (Cope et al. 1992; Trapani et al. 1999; Hemauer et al. 1999; Hsu and Tsay 2001; Diaz et al. 2002).

Table 2. Autoimmune diseases that are reported in association with parvovirus B19 infection

Involved organs	Disease category	Disease
Joints	Arthralgias Arthritis	Monoarthritis Oligoarthritis Polyarthritis Rheumatoid arthritis? Juvenile idiopathic arthritis
Connective tissue/vessels	Systemic lupus erythematosus (SLE) Vasculitis	Leucoclastic vasculitis Purpura Henoch-Schönlein Papular-purpuric gloves and socks syndrome (PPGSS) Kawasaki disease? Giant cell arteritis (GCA) Polyarteritis nodosa Wegener's granulomatosis
	Dermatomyositis	
Blood cells	Autoimmune neutropenia Autoimmune thrombocytopenia Idiopathic thrombocytopenic purpura (ITP) Autoimmune haemolytic anaemia Virus-associated haemophagocytic syndrome (VAHS)	

Patients with recent B19 infection may develop leucocytoclastic vasculitis (Chakravarty and Merry 1999). Vasculitis-like syndromes in association with a B19 infection are also found in patients with Henoch-Schönlein purpura, papular-purpuric gloves- and socks-syndrome (PPGSS) and Kawasaki disease (Smith et al. 1998; Grilli et al. 1999; Cohen 1994; Cioc et al. 2002).

In patients with giant cell arteritis (GCA), a statistically significant association between histological evidence of GCA and the presence of B19 DNA in temporal artery biopsies has been described (Gabriel et al. 1999). Detection of parvoviral RNA by in situ RT-PCR demonstrated that the endothelial cells and surrounding mononuclear cells were the viral targets in various connective tissue diseases. As revealed by immunohistochemistry, viral protein showed an equivalent histological distribution in these tissue specimens (Magro et al. 2002). Polyarteritis nodosa has been linked to persistent B19 infection and may be cured by immunoglobulin treatment (Corman and Dolson 1992; Viguier et al. 2001). Furthermore, viral genomes have been detected in muscle biopsies of patients with dermatomyositis (Chevrel et al. 2000). However, this seems to be a rare event.

Besides the direct destruction of erythrocytes during acute and persistent B19 infection, autoimmune haemolytic anaemia is also known to be associated with the virus. Autoimmune cytopenias are well known haematological disorders that may affect any bone-marrow cell lineage. In the vast majority of cases only one of the cell lines is affected. However, concurrent autoimmune-mediated destruction of neutrophil granulocytes and thrombocytes due to persistent parvovirus B19 infection is known (Scheurlen et al. 2001). Parvovirus B19 has been identified as a possible trigger in some cases of immune thrombocytopenia (Hanada et al. 1989; Lefrere et al. 1989; Murray et al. 1994; Hida et al. 2000). The results of an examination of the bone marrow of children with neutropenia for B19 DNA indicated that B19 infection may be a common cause of immune-mediated neutropenia in childhood (15 of 19 patients) (McClain et al. 1993).

5.4 Parvovirus B19 and Cardiovascular Disease

In fetal infection associated with hydrops fetalis, parvovirus B19 DNA could be shown to be present in the nuclei of myocytes, indicating a cardiac tropism of the infection (Berry et al. 1992; Morey et al. 1992a,b; Porter et al. 1988; Naides and Weiner 1989). The infection of myocardial cells is thought to contribute to the development of hydrops fetalis. In paediatric cardiac transplant patients, parvovirus B19 infection has been reported to cause generalised disease as well as possible myocarditis (Nour et al. 1993; Janner et al. 1994; Enders et al. 1998; Murry et al. 2001; Dettmeyer et al. 2003). In addition, several groups have described that acute parvovirus B19 infection may be associated with myocarditis both in children and in adults (Lamparter et al. 2003; Munro et al. 2003; Papadogiannakis et al. 2002; Bültmann et al. 2003a). In these cases, high amounts of viral DNA can be detected in the myocard and the peripheral blood, indicating a highly productive acute infection.

During recent years, parvoviral genome sequences have been shown to be present in a high number of myocardial biopsies as well of patients with dilated myocardiopathy, such as in healthy transplant donors and patients undergoing myocard biopsy from other reasons. The frequency of DNA detection ranged from 0% to 40% in controls and from 11% to 42% in the patient groups (Bültmann et al. 2003a; Lotze et al. 2004; Pankuweit et al. 2003; Donoso Manke et al. 2004; Klein et al. 2004). In situ hybridisation revealed that the viral DNA could be detected in endothelial cells. In general the virus load was rather low. These findings may be used as an indication that the genomes of parvovirus B19 may persist in a latency stage in myocardial cells. Similar observations have been made analysing skin, liver, synovial tissue and bone marrow samples. The question as to whether viral proteins are produced in this setting has not yet been answered. Since it is clear that parvovirus B19 infection is common and the development of dilated cardiomyopathy is a rare event, the involvement of important, but as-yet-unknown factors are involved in the causation of parvovirus B19-associated heart disease.

5.5 Conclusion

Parvovirus B19 infection is associated with a board spectrum of symptoms. The acute infection is characterised by the destruction of erythroid precursor cells and high viraemia, up to 10^{13} particles per millilitre of blood. Persistent infections with ongoing virus production are observed in about 20% of the patients, preferentially but not exclusively in immunosuppressed individuals. These persistent infections may induce or trigger various forms of autoimmune reactions resulting in different forms of rheumatic diseases.

Furthermore, the viral genome may be detected in various tissues and organs in individuals with past B19 infection. In general, this stage of latency has not been found to be associated with virus production, viraemia or distinct disease. Therefore, also in symptomatic patients, the mere presence of viral DNA does not prove that the present disease is caused by parvovirus B19. With respect to the association of parvovirus B19 with dilated cardiomyopathy, several open questions remain to be answered:

1. Do the patients produce virus or viral proteins in the myocardial tissue or in the cells of other organs?
2. Do the patients display serological markers for recent or persistent B19 infection?
3. Is dilated cardiomyopathy associated with reactivations of the latent viral genome observed in the patients?
4. Do patients with dilated cardiomyopathy develop autoimmune reactions as a consequence of parvovirus B19 infection?

Until these basic virological problems have been addressed, the causal association between parvovirus B19 infection and the development of dilated cardiomyopathy remains speculative.

References

Barah F, Vallely PJ, Chiswick ML, Cleator GM, Kerr JR (2001) Human parvovirus B19 infection associated with acute meningoencephalitis. Lancet 358:729–730

Berry PJ, Gray ES, Porter HJ, Burton PA (1992) Parvovirus infection of the human fetus and newborn. Semin Diagn Pathol 9:4–12

Bousvaros A, Sundel R, Thorne GM, McIntosh K, Cohen M, Erdman DD, Perez-Atayde A, Finkel TH, Colin AA (1998) Parvovirus B19-associated interstitial lung disease, hepatitis, and myositis. Pediatr Pulmonol 26:365–369

Bültmann B, Klingel K, Sotlar K, Bock CT, Baba HA, Sauter M, Kandolf R (2003b) Fatal parvovirus B19-associated myocarditis clinically mimicking ischemic heart disease: an endothelial cell-mediated disease. Hum Pathol 34:92–95

Bültmann BD, Klingel K, Sotlar K, Bock CF, Kandolf R (2003a) Parvovirus B19: a pathogen responsible for more than hematologic disorders. Virchows Arch 442:8–17

Caper E, Kurtzman GJ (1996) Human parvovirus B19 infection. Curr Opin Hematol 3:111–117

Chakravarty K, Merry P (1999) Systemic vasculitis and atypical infections: report of two cases. Postgrad Med J 75:544–546

Cherry JD (1994) Parvovirus infections in children and adults. Adv Pediatr 46:245–269
Chevrel G, Calvet A, Belin V, Miossec P (2000) Dermatomyositis associated with the presence of parvovirus B19 DNA in muscle. Rheumatology (Oxford) 39:1037–1039
Cioc AM, Sedmak DD, Nuovo GJ, Dawood MR, Smart G, Magro CM (2002) Parvovirus B19 associated adult Henoch Schonlein purpura. J Cutan Pathol 29:602–607
Cohen BJ (1994) Human parvovirus B19 infection in Kawasaki disease. Lancet 344:59
Cohen BJ, Buckley MM (1988) The prevalence of antibody to human parvovirus B19 in England and Wales. J Med Microbiol 25:151–153
Cope AP, Jones A, Brozovic M, Shafi MS, Maini RN (1992) Possible induction of systemic lupus erythematosus by human parvovirus. Ann Rheum Dis 51:803–804
Corman L, Dolson DJ (1992) Polyarteritis nodosa and parvovirus B19 infection. Lancet 339:491
Dettmeyer R, Kandolf R, Banaschak S, Eis-Hübinger AM, Madea B (2003) Fatal parvovirus B19 myocarditis in an 8-year old boy. J Forensic Sci 48:183–186
Diaz F, Collazos J, Mendoza F, de la Viuda JM, Cazallas J, Urkijo JC, Flores M (2002) Systemic lupus erythematosus associated with acute parvovirus B19 infection. Clin Microbiol Infect 8 :115–117
Diss TC, Pan Lx, Du MQ, Peng HZ, Kerr JR (1999) Parvovirus B19 is associated with benign testes as well as testicular germ cell tumours. Mol Pathol 52:349–352
Donoso Mantke O, Nitsche A, Meyer R, Klingel K, Niedrig M (2004) Analysing myocardial tissue from explanted hearts of heart transplant recipients and multi-organ donors for the presence of parvovirus B19 DNA. J Clin Virol 31:32–39
Dorsch S, Kaufmann B, Schaible U, Prohaska E, Wolf H, Modrow S (2001) The VP1-unique region of parvovirus B19: amino acid variability and antigenic stability. J Gen Virol 82 :191–199
Dorsch S, Liebisch G, Kaufmann B, Von Landenberg P, Hoffmann JH, Drobnik W, Modrow S (2002) The VP1-unique region of parvovirus B19 and its constituent phospholipase A2-like activity. J Virol 76:2014–2018
Eis-Hübinger AM, Reber U, Abdul-Nour T, Glatzel U, Lauschke H, Putz U (2001) Evidence for persistence of parvovirus B19 DNA in livers of adults. J Med Virol 65:395–401
Enders G, Dotsch J, Bauer J, Nutzenadel W, Hengel H, Haffner D, Schalasta G, Searle K, Brown KE (1998) Life threatening parvovirus B19 associated myocarditis and cardiac transplantation as possible therapy: two case reports. Clin Infect Dis 26 :355–358

Erdman DD, Durigen EL, Wang QY, Anderson LJ (1996) Genetic diversity of human parvovirus B19: Sequence analysis of the VP1/VP2 gene from multiple isolates. J Gen Virol 77:2767–2774

Fu Y, Ishii KK, Munakata Y, Saitoh T, Kaku M, Sasaki T (2002) Regulation of tumor necrosis factor alpha promoter by human parvovirus B19 NS1 through activation of AP-1 and AP-2. J Virol 76:5395–5403

Gabriel SE, Espy M, Erdman DD, Björnsson J, Smith TF, Hunder GG (1999) Characterization of cis-acting and NS1-responsive elements in the p6 promoter of parvovirus B19. Arthritis Rheum 42:1255–1262

Gareus R, Gigler A, Hemauer A, Leruez-Ville M, Morinet F, Wolf H, Modrow S (1998) Characterization of cis-acting and NS1-responsive elements in the p6 promoter of parvovirus B19. J Virol 72:609–616

Grilli R, Izquirdo MJ, Farina MC, Kutzner H, Gadea I, Martin L, Requena L (1999) Papular-purpuric 'gloves and socks' syndrome: polymerase chain reaction demonstration of parvovirus B19 DNA in cutaneous lesions and sera. J Am Acad Dermatol 41:793–796

Hanada T, Koike K, Hirano C, Takeya T, Suzuki T, Matsunaga Y, Takita H (1989) Childhood transient erythroblastopenia complicated by thrombocytopenia and neutropenia. Eur J Haematol 42:77–80

Haseyama K, Kudoh T, Yoto Y, Suzuki N, Chiba S (1997) Detection of human parvovirus B19 DNA in cerebrospinal fluid. Pediatr Infect Dis J 16 :324–326

Heegaard ED, Brown KE (2002) Human parvovirus B19. Clin Microbiol Rev 15:485–505

Heegaard ED, Qvortrup K, Christensen J (2002a) Baculovirus expression of erythrovirus V9 capsids and screening by ELISA: Serologic cross-reactivity with erythrovirus B19. J Med Virol 66:246–252

Heegaard ED, Petersen BL, Heilmann CJ, Hornsleth A (2002b) Prevalence of parvovirus B19 and parvovirus V9 DNA and antibodies in paired bone marrow and serum samples from healthy individuals. J Clin Microbiol 40:933–936

Hemauer A, von Poblotzki A, Gigler A, Cassinotti P, Siegl G, Wolf H, Modrow S (1996) Sequence variability among different parvovirus B19 isolates. J Gen Virol 77:1781–1785

Hemauer A, Beckenlehner K, Wolf H, Lang B, Modrow S (1999) Acute parvovirus B19 infection in connection with a flare of systemic lupus erythematodes in a female patient. J Clin Virol 14:73–77

Hida M, Shimamura J, Ueno E, Watanabe J (2000) Childhood idiopathic thrombocytopenic purpura associated with human parvovirus B19 infection. Pediatr Int 42:708

Hokynar K, Söderlund-Venermo M, Pesonen M, Ranki A, Kiviluoto O, Partio EK, Hedman K (2002) A new parvovirus genotype persistent in human skin. Virology 302:224–228

Hsu T-C, Tsay GJ (2001) Human parvovirus B19 infection in patients with systemic lupus erythematodes. Rheumatology 40:152–157
Janner D, Bork J, Baum M, Chinnock R (1994) Severe pneumonia after heart transplantation as a result of human parvovirus B19. J Heart Lung Transplant 13:336–338
Kajigaya S, Anderson S, Young NS, Field A, Warrener P, Bansal G, Collett MS (1992) Unique region of the minor capsid protein of human parvovirus B19 is exposed on the virion surface. J Clin Invest 89:2023–2029
Kaufmann B, Simpson AA, Rossmann MG (2004) The structure of human parvovirus B19. Proc Natl Acad Sci U S A 101:11628–11633
Kawase M, Momoeda M, Young NS, Kajigaya S (1995) Most of the VP1-unique region of parvovirus B19 is on the capsid surface. Virology 211:359–366
Kelly HA, Siebert D, Hammond R, Leydon J, Kiely P, Maskill W (2000) The age-specific prevalence of human parvovirus immunity in Victoria, Australia compared with other parts of the world. Epidemiol Infect 124:449–457
Kerr JR, Modrow S (2005) Human and primate parvovirus infections and associated disease. In: Berns K, et al. (eds) Parvoviruses. Arnold publishers, Hodder (in press)
Kerr JR, Cartron JP, Curran MD, Moore JE, Elliott JRM, Mollan RAB (1995) A study of the role of parvovirus B19 in rheumatoid arthritis. Br J Rheumatol 34:809–813
Kerr JR, Kane D, Crowley B, Leonard N, O'Briain S, Coyle PV, Mulcahy F, et al. (1997) Parvovirus B19 infection in AIDS patients. Int J STD AIDS 8:184–186
Kerr JR, Barah F, Mattey DL, Laing I, Hopkins SJ, Hutchinson IV, Tyrrell DA (2001) Circulating tumour necrosis factor-alpha and interferon-gamma are detectable during acute and convalescent parvovirus B19 infection and are associated with prolonged and chronic fatigue. J Gen Virol 82:3011–3019
Klein RM, Jiang H, Niederacher D, Adams O, Du M, Horlitz M, Schley P, Marx R, Lankisch MR, Brehm MU, Strauer BE, Gabbert HE, Scheffold T, Gulker H (2004) Frequency and quantity of the parvovirus B19 genome in endomyocardial biopsies from patients with suspected myocarditis or idiopathic left ventricular dysfunction. Z Kardiol 93:300–309
Koduri PR (1998) Novel cytomorphology of the giant proerythroblasts of parvovirus B19 infection. Am J Hematol 58:95–99
Kurtzman GJ, Ozawa K, Cohen B, Hanson G, Oseas R, Blase RM, Young NS (1987) Parvovirus B19 as a possible causative agent of fulminant liver failure and associated aplastic anaemia. N Engl J Med 317:287–294
Lamparter S, Schoppet M, Pankuweit S, Maisch B (2003) Acute parvovirus B19 infection associated with myocarditis in an immunocompetent adult. Hum Pathol 34:725–728

Langnas AN, Markin RS, Cattral MS, Naides SJ (1995) Parvovirus B19 as a possible causative agent of fulminant liver failure and associated aplastic anaemia. Hepatology 22:1661–1665

Lefrere JJ, Courouce AM, Kaplan C (1989) Parvovirus and idiopathic thrombocytopenic purpura. Lancet 1:279

Lehmann HW, Modrow S (2003) Parvovirus B19 infection – cause or trigger of rheumatic disease. Recent Results Dev Virol 5:197–211

Lehmann HW, Knöll A, Küster RM, Modrow S (2003) Frequent infection with a viral pathogen, parvovirus B19, in rheumatic diseases of childhood. Arthritis Rheum 48:1631–1638

Liefeldt L, Plentz A, Klempa B, Kershaw O, Endres AS, Raab U, Neumayer HH, Meisel H, Modrow S (2005) Recurrent high level parvovirus B19/genotype 2 viremia in a renal transplant recipient analyzed by real-time PCR for simultaneous detection of genotypes 1 to 3. J Med Virol 75:161–169

Loizou S, Cazabon JK, Walport MJ, Tait D, So AK (1997) Similarities of specificity and cofactor dependence in serum antiphospholipid antibodies from patients with human parvovirus B19 infection and from those with systemic lupus erythematosus. Arthritis Rheum 40:103–108

Lotze U, Egerer R, Tresselt C, Gluck B, Dannberg G, Stelzner A, Figulla HR (2004) Frequent detection of parvovirus B19 genome in the myocardium of adult patients with idiopathic dilated cardiomyopathy. Med Microbiol Immunol (Berl) 193:75–82

Magro CM, Crowson AN, Dawood M, Nuovo GJ (2002) Parvoviral infection of endothelial cells and its possible role in vasculitis and autoimmune diseases. J Rheumatol 29 :1227–1235

McClain K, Estrov Z, Chen H, Mahoney DH jr (1993) Chronic neutropenia of childhood: frequent association with parvovirus infection and correlations with bone marrow culture studies. Br J Haematol 85:57

Miller E, Fairley CK, Cohen BJ, Seng C (1998) Immediate and long term outcome of human parvovirus B19 infection in pregnancy. Br J Obstet Gynaecol 105:174–178

Mitchell LA (2002) Parvovirus B19 nonstructural (NS1) protein as a transactivator of interleukIn-6 synthesis: common pathway in inflammatory sequelae of human parvovirus infections? J Med Virol 67:267–274

Moffatt S, Tanaka N, Tada K, Nose M, Nakamura M, Muraoka O (1996) A cytotoxic nonstructural protein, NS1, of human parvovirus B19 induced activation of interleukin 6-gene expression. J Virol 70:8485–8491

Moffatt S, Yaegashi N, Tada K, Tanaka N, Sugamura K (1998) Human parvovirus B19 nonstructural (NS1) protein induces apoptosis in erythroid lineage cells. J Virol 72:3018–3028

Moore TL (2000) Parvovirus-associated arthritis. Curr Opin Rheumatol 12:289–294
Morey AL, Keeling JW, Porter JH, Fleming KA (1992a) Clinical and histopathological features of parvovirus B19 infection in the human fetus. Br J Obstet Gynaecol 99:566–574
Morey AL, Porter HJ, Keeling JW, Fleming KA (1992b) Non-isotopic in situ hybridisation and immunophenotyping of infected cells in the investigation of human fetal parvovirus infection. J Clin Pathol 45:673–678
Morey AL, Ferguson DJ, Fleming KA (1993) Ultrastructural features of fetal erythroid precursors infected with parvovirus B19 in vitro: evidence of cell death by apoptosis. J Pathol 169:213–220
Morita E, Nakashima A, Asao H, Sato H, Sugamura K (2003) Human parvovirus B19 nonstructural protein (NS1) induces cell cycle arrest at G(1) phase. J Virol 77:2915–2921
Munro K, Croxson MC, Thomas S, Wilson NJ (2003) Three cases of myocarditis in childhood associated with human parvovirus (B19 virus). Pediatr Cardiol 24:473–475
Murai C, Munakata Y, Takahashi Y, Ishii T, Shibata S, Muryoi T, Funato T, Nakamura M, Sugamura K, Sasaki T (1999) Rheumatoid arthritis after human parvovirus B19 infection. Ann Rheum Dis 58:130–132
Murray JC, Kelley PK, Hogrefe WR, McClain KL (1994) Childhood idiopathic thrombocytopenic purpura: association with human parvovirus B19 infection. Am J Pediatr Hematol Oncol 16:314–319
Murry CE, Keith RJ, Reichenbach DD (2001) Fatal parvovirus myocarditis in a 5 year old girl. Hum Pathol 32:342–345
Naides SJ, Weiner CP (1989) Antematal diagnosis and palliative treatment of non-immune hydrops fetalis secondary to fetal parvovirus B19 infection. Prenat Diagn 9:105–114
Naides SJ, Scharosch LL, Foto F, Howard EJ (1990) Rheumatologic manifestations of human parvovirus B19 infection in adults. Initial two-year clinical experience. Arthritis Rheum 33:1297–1309
Narvaez Garcia FJ, Domingo-Domenech E, Castro-Bohorquez FJ, Biosca M, Garcia-Quintana A, Perez-Vega C, Vilaseca-Momplet J (2001) Lupus-like presentation of parvovirus B19 infection. Am J Med 111:573–575
Negro A, Regolisti G, Perazzoli F, Coghi P, Tumiati B, Rossi E (2001) Human parvovirus B19 infection mimicking systemic lupus erythematosus in an adult patient. Ann Ital Med Int 16:125–127
Nguyen QT, Sifer C, Schneider V, Allaume X, Servant A, Bernaudin F, Auguste V, Garbarg-Chenon A (1999) Novel human erythrovirus associated with transient aplastic anemia. J Clin Microbiol 37:2483–2487

Nguyen QT, Wong S, Heegaard ED, Brown KE (2002) Identification and characterization of a second novel human erythrovirus variant, A6. Virology 301:374–380

Nikkari S, Luukkainen R, Möttönen T, Meurman O, Hannonen P, Skurnik M, Toivanen P (1994) Does parvovirus B19 have a role in rheumatoid arthritis? Ann Rheum Dis 53:106–111

Nikkari S, Roivainen A, Hannonen P, Mottonen T, Luukkainen R, Yli-Jama T, Toivanen P (1995) Persistence of parvovirus B19 in synovial fluid and bone marrow. Ann Rheum Dis 54:597–600

Nikkari S, Vuorinen T, Kotilainen P, Lammintausta K (1999) Presence of parvovirus B19 DNA in the skin of healthy individuals. Arthritis Rheum 42:S338

Nour B, Green M, Michaels M, Reyes J, Tzakis A, Gartner JC, McLoughlin L, Starzl TE (1993) Parvovirus B19 infection in pediatric transplant patients. Transplantation 56:835–838

Nyman M, Tolfvenstam T, Petersson K, Krassny C, Skjoldebrand-Sparre L, Broliden K (2002) Detection of human parvovirus B19 infection in first-trimester fetal loss. Obstet Gynecol 99:795

Ozawa K, Kurtzman GJ, Young N (1986) Replication of the B19 parvovirus in human bone marrow cultures. Science 233:883–886

Ozawa K, Kurtzman G, Young N (1987) Productive infection by B19 parvovirus of human erythroid bone marrow cells in vitro. Blood 70:384–391

Pallier C, Greco A, Le Junter J, Saib A, Vassias I, Morinet F, et al. (1997) The 3′ untranslated region of the B19 parvovirus capsid protein mRNAs inhibits its own mRNA translation in non-permissive cells. J Virol 71:9482–9489

Pankuweit S, Moll R, Baandrup U, Portig I, Hufnagel G, Maisch B (2003) Prevalence of the parvovirus B19 genome in endomyocardial biopsy specimens. Hum Pathol 34:497–503

Papadogiannakis N, Tolfvenstam T, Fischler B, Norbeck O, Broliden K (2002) Active, fulminant, lethal myocarditis associated with parvovirus B19 infection in an infant. Clin Infect Dis 35 :1027–1031

Pont J, Puchhammer-Stöckl E, Chott A, Popow-Kraupp T, Kienzer I, Postner G, Honetz N (1992) Recurrent granulocytic aplasia as clinical presentation of a persistent parvovirus B19 infection. Br J Haematol 80:160–165

Porter HJ, Khong TY, Evans MF, Chan VT, Fleming KA (1988) Parvovirus as a cause of hydrops fetalis: detection by in situ DNA hybridisation. J Clin Pathol 41 :381–383

Raab U, Bauer B, Gigler A, Beckenlehner K, Wolf H, Modrow S (2001) Cellular transcription factors that interact with the p6 promoter elements of parvovirus B19. J Gen Virol 82:1473–1480

Raab U, Beckenlehner K, Lowin T, Niller HH, Doyle S, Modrow S (2002) NS1-protein of parvovirus B19 interacts directly with DNA sequences of the p6-promoter and with cellular transcription factors Sp1/Sp3. Virology 293:86–93
Rogers BB, Singer DB, Mak SK, Gary GW, Fikrig MK, McMillan PM (1993) Detection of human parvovirus B19 in early spontaneous abortuses using serology, histology, electron microscopy, in situ hybridization, and the polymerase chain reaction. Obstet Gynecol 81:402–408
Rosenfeld SJ, Yoshimoto K, Kajigaya S, Anderson S, Young NS, Field A, Warrener P, Bansal G, Collett MS (1992) Unique region of the minor capsid protein of human parvovirus B19 is exposed on the virion surface. J Clin Invest 89:2023–2029
Saal JG, Stendle M, Einsele H, Muller CA, Fritz P, Zacher J (1992) Persistence of B19 parvovirus in synovial membranes of patients with rheumatoid arthritis. Rheumatology 12:147–151
Scheurlen W, Ramasubbu K, Wachowski O, Hemauer A, Modrow S (2001) Chronic autoimmune thrombocytopenia/neutropenia in a boy with persistent parvovirus B19 infection. J Clin Virol 20:173–178
Schneider B, Becker M, Brackmann HH, Eis-Hübinger AM (2004) Contamination of coagulation factor concentrates with human parvovirus B19 genotype 1 and 2. Thromb Haemost 2004 92:838–845
Servant A, Laperche S, Lallemand F, Marinho V, De Saint Maur G, Meritet JF, Garbarg-Chenon A (2002) Genetic diversity within human erythroviruses: identification of three genotypes. J Virol 76:9124–9134
Shade RO, Blundell MC, Cotmore SF, Tattersall P, Astell CR (1986) Nucleotide sequence and genome organisation of human parvovirus B19 isolated from the serum of a child during aplastic crisis. J Virol 58:921–936
Smith PT, Landry ML, Carey H, Krasnoff J, Cooney E (1998) Papular-purpuric gloves and socks syndrome associated with acute parvovirus B19 infection: case report and review. Clin Infect Dis 27:164–168
Söderlund M, von Essen R, Haapasaari J, Kiistala U, Kiviluoto O, Hedman K (1997) Persistence of parvovirus B19 DNA in synovial membranes of young patients with and without chronic arthropathy. Lancet 349:1063–1065
Sol N, Le Junter J, Vassias I, Freyssinier JM, Thomas A, Prigent AF, Morinet F (1999) Possible interaction between the NS-1 protein and tumor necrosis factor alpha pathways in erythroid cell apoptosis induced by human parvovirus B19. J Virol 73:8762–8770
Stahl HD, Pfeiffer R, Von Salis-Soglio G, Emmrich F (2000) Parovirus B19-associated mono- and oligoarticular arthritis may evolve into a chronic inflammatory arthropathy fulfilling criteria for rheumatoid arthritis or spondylarthropathy. Clin Rheumatol 19:510–511

Takahashi T, Ozawa K, Takahashi K, Asano S, Takaku F (1990) Susceptibility of human erythropoietic cells to B19 parvovirus in vitro increases with differentiation. Blood 75:603–610
Takahashi Y, Murai C, Shibata S, Munakata Y, Ishii T, Ishii K, Saitoh T, Sawai T, Sugamura K, Sasaki T (1998) Human parvovirus as a causative agent for rheumatoid arthritis. Proc Natl Acad Sci USA 95:8227–8232
Tovari E, Mezey I, Hedman K, Czirjak L (2002) Self limiting lupus-like symptoms in patients with parvovirus B19 infection. Ann Rheum Dis 61:662–663
Trapani S, Ermini M, Falcini F (1999) Human parvovirus B19 infection: its relationship with systemic lupus erythematosus. Semin Arthritis Rheum 28:319–325
Tsujimura M, Matsushita K, Shiraki H, Sato K, Okochi K, Maeda Y (1995) Human parvovirus B19 infection in blood donors. Vox Sang 69:206–212
Vassias I, Hazan U, Michel Y, Sawa C, Handa H, Gouya L, Morinet F (1998) Regulation of human B19 parvovirus promoter expression by hGABP (E4TF1) transcription factor. J Biol Chem 273:8287–8293
Viguier M, Guillevin L, Laroche L (2001) Treatment of parvovirus B19-associated polyarteritis nodosa with intravenous immune globulin. N Engl J Med 344:1481–1482
Von Landenberg P, Lehmann HW, Knoll A, Dorsch S, Modrow S (2003) Antiphospholipid antibodies in pediatric and adult patients with rheumatic disease are associated with parvovirus B19 infection. Arthritis Rheum 48:1939–1947
Wardeh A, Marik P (1998) Acute lung injury due to parvovirus pneumonia. J Intern Med 244:257–260
Woolf AD, Campion GV, Chisbick A, Wise S, Cohen BJ, Klouda PT, Caul O, Dieppe PA (1989) Clinical manifestation of human parvovirus B19 infection in adults. Arch Intern Med 149:1153–1156
Yaegashi N, Niinuma T, Chisaka H, Watanabe T, Uehara S, Okamura K, Moffat S, Sugamura K, Yajima A (1998) The incidence of, and factors leading to, parvovirus B19-related hydrops fetalis following maternal infection; report of 10 cases and meta-analysis. J Infect 37:28–35
Yoto Y, Kudoh T, Haseyama K, Tsutsumi H (2001) Human parvovirus B19 and meningoencephalitis. Lancet 358:2168
Young N (1988) Hematological and hematopeitic consequences of parvovirus B19 infection. Semin Hematol 25:159–172
Zadori Z, Szelei J, Lacoste MC, Li Y, Gariepy S, Raymond P, Alliare IR, Tijssen P (2001) A viral phospholipase A2 is required for parvovirus infectivity. Dev Cell 1:291–302

6 Parvovirus B19: A New Emerging Pathogenic Agent of Inflammatory Cardiomyopathy

C.-T. Bock

Abstract. The human parvovirus B19 (PVB19), an erythrovirus causing diverse clinical manifestations ranging from asymptomatic or mild to more severe outcomes such as hydrops fetalis, is the only currently known human pathogenic parvovirus. Recently, PVB19 has been identified as a causative agent of pediatric and adult inflammatory cardiac diseases. The first hints for a possible etiopathogenetic role of the PVB19 infection and the development of cardiac dysfunction were demonstrated by molecular biology methods such as in situ hybridization (ISH) and polymerase chain reaction (PCR). In this regard, PVB19-associated inflammatory cardiomyopathy is characterized by infection of endothelial cells of small intracardiac arterioles and venules, which may be associated with endothelial dysfunction, impairment of myocardial microcirculation, and penetration of inflammatory cells in the myocardium.

6.1 Introduction

Although enteroviruses (EV) have long been considered the most common cause of inflammatory cardiomyopathy, human parvovirus B19 (PVB19) is emerging as a new important candidate. In this regard, recent studies have indicated an association of PVB19 with acute myocarditis in children and adults. Whether or not PVB19 has an impact on ongoing inflammatory cardiomyopathy in adult patients is still unclear. Using in situ hybridization (ISH) techniques, we have recently demonstrated that endothelial cells, but not cardiac myocytes, are the PVB19-specific target cells in PVB19-associated myocarditis. Furthermore, PVB19 genomes were found in patients with unexplained isolated diastolic dysfunction. A strong association of PVB19 infection with the incidence of endothelial dysfunction appears to be obvious and consistent with the hypothesis that PVB19-induced endothelial dysfunction may be a possible pathomechanism underlying diastolic dysfunction.

6.2 Classification, Structure, and Organization of PVB19

Yvonne Cossart, an Australian virologist working in London in the mid-1970s, discovered PVB19 while investigating laboratory assays for hepatitis B. The peculiar name of this virus was derived from the blood bank code of one of the donors whose serum was number 19 in panel B (Cossart et al. 1975).

The human parvovirus B19 is a member of the erythroviruses within the family of Parvoviridae (Heegaard and Brown 2002). The classification of the family Parvoviridae relies on morphology and functional characteristics. Parvoviruses are among the smallest DNA-containing viruses able to infect mammalian cells (for review see Berns 2001).

Parvovirus B19 is a non-enveloped virus of about 22–24 nm in diameter. The morphology of PVB19 comprises an icosahedral symmetry.

Using electron microscopic techniques, DNA-containing capsids as well as empty capsids are detectable in the sera of infected patients (Heegaard and Brown 2002). The linear single-stranded DNA genome of PVB19 contains about 5,600 nucleotides (nt), composed of an internal coding region flanked by terminal repeat sequences of 383 nt that are essential for viral replication (Morinet 1992; Zhi et al. 2004).

The viral genome contains two large open reading frames (ORFs). The first ORF is located at the 5′-half of the genome and encodes the non-structural protein NS1 with a molecular mass of 77 kDa. The NS1 protein represents site-specific DNA-binding, and is associated with transcriptional and helicase activities (Cotmore et al. 1986; Doerig et al. 1987; Heegaard and Brown 2002). The NS1 protein acts as a transactivator on its own P6 promoter as well as on cellular [e.g., interleuking (IL)-6] and viral [e.g. human immunodeficiency virus (HIV)] promoters (Mitchell 2002; Sol et al. 1993). It has been demonstrated that the cytotoxicity of PVB19 is closely related to cellular interference mechanisms of NS1 with factors (e.g., p21/WAF, caspase 3) of cellular signaling pathways leading to apoptosis of the host cell (Hsu et al. 2004; Moffatt et al. 1998; Sol et al. 1999).

The second ORF is located in the 3′-half of the PVB19 genome and encodes the major VP1 and minor VP2 structural capsid proteins (Cotmore et al. 1986). VP1 consists of a unique sequence of 227 amino acids (VP1u; 84 kDa) and is followed by the entire VP2 sequence (554 amino acids; 58 kDa). Both structural proteins have also a variety of other functions that are important for the viral life cycle, including the induction of cytokines. Noteworthy is that the VP1 protein comprises a viral phospholipase A2 activity which is an intermediate during the synthesis of eicosanoids, prostaglandins, and leukotrienes that play an important role in inflammatory reactions, but may also lead to host cell dysfunction (Dorsch et al. 2002). Three additional ORFs encoding small proteins (7.5, 9, and 11 kDa) with as-yet-unknown functions have also been described (Heegaard and Brown 2002; St Amand and Astell 1993).

Transcription produces at least nine overlapping mRNA transcripts initiated by the single PVB19 P6 promoter (Deiss et al. 1990; Luo and Astell 1993; Ozawa et al. 1988). Transcription studies of permissive and non-permissive cell lines with plasmids containing PVB19 genomes suggest that, in non-permissive cells, there may be a block in full-length transcription leading to an overexpression of the cytotoxic NS1 protein and reduced or lack of expression of VP1 and VP2 proteins (Liu et al. 1992). These data suggested that the NS1 protein is responsible for the death of erythroid progenitors and endothelial cells by apoptosis, and, additionally, some cells, such as megakaryocytes, may be lysed by restricted expression of viral proteins in the absence of complete viral propagation.

The genetic diversity among PVB19 isolates has been reported to be very low, with less than 1% to 2% nucleotide divergence in the whole genome, although full-length sequences are available only for a limited number of isolates (Hicks et al. 1996; Hokynar et al. 2002; Nguyen et al. 1999; Shade et al. 1986). However, some isolates obtained from patients with persistent PVB19 infection have been reported to exhibit a higher degree of variability in some parts of the genome, with the VP1 region being the most variable at both the DNA and protein levels with up to 4% and 8% divergence, respectively (Hemauer et al. 1996). Three erythrovirus genotypes (genotype 1, 2 and 3) have been described so far based on restriction and sequence analysis of the PVB19 genome (Mori et al. 1987; Morinet et al. 1986; Nguyen et al. 1999; Umene and Nunoue 1991). PVB19 (isolate PVBaua) represents the prototype virus of erythrovirus genotype 1. The isolate LaLi representing the genotype 2 was first detected to be persistently present in human skin (Hokynar et al. 2002), and the isolate V9 (or D91.1), which was associated with transient aplastic anemia, is the prototype virus of genotype 3 (Nguyen et al. 1999).

6.3 The PVB19 Life Cycle

The PVB19 life cycle includes binding of the virus to the host cell receptor, penetration of permissive cells by endocytosis, DNA replication, RNA transcription, translation of the viral proteins, assembly of virions and finally cell lysis with the release of infectious virions (for review see Heegaard and Brown 2002; Berns 2001).

Globoside, a neutral glycolipid that acts as a cellular receptor, accounts for the tropism of the virus for erythroid cells (Kolmos 1994). The presence of globoside on the cell surface has been demonstrated for erythroid precursor cells, red blood cells, megakaryocytes, and endothelial cells (Brown et al. 1993). Globoside is also known as erythrocyte P antigen. Rare persons of the P phenotype blood group, whose erythrocytes lack P antigen, are not susceptible to infection with PVB19 (Brown et al. 1994a). These patients have no serologic evidence of prior infection, and their marrow erythroid progenitors proliferate normally in the presence of high concentrations of the virus. Recent studies have shown that the erythrocyte P antigen is necessary but not sufficient for successful infection of human hematopoietic cells (Weigel-Kelley et al. 2001). These studies further suggest the existence of a putative cellular coreceptor for efficient entry of parvovirus B19 into human cells. The erythroid progenitor cell line K562 becomes adherent and permissive for PVB19 entry after induction of $\alpha_5\beta_1$-integrin (Weigel-Kelley et al. 2003). In this regard a cell line permissive for PVB19 infection must express the P antigen and the co-receptor $\alpha_5\beta_1$-integrin. Important to note is that cardiac endothelial cells, which are putative target cells for PVB infection, comprise both the P antigen and $\alpha_5\beta_1$-integrin (Szekanecz et al. 1992; Urbich et al. 2002; Weigel-Kelley et al. 2003).

6.4 PVB19 Infection Is Associated with Various Diseases

The prevalence of specific immunoglobulin G (IgG) antibodies is approximately 15% in young children, 60% in adults, and more than 85% in those 70 years or older (Brown et al. 1994b). PVB19 entered the medical curriculum as an agent of human disease when its association with erythema infectiosum (fifth disease), hydrops fetalis, and transient aplastic anemia was demonstrated in the 1980s (Anderson et al. 1983, 1985; Kariyawasam et al. 2000; Pamidi et al. 2000; Pattison 1987). During the last few years, a growing number of reports have been published demonstrating an association between PVB19 and many other clinical diseases (Koch 2001; Lehmann et al. 2003; Pattison 1988; Trapani et al. 1999), like arthritis (Moore 2000; Stahl et al. 2000; Takahashi et al. 1998), myocarditis (Brown et al. 1994a; Enders et al. 1998; Malm et al. 1993; Murry et al. 2001; Nigro et al. 2000; Orth et al. 1997; Schowengerdt et al. 1997), various vasculitic syndromes (Corman and Dolson 1992; Dingli et al. 2000; Finkel et al. 1994; Trapani et al. 1999), hepatitis

(Drago et al. 1999; Hillingso et al. 1998; Karetnyi et al. 1999; Sokal et al. 1998; Yoto et al. 1996), and neurological disorders (Barah et al. 2001; Yoto et al. 1994). A growing number of reports have suggested an association between PVB19 infection with acute and chronic cardiac diseases (Bultmann et al. 2003a; Enders et al. 1998; Malm et al. 1993; Murry et al. 2001; Nigro et al. 2000; Orth et al. 1997; Salimans et al. 1989; Schowengerdt et al. 1997).

6.5 PVB19: A Cardiac Pathogenic Agent in Inflammatory Cardiomyopathy

Fetal myocardial cells express the cellular receptor (P antigen) for PVB19 (Brown et al. 1993). However, P antigen has not been demonstrated on cardiac myocytes of older children or adults. In addition, given that PVB19 requires the host S-phase enzymes for viral replication, cytotoxic virus replication seems unlikely to occur in post-partum myocytes, because these cells are withdrawn from the cell cycle around the time of birth (Murry et al. 2001).

The first hints for a possible etiopathogenetic role of the PVB19 infection and the development of cardiac dysfunctions were demonstrated by ISH experiments (Bultmann et al. 2003a,b; Kandolf 2004; Klingel et al. 2004). Radioactive ISH detected viral genomes in endothelial cells (ECs) of the myocardium predominantly in the venular compartment and in small arteries and arterioles, but not in cardiac myocytes or other myocardial tissue components in patients with acute inflammatory cardiomyopathy.

Concomitant with EC infection, marked expression of the adhesion molecule E-selectin was noted, accompanied by margination, adherence, penetration, and perivascular infiltration of macrophages and T lymphocytes. Due to the high viral load in cardiac ECs, which was confirmed by application of quantitative real-time PCR, PVB19 infection of endothelial cells appears to be capable of inducing impaired coronary microcirculation with secondary cardiac myocyte necrosis (Bultmann et al. 2003b; Kandolf 2004).

In contrast to acute PVB19-associated inflammatory cardiomyopathy, virus persistence in chronic disease is predominantly observed in endothelial cells of the close-meshed capillary system of the myocardium in association with a chronic type of inflammation predominated by macrophages (Klingel et al. 2004). It is important to note that persistent PVB19 infection of endothelial cells is only of etiopathogenic significance in the presence of immunohistologically proven chronic inflammation. By contrast, the latent type of PVB19 persistence as observed in non-inflammatory myocardial tissue does not explain myocardial dysfunction.

Additionally, regarding the pathogenesis, high cytokine levels (IFN-γ, TNF-α, IL-6, and IL-8) have been described in PVB19-associated acute and chronic myocarditis and were made responsible for induction of myocardial necrosis and inflammation (Nigro et al. 2000).

From the presence of PVB19 genomes in the cardiac allografts without concomitant viremia, it was concluded that PVB19 can persist in cardiac tissue (Schowengerdt et al. 1997). Recently, PVB19 persistence in consecutive endomyocardial biopsies (EMBs) for more than 3 years has been described in a young immunocompetent girl clinically and histologically characterized by ongoing myocarditis (Nigro et al. 2000). Preliminary retrospective analysis of EMBs disclosed PVB19 persistence with low viral load in consecutive EMBs of eight immunocompetent patients suffering from chronic myocarditis or dilated cardiomyopathy over a period of 6–18 month (Bultmann et al. 2003b).

6.6 PVB19 Infection and Isolated Endothelial and Diastolic Dysfunction: Molecular Diagnosis of Myocardial PVB19 Infection

Besides acute and chronic forms of inflammatory cardiomyopathy, a recent collaborative clinical study revealed an association of PVB19 with isolated endothelial and diastolic dysfunction (Tschöpe et al. 2005). EMBs obtained from a series of patients presenting with isolated endothelial and diastolic dysfunction were analyzed for the presence of PVB19 genomes. As a result, in 84% of these patients, PVB19 genomes were detectable by molecular methods (see also Fig. 1). The qualitative detection of PVB19 DNA isolated from endomyocardial biopsies of 14 representative patients was performed by nested PCR (nPCR)

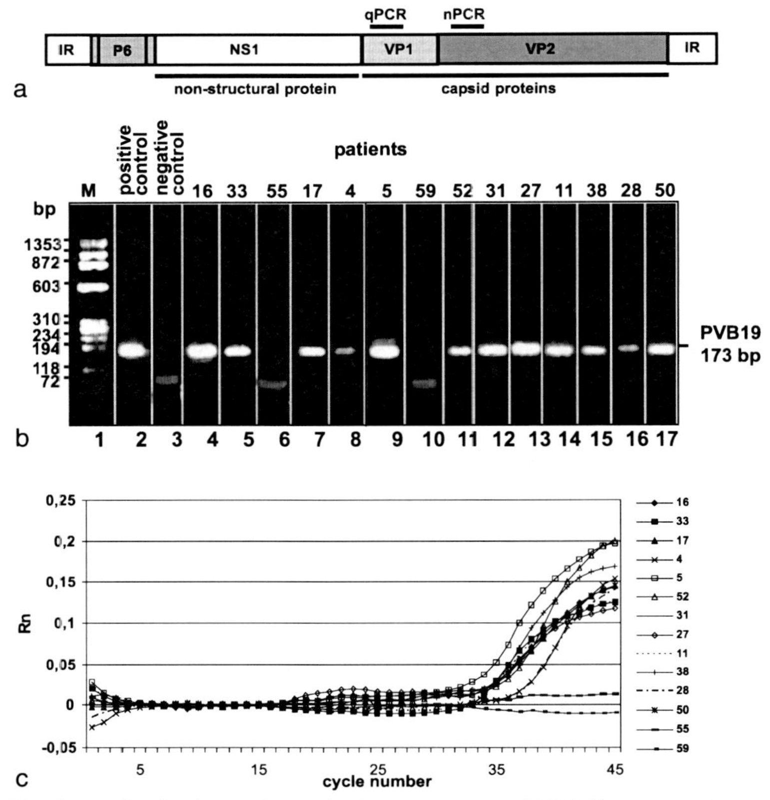

Fig. 1a–c. Qualitative and quantitative assessment of PVB19 genomes in endomyocardial biopsies. **a** Schematic presentation of the PVB19 genome. The positions of the nested PCR (*nPCR*) and quantitative PCR (*qPCR*) amplicons are indicated. *IR*, inverted repeat; *P6*, P6 promoter; note that the promoter region overlaps with the IR-and NS1 region; *NS1*, non-structural protein 1; *VP*, capsid protein **b** Qualitative detection of PVB19 DNA isolated from endomyocardial biopsies of 14 representative patients. PCR products were separated on 1.8% agarose gels and visualized by ethidium bromide staining. A PhiX174 BsuRI DNA marker is seen in *lane 1*. **c** Quantitative real-time PCR (*qPCR*) amplification plot of PVB19 DNA loads in endomyocardial biopsies of the representative patients (depicted at the *right*). The fluorescence reporter signal (*Rn*) emitted by the TaqMan probe was measured and plotted against cycle number

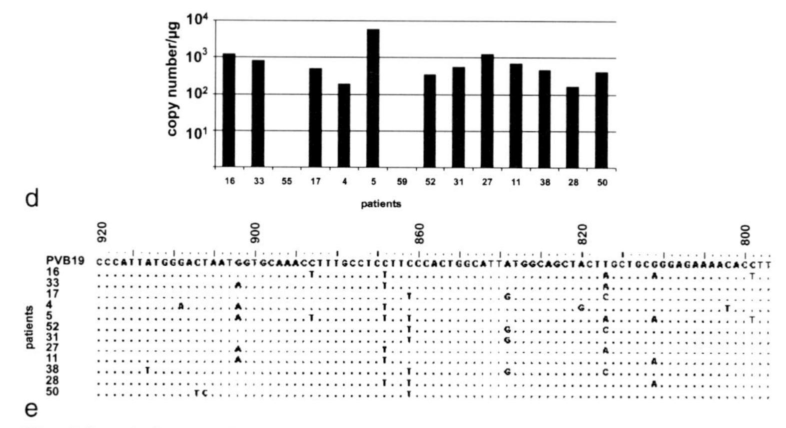

Fig. 1d,e. d Graph of the PVB19 DNA copy numbers per microgram of isolated nucleic acid calculated from the respective amplification plot. **e** Alignment of PVB19 sequences of the virus-positive patients with reference PVB19 sequences (accession No. U38509). Sequence homologies are denoted as *dots*; point mutations are shown in *letters*. *Ruler on top* was numbered according to the VP1 sequences (VP1 ATG = 1, minus strand)

using PVB19-VP2-specific primers generating a 173-bp PVB19-VP2 fragment (Fig. 1a and b). PVB19-specific DNA fragments are shown for 12 PVB19-positive and 2 PVB19-negative patients (Fig. 1b; lanes 4 to 17). Note that no PVB19 DNA was amplified from patients 55 and 59 (Fig. 1b; lanes 6 and 10). PVB19 DNA loads in endomyocardial biopsies of the 14 representative patients are demonstrated by quantitative real-time PCR amplification (qPCR; TaqMan PCR; Fig. 1c and d). The amplification plot of the quantitative real-time PCR (Fig. 1c) and the calculation of viral load (Fig. 1d) demonstrated low viral titers of myocardial PVB19 genomes of approximately 10^2-10^3 PVB19 genome equivalents/μg isolated nucleic acid, which is consistent with a persistent type of infection. The sequences of the amplified PVB19 VP2 regions of virus-positive patients were aligned with reference PVB19 sequences (accession No. U38509) and confirmed the positive PVB19 PCR results (Fig. 1e). Sequence analyses of the PVB19 amplicons revealed randomly distributed point mutations in the VP1/2 region of the PVB19 isolates. With respect to quality control of the PCR assays, it is noteworthy that the PVB19 isolates showed sequence differences when compared to each other. Therefore, cross-contaminations of the patient's samples with PVB19 as may occur during nPCR procedures can be excluded. Within the scope of the sequence analysis of the PVB19 genomes, the hypothesis arises whether it is likely that different PVB19 genotypes and virus variants may account for the highly variable courses of myocarditis with respect to endothelial and myocardial dysfunction. In addition, it has to be determined whether or not PVB19 variants (genotypes or mutants) show different susceptibility to therapeutic treatments, especially with respect to interferon sensitivity or resistance.

6.7 Conclusion

PVB19 infection of the cardiac endothelium may cause endothelial dysfunction in the absence of coronary artery disease. Endothelial dysfunction appears to reach a very high level during acute infection of susceptible patients, mimicking the clinical phenotype of myocardial infarction (Bultmann et al. 2003a; Kuhl et al. 2003). In addition, our data suggest a possible pathogenic role of PVB19 in a considerable proportion of patients with progressive ventricular dysfunction as well as with isolated diastolic dysfunction. As to whether different cardiotropic viral genotypes in addition to host- specific factors may explain the variable clinical phenotypes remains to be elucidated. So far, a strong association of PVB19 with the incidence of endothelial dysfunction is obviously suggesting PVB19-induced impairment of the coronary microcirculation in acute and chronic cardiac disease. PVB19 should therefore be considered as a potential pathogenic factor in patients with inflammatory cardiomyopathy.

References

Anderson MJ, Jones SE, Fisher-Hoch SP, Lewis E, Hall SM, Bartlett CL, Cohen BJ, Mortimer PP, Pereira MS (1983) Human parvovirus, the cause of erythema infectiosum (fifth disease)? Lancet 1:1378

Anderson MJ, Higgins PG, Davis LR, Willman JS, Jones SE, Kidd IM, Pattison JR, Tyrrell DA (1985) Experimental parvoviral infection in humans. J Infect Dis 152:257–265
Barah F, Vallely PJ, Chiswick ML, Cleator GM, Kerr JR (2001) Association of human parvovirus B19 infection with acute meningoencephalitis. Lancet 358:729–730
Berns KI (2001) Parvoviridae: the viruses and their replication. In: Fields BN, Knipe PM, Howley PM, et al (eds) Fields virology, 3rd edn. Lippincott-Raven, Philadelphia, pp 2173–2197
Brown KE, Anderson SM, Young NS (1993) Erythrocyte P antigen: cellular receptor for B19 parvovirus. Science 262:114–117
Brown KE, Hibbs JR, Gallinella G, Anderson SM, Lehman ED, McCarthy P, Young NS (1994a) Resistance to parvovirus B19 infection due to lack of virus receptor (erythrocyte P antigen). N Engl J Med 330:1192–1196
Brown KE, Young NS, Liu JM (1994b) Molecular, cellular and clinical aspects of parvovirus B19 infection. Crit Rev Oncol Hematol 16:1–31
Bultmann BD, Klingel K, Sotlar K, Bock CT, Baba HA, Sauter M, Kandolf R (2003a) Fatal parvovirus B19-associated myocarditis clinically mimicking ischemic heart disease: an endothelial cell-mediated disease. Hum Pathol 34:92–95
Bultmann BD, Klingel K, Sotlar K, Bock CT, Kandolf R (2003b) Parvovirus B19: a pathogen responsible for more than hematologic disorders. Virchows Arch 442:8–17
Corman LC, Dolson DJ (1992) Polyarteritis nodosa and parvovirus B19 infection. Lancet 339:491
Cossart YE, Field AM, Cant B, Widdows D (1975) Parvovirus-like particles in human sera. Lancet 1:72–73
Cotmore SF, McKie VC, Anderson LJ, Astell CR, Tattersall P (1986) Identification of the major structural and nonstructural proteins encoded by human parvovirus B19 and mapping of their genes by procaryotic expression of isolated genomic fragments. J Virol 60:548–557
Deiss V, Tratschin JD, Weitz M, Siegl G (1990) Cloning of the human parvovirus B19 genome and structural analysis of its palindromic termini. Virology 175:247–254
Dingli D, Pfizenmaier DH, Arromdee E, Wennberg P, Spittell PC, Chang-Miller A, Clarke BL (2000) Severe digital arterial occlusive disease and acute parvovirus B19 infection. Lancet 356:312–314
Doerig C, Beard P, Hirt B (1987) A transcriptional promoter of the human parvovirus B19 active in vitro and in vivo. Virology 157:539–542

Dorsch S, Liebisch G, Kaufmann B, Von LP, Hoffmann JH, Drobnik W, Modrow S (2002) The VP1 unique region of parvovirus B19 and its constituent phospholipase A2-like activity. J Virol 76:2014–2018

Drago F, Semino M, Rampini P, Rebora A (1999) Parvovirus B19 infection associated with acute hepatitis and a purpuric exanthem. Br J Dermatol 141:160–161

Enders G, Dotsch J, Bauer J, Nutzenadel W, Hengel H, Haffner D, Schalasta G, Searle K, Brown KE (1998) Life-threatening parvovirus B19-associated myocarditis and cardiac transplantation as possible therapy: two case reports. Clin Infect Dis 26:355–358

Finkel TH, Torok TJ, Ferguson PJ, Durigon EL, Zaki SR, Leung DY, Harbeck RJ, Gelfand EW, Saulsbury FT, Hollister JR (1994) Chronic parvovirus B19 infection and systemic necrotising vasculitis: opportunistic infection or aetiological agent? Lancet 343:1255–1258

Heegaard ED, Brown KE (2002) Human parvovirus B19. Clin Microbiol Rev 15:485–505

Hemauer A, von PA, Gigler A, Cassinotti P, Siegl G, Wolf H, Modrow S (1996) Sequence variability among different parvovirus B19 isolates. J Gen Virol 77:1781–1785

Hicks KE, Cubel RC, Cohen BJ, Clewley JP (1996) Sequence analysis of a parvovirus B19 isolate and baculovirus expression of the non-structural protein. Arch Virol 141:1319–1327

Hillingso JG, Jensen IP, Tom-Petersen L (1998) Parvovirus B19 and acute hepatitis in adults. Lancet 351:955–956

Hokynar K, Soderlund-Venermo M, Pesonen M, Ranki A, Kiviluoto O, Partio EK, Hedman K (2002) A new parvovirus genotype persistent in human skin. Virology 302:224–228

Hsu TC, Wu WJ, Chen MC, Tsay GJ (2004) Human parvovirus B19 nonstructural protein (NS1) induces apoptosis through mitochondria cell death pathway in COS-7 cells. Scand J Infect Dis 36:570–577

Kandolf R (2004) [Virus etiology of inflammatory cardiomyopathy]. Dtsch Med Wochenschr 129:2187–2192

Karetnyi YV, Beck PR, Markin RS, Langnas AN, Naides SJ (1999) Human parvovirus B19 infection in acute fulminant liver failure. Arch Virol 144:1713–1724

Kariyawasam HH, Gyi KM, Hodson ME, Cohen BJ (2000) Anaemia in lung transplant patient caused by parvovirus B19. Thorax 55:619–620

Klingel K, Sauter M, Bock CT, Szalay G, Schnorr JJ, Kandolf R (2004) Molecular pathology of inflammatory cardiomyopathy. Med Microbiol Immunol (Berl) 193:101–107

Koch WC (2001) Fifth (human parvovirus) and sixth (herpesvirus 6) diseases. Curr Opin Infect Dis 14:343–356

Kolmos HJ (1994) [Blood group P antigen is the cellular receptor for human parvovirus B 19]. Ugeskr Laeger 156:6227

Kuhl U, Pauschinger M, Bock T, Klingel K, Schwimmbeck CP, Seeberg B, Krautwurm L, Poller W, Schultheiss HP, Kandolf R (2003) Parvovirus B19 infection mimicking acute myocardial infarction. Circulation 108:945–950

Lehmann HW, Knoll A, Kuster RM, Modrow S (2003) Frequent infection with a viral pathogen, parvovirus B19, in rheumatic diseases of childhood. Arthritis Rheum 48:1631–1638

Liu JM, Green SW, Shimada T, Young NS (1992) A block in full-length transcript maturation in cells nonpermissive for B19 parvovirus. J Virol 66:4686–4692

Luo W, Astell CR (1993) A novel protein encoded by small RNAs of parvovirus B19. Virology 195:448–455

Malm C, Fridell E, Jansson K (1993) Heart failure after parvovirus B19 infection. Lancet 341:1408–1409

Mitchell LA (2002) Parvovirus B19 nonstructural (NS1) protein as a transactivator of interleukin-6 synthesis: common pathway in inflammatory sequelae of human parvovirus infections? J Med Virol 67:267–274

Moffatt S, Yaegashi N, Tada K, Tanaka N, Sugamura K (1998) Human parvovirus B19 nonstructural (NS1) protein induces apoptosis in erythroid lineage cells. J Virol 72:3018–3028

Moore TL (2000) Parvovirus-associated arthritis. Curr Opin Rheumatol 12:289–294

Mori J, Beattie P, Melton DW, Cohen BJ, Clewley JP (1987) Structure and mapping of the DNA of human parvovirus B19. J Gen Virol 68:2797–2806

Morinet F (1992) [Parvovirus B19 infection]. Pathol Biol (Paris) 40:621–622

Morinet F, Tratschin JD, Perol Y, Siegl G (1986) Comparison of 17 isolates of the human parvovirus B 19 by restriction enzyme analysis. Brief report. Arch Virol 90:165–172

Murry CE, Jerome KR, Reichenbach DD (2001) Fatal parvovirus myocarditis in a 5-year-old girl. Hum Pathol 32:342–345

Nguyen QT, Sifer C, Schneider V, Allaume X, Servant A, Bernaudin F, Auguste V, Garbarg-Chenon A (1999) Novel human erythrovirus associated with transient aplastic anemia. J Clin Microbiol 37:2483–2487

Nigro G, Bastianon V, Colloridi V, Ventriglia F, Gallo P, D'Amati G, Koch WC, Adler SP (2000) Human parvovirus B19 infection in infancy associated with acute and chronic lymphocytic myocarditis and high cytokine levels: report of 3 cases and review. Clin Infect Dis 31:65–69

Orth T, Herr W, Spahn T, Voigtlander T, Michel D, Mertens T, Mayet WJ, Dippold W, Meyer zum Buschenfelde KH (1997) Human parvovirus B19 infection associated with severe acute perimyocarditis in a 34-year-old man. Eur Heart J 18:524–525

Ozawa K, Ayub J, Young N (1988) Functional mapping of the genome of the B19 (human) parvovirus by in vitro translation after negative hybrid selection. J Virol 62:2508–2511

Pamidi S, Friedman K, Kampalath B, Eshoa C, Hariharan S (2000) Human parvovirus B19 infection presenting as persistent anemia in renal transplant recipients. Transplantation 69:2666–2669

Pattison JR (1987) B19 virus–a pathogenic human parvovirus. Blood Rev 1:58–64

Pattison JR (1988) Diseases caused by the human parvovirus B19. Arch Dis Child 63:1426–1427

Salimans MM, van de Rijke FM, Raap AK, van Elsacker-Niele AM (1989) Detection of parvovirus B19 DNA in fetal tissues by in situ hybridisation and polymerase chain reaction. J Clin Pathol 42:525–530

Schowengerdt KO, Ni J, Denfield SW, Gajarski RJ, Bowles NE, Rosenthal G, Kearney DL, Price JK, Rogers BB, Schauer GM, Chinnock RE, Towbin JA (1997) Association of parvovirus B19 genome in children with myocarditis and cardiac allograft rejection: diagnosis using the polymerase chain reaction. Circulation 96:3549–3554

Shade RO, Blundell MC, Cotmore SF, Tattersall P, Astell CR (1986) Nucleotide sequence and genome organization of human parvovirus B19 isolated from the serum of a child during aplastic crisis. J Virol 58:921–936

Sokal EM, Melchior M, Cornu C, Vandenbroucke AT, Buts JP, Cohen BJ, Burtonboy G (1998) Acute parvovirus B19 infection associated with fulminant hepatitis of favourable prognosis in young children. Lancet 352:1739–1741

Sol N, Morinet F, Alizon M, Hazan U (1993) Trans-activation of the long terminal repeat of human immunodeficiency virus type 1 by the parvovirus B19 NS1 gene product. J Gen Virol 74:2011–2014

Sol N, Le JJ, Vassias I, Freyssinier JM, Thomas A, Prigent AF, Rudkin BB, Fichelson S, Morinet F (1999) Possible interactions between the NS-1 protein and tumor necrosis factor alpha pathways in erythroid cell apoptosis induced by human parvovirus B19. J Virol 73:8762–8770

St Amand J, Astell CR (1993) Identification and characterization of a family of 11-kDa proteins encoded by the human parvovirus B19. Virology 192:121–131

Stahl HD, Seidl B, Hubner B, Altrichter S, Pfeiffer R, Pustowoit B, Liebert UG, Hofmann J, von Salis-Soglio G, Emmrich F (2000) High incidence of parvovirus B19 DNA in synovial tissue of patients with undifferentiated mono- and oligoarthritis. Clin Rheumatol 19:281–286

Szekanecz Z, Humphries MJ, Ager A (1992) Lymphocyte adhesion to high endothelium is mediated by two beta 1 integrin receptors for fibronectin, alpha 4 beta 1 and alpha 5 beta 1. J Cell Sci 101:885–894
Takahashi Y, Murai C, Shibata S, Munakata Y, Ishii T, Ishii K, Saitoh T, Sawai T, Sugamura K, Sasaki T (1998) Human parvovirus B19 as a causative agent for rheumatoid arthritis. Proc Natl Acad Sci U S A 95:8227–8232
Trapani S, Ermini M, Falcini F (1999) Human parvovirus B19 infection: its relationship with systemic lupus erythematosus. Semin Arthritis Rheum 28:319–325
Tschöpe C, Bock CT, Kasner M, Noutsias M, Westermann D, Schwimmbeck PL, Pauschinger M, Poller WC, Kuhl U, Kandolf R, Schultheiss HP (2005) High prevalence of cardiac parvovirus B19 infection in patients with isolated left ventricular diastolic dysfunction. Circulation 111:879–886
Umene K, Nunoue T (1991) Genetic diversity of human parvovirus B19 determined using a set of restriction endonucleases recognizing four or five base pairs and partial nucleotide sequencing: use of sequence variability in virus classification. J Gen Virol 72:1997–2001
Urbich C, Dernbach E, Reissner A, Vasa M, Zeiher AM, Dimmeler S (2002) Shear stress-induced endothelial cell migration involves integrin signaling via the fibronectin receptor subunits alpha(5) and beta(1). Arterioscler Thromb Vasc Biol 22:69–75
Weigel-Kelley KA, Yoder MC, Srivastava A (2001) Recombinant human parvovirus B19 vectors: erythrocyte P antigen is necessary but not sufficient for successful transduction of human hematopoietic cells. J Virol 75:4110–4116
Weigel-Kelley KA, Yoder MC, Srivastava A (2003) Alpha5beta1 integrin as a cellular coreceptor for human parvovirus B19: requirement of functional activation of beta1 integrin for viral entry. Blood 102:3927–3933
Yoto Y, Kudoh T, Asanuma H, Numazaki K, Tsutsumi Y, Nakata S, Chiba S (1994) Transient disturbance of consciousness and hepatic dysfunction associated with human parvovirus B19 infection. Lancet 344:624–625
Yoto Y, Kudoh T, Haseyama K, Suzuki N, Chiba S (1996) Human parvovirus B19 infection associated with acute hepatitis. Lancet 347:868–869
Zhi N, Zadori Z, Brown KE, Tijssen P (2004) Construction and sequencing of an infectious clone of the human parvovirus B19. Virology 318:142–152

7 Role of Hepatitis C Virus in Cardiomyopathies

A. Matsumori

Abstract. Virus infection was conventionally considered to cause myocarditis, which resulted in development of dilated cardiomyopathy. Recent studies suggest that hepatitis C virus (HCV) is involved in the development of dilated cardiomyopathy, hypertrophic cardiomyopathy and arrhythmogenic right ventricular cardiomyopathy in addition to myocarditis. Furthermore, left ventricular aneurysm represents the same morbid state not only after myocardial infarction but also after myocarditis. There were wide variations in the frequency of detection of HCV genomes in cardiomyopathies in different regions or in different populations. Major histocompatibility complex class II genes may play a role in the susceptibility to HCV infection, and may influence the development of different phenotypes of cardiomyopathies. If it is the fact that the myocardial damage is caused by HCV, it might be expected that interferon (IFN) treatment would be useful for its treatment. Patients receiving IFN treatment of hepatitis were screened by thallium myocardial scintigraphy, and an abnormality was discovered in half of patients. Treatment with IFN resulted in disappearance of the image abnormality. It has thus been suggested that mild myocarditis and myocardial damage may be cured with IFN. We have recently found that high concentrations of circulating cardiac troponin T are a specific marker of cardiac involvement in HCV infection. By measuring cardiac troponin T in patients with HCV infection, the prevalence of cardiac involvement in hepatitis C virus infection will be clarified. We are proposing a collaborative work on global network on myocarditis/cardiomyopathies due to HCV infection.

7.1 Introduction

Cardiomyopathies may present as idiopathic dilated, hypertrophic or restrictive disease, arrhythmogenic right ventricular cardiomyopathy or several other distinct disorders of the heart muscle (Richardson et al. 1996). Dilated cardiomyopathy, hypertrophic cardiomyopathy and restrictive cardiomyopathy are heterogeneous myocardial disorders of multifactorial aetiologies, including genetic anomalies and acquired immune pathogenetic factors, such as viral infections (Matsumori 1997). Dilated cardiomyopathy is a relatively common myocardial disorder, which may lead to severe heart failure. Along with ischaemic heart disease, it represents the main antecedent of heart transplantation in Western countries, where epidemiological studies performed a decade ago have measured 5-year survival rates as low as 30% to 40% after its initial diagnosis. In contrast, few large-scale studies have been conducted to examine the prevalence, prognosis and management patterns of cardiomyopathies in Asian populations.

Recently, nationwide clinico-epidemiological surveys of cardiomyopathies were performed in Japan (Miura et al. 2002; Matsumori et al. 2002). The total number of patients was estimated at 17,700 (prevalence; 14.0 per 100,000) for dilated cardiomyopathy, 21,900 (17.3 per 100,000) for hypertrophic cardiomyopathy, 300 (0.2 per 100,000) for restrictive cardiomyopathy and 520 (0.4 per 100,000) for arrhythmogenic right ventricular cardiomyopathy. The prevalence of dilated cardiomyopathy and hypertrophic cardiomyopathy was higher in men than women: the men-to-women ratios were 2.6 and 2.3 for dilated cardiomyopathy and hypertrophic cardiomyopathy, respectively. The occurrence of cardiomyopathies was most frequent in the age range of 60–69 years both in dilated cardiomyopathy and in hypertrophic cardiomyopathy.

7.2 Role of Viruses in the Pathogenesis of Cardiomyopathies

The myocardium is involved in a wide range of viral infections. In some cases, myocarditis may be the primary disorder; in others, it may occur as part of a systemic disease. Myocarditis is thought to be most commonly caused by enteroviruses, particularly coxsackievirus B. However, in many cases, when myocarditis has been diagnosed on the basis of clinical characteristics, no definite confirmation of viral origin is obtained, despite extensive laboratory investigations. The evidence is often only circumstantial and a direct, conclusive proof of cardiac involvement is not available (Kawai et al. 1987; Abelmann and Lorell 1989; Olinde and O'Connell 1994). However, accumulating evidence links viral myocarditis with the eventual development of dilated cardiomyopathy (Johnson and Palacios 1982; Matsumori and Kawai 1982a,b; Caforio et al. 1990; Matsumori 1993; Feldman and McNamara 2000; Liu and Mason 2001).

The clinical presentation of viral myocarditis is variable. When myocardial necrosis occurs diffusely, congestive heart failure develops and, later, dilated cardiomyopathy. If myocardial lesions are localized, a ventricular aneurysm may form. When complicated with arrhythmias, myocarditis presents as arrhythmogenic right ventricular cardiomyopathy (Matsumori 1993). When myocardial necrosis is localized to the subendocardium, restrictive cardiomyopathy may develop. While it has not been established that hypertrophic cardiomyopathy may be a complication of viral myocarditis, asymmetrical septal hypertrophy has, in fact, sometimes been observed in patients with myocarditis (Fig. 1; Kawano et al. 1994).

The myocardium may be the target of several types of viral infections. Recently, the importance of hepatitis C virus (HCV) has been noted in patients with hypertrophic cardiomyopathy, dilated cardiomyopathy and myocarditis and myocarditis (Matsumori et al. 1995, 1996, 1998a,b,

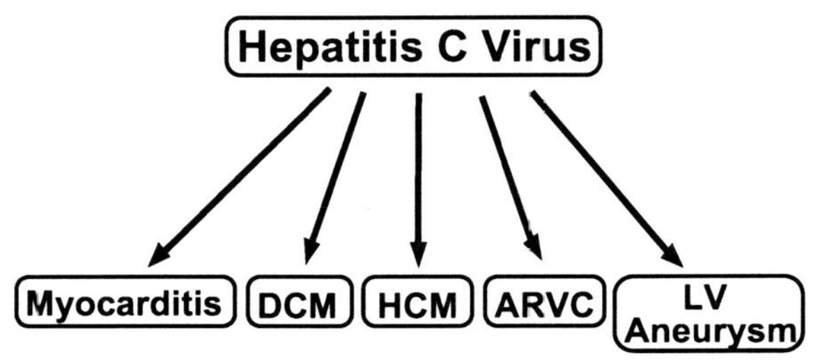

Fig. 1. Hepatitis C virus causes various heart diseases. *ARVC*, arrhythmogenic right ventricular cardiomyopathy; *DCM*, dilated cardiomyopathy; *HCM*, hypertrophic cardiomyopathy; *LV*, left ventricle

1999, 2000; Matsumori 1997, 2001, 2003, 2005; Okabe et al. 1997; Takeda et al. 1999; Ooyake et al. 1999; Sato et al. 2000; Nakamura et al. 2000). In a collaborative research project of the Committees for the Study of Idiopathic Cardiomyopathy in Japan, HCV antibody was found in 74 of 697 patients (10.6%) with hypertrophic cardiomyopathy and in 42 of 663 patients (6.3%) with dilated cardiomyopathy; this prevalence in patients with cardiomyopathies was significantly higher than in age-matched volunteer blood donors in Japan (2.4%).

The global prevalence of HCV carriers is estimated to average 3%, ranging from 0.1% to 10% or more in different countries (Cohen 1999). In Europe, the overall prevalence is 1% with a north-south gradient, ranging from 0.5% in northern countries to 2% in Mediterranean countries. Recent studies have shown high prevalence in Eastern Europe, ranging from 0.7% to 5%. There are 170 million chronic HCV carriers throughout the world, of whom an estimated 2 million are in Japan, 2.7 million in the United State and 5 million in Western Europe.

Until relatively recently, blood transfusion posed a major risk of HCV infection in developed countries. The introduction in 1989 and 1992 of improved blood-screening tests by the detection of anti-HCV infection (Beld et al. 2000) changed this.

7.3 HCV Infection and Cardiomyopathies

We first evaluated 31 patients with cardiomyopathy and myocarditis by polymerase chain reaction (PCR) for the presence of RNA viruses such as enterovirus, cardiovirus, hepatitis A virus, human immunodeficiency viruses (HIV)1 and 2, human T lymphocytic leukaemia virus (HTLV) I, influenza A and B viruses, and reovirus. We also evaluated patients with cardiomyopathy and myocarditis for DNA viruses such as adenovirus, cytomegalovirus, Epstein-Barr virus, hepatitis B virus, human herpesvirus 6, varicella-zoster virus, and herpes simplex virus types 1 and 2. However, enterovirus RNA was detected in only one (3.2%) patient with dilated cardiomyopathy, and no other virus genomes were found. On the other hand, we found HCV RNA in six patients (19.4%) with dilated cardiomyopathy (Matsumori et al. 1995; Table 1).

7.3.1 HCV Infection and Dilated Cardiomyopathy

Over a 10-year period, we identified 19 of 191 patients (9.9%) with dilated cardiomyopathy who had evidence of HCV infection on the basis of a positive immunoradiometric assay, whereas only 1 of 40 patients (2.5%) of those with ischaemic heart disease was positive for the HCV antibody (Matsumori 2005). Since the prevalence of positive HCV antibody in voluntary blood donors in Japan was 2.4% in subjects 55 to

Table 1. PCR analysis for RNA viruses in the hearts of patients with myocarditis and cardiomyopathy

Virus	Positive (n)/total $n = 31$ (%)
Cardiovirus	0
Enterovirus (coxsackievirus B)	1 (3.2%)
Hepatitis A	0
Hepatitis C	6 (19.4%)
HIV 1 and 2	0
HTLV-1	0
Influenza A	0
Influenza B	0
Reovirus	0

59 years of age, the difference was statistically significant (Table 2). None of the patients with HCV antibody had a history of intravenous drug use. Mildly elevated levels of serum aminotransferase were found in some of the patients who had blood transfusions. Three patients had a history of hepatitis, and mildly elevated serum aminotransferase was measured in 10 patients. The primary findings at presentation were heart failure and cardiac arrhythmias. Of the 19 patients with HCV antibodies, 10 patients had HCV RNA in the serum, and all 6 patients had type 1b HCV (Table 3).

Quantitative analysis of HCV RNA showed that the copy number in the serum was 8×10^2 to 1.2×10^6 copies/ml. HCV RNA was found in the heart of 8 patients. Negative strands of HCV RNA were detected in the heart of 2 patients. Because negative RNA molecules are considered to be intermediates in the replication of the HCV genome, it is supposed that HCV replicates in myocardial tissues.

Table 2. Frequency of HCV antibody in patients with cardiomyopathies at Kyoto University

Positive (*n*)		Total (*n*)	Frequency
Dilated cardiomyopathy	19	191*	9.9%
Hypertrophic cardiomyopathy	16	113*	14.1%
ARVC	1	2	50.0%

ARVC, arrhythmogenic right ventricular cardiomyopathy
*$p < 0.0001$ vs volunteer blood donors; 2.4% (24 of 1,039)

Table 3. Clinical features of dilated cardiomyopathy associated with HCV infection

Cases	$n = 19$
Age, sex	62 ± 15yo (17–83) M:10, F:9
PH	Hepatitis 3
	Elevated aminotransferase 10
HCV type	1b, 7; 2, 3
Titre	8×10^2–1.2×10^6 copies/ml
HCV RNA in the heart	(+) Strands 8/13, (–) strands 2/8

7.3.2 HCV Infection and Hypertrophic Cardiomyopathy

Over a 10-year period, 16 of 113 patients (14.1%) with hypertrophic cardiomyopathy were identified who had evidence of HCV infection on the basis of positive HCV antibody (Table 2). When compared with the prevalence of positive HCV antibody in voluntary blood donors, the difference was statistically significant. Of these 16 patients, none of the patients had a family history of hypertrophic cardiomyopathy. Seven patients had hepatoma, 4 patients had blood transfusions, and mildly elevated serum aminotransferases were measured in 10 patients. Nine patients had ace of spade-shaped deformities of the left ventricle with a ratio of apical thickness to middle anterior free wall thickness exceeding 1.3, and were diagnosed as apical hypertrophic cardiomyopathy (Matsumori 2005). None had angiographically visible coronary artery disease.

Histopathological studies showed mild to severe degrees of myocyte hypertrophy in the right or left ventricle, and mild to moderate fibrosis, and mild cellular infiltration was seen. Type 1b HCV RNA was detected in the serum of 7 patients. Quantitative analysis of HCV RNA showed that the copy number in the serum was 5.5×10^3 to 4×10^6 copies/ml. HCV RNA was found in the biopsy specimens of 6 patients. Negative strands of hepatitis C virus RNA were found in the hearts of 2 patients (Table 4). Analysis by fluorescent single-stand conformation

Table 4. Clinical features of hypertrophic cardiomyopathy associated with HCV infection

Cases	$n = 16$
Age, sex	65 ± 10yo (50–81) M:9, F:7
PH	Hepatoma 7, blood transfusion 4
	Elevated aminotransferase 10
Type	ASH 9, APH 5, obstructive 3
HCV type	1b, 7; 2, 3
Titre	5.5×10^3–4×10^6 copies/ml
HCV RNA in the heart	(+) Strands 6/8, (–) strands 2/6

APH, apical hypertrophy; ASH, asymmetrical septal hypertrophy; PH, past history

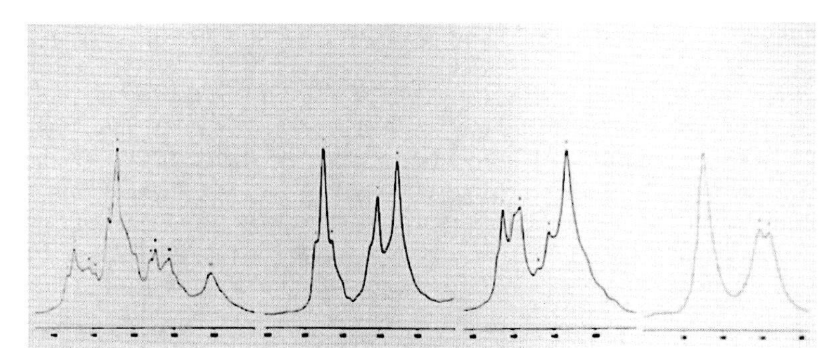

Fig. 2. Analysis by fluorescent single-stranded conformation polymorphism of sera from 4 patients with cardiomyopathy. The presence of multiple clones in the sera is shown

polymorphism showed the presence of multiple clones in the sera of patients with hypertrophic cardiomyopathy (Fig. 2).

Recently, Teragaki and co-workers studied the prevalence of HCV infection in 80 Japanese patients with hypertrophic cardiomyopathy, and found that 22.5% of the patients were positive for HCV antibody. When compared with controls, the difference was statistically significant. In their study, 7 patients had HCV type 1b, and 5 patients had type 2a (Teragaki et al. 2003). The prevalence of HCV infection in their study was more frequent than that of our study.

7.3.3 HCV Infection and Heart Diseases: A Multicentre Study in Japan

As shown above, hypertrophic and dilated cardiomyopathies were both associated with significantly higher prevalences of positive antibodies than was measured among blood donors. In addition, positive HCV antibody was more prevalent in patients with hypertrophic cardiomyopathy than in those with dilated cardiomyopathy. Positive HCV antibody was detected in 650 of 11,967 patients (5.4%) seeking care at five university hospitals, a significantly higher prevalence than in volunteer blood donors. Of the cardiac abnormalities observed in these patients with positive HCV antibody, arrhythmias were the most frequent (21.5%).

Electrocardiographic abnormalities were found in 130 of 349 patients tested (62.8%), most often in the form of arrhythmias or conduction disturbances (Matsumori et al. 1998a). Echocardiographic examination suggested that HCV infection was associated with left ventricular hypertrophy in over one half of the patients, ventricular dilatation in 40% and decreased left ventricular systolic function in 34%.

The study suggests that several cardiac abnormalities other than cardiomyopathic disorders (e.g. arrhythmias) may result from HCV infection, which may be a risk factor for such conditions (hypertension, myocardial infarction, etc.; Table 5), although further study is necessary to confirm these associations. More recently, a possible role of HCV infection in the pathogenesis of atherosclerosis has been reported (Ishizaka et al. 2002).

Table 5. Clinical diagnoses of patients with HCV antibody

Clinical diagnoses	n (total n = 349)	Frequency (%)
Arrhythmia	75	21.5
Hypertension	71	20.3
Myocardial infarction	57	16.3
Diabetes mellitus	50	14.3
Angina pectoris	45	12.9
Renal disease	41	11.7
Valvular heart disease	33	9.5
Congestive heart failure	28	8.0
Hypertrophic cardiomyopathy	23	6.6
Post valvular replacement and/or CABG	22	6.3
Dilated cardiomyopathy	20	5.7
Cerebrovascular disease	7	2.0
Unclassified cardiomyopathy	2	0.6
Myocarditis	1	0.3

CABG, coronary artery bypass grafting
(Reproduced from Matsumori et al. 1998a with permission of Japan Circulation Society)

7.3.4 HCV Genomes from Paraffin Heart Sections

A collaborative multicentre study was performed by the Scientific Council on Cardiomyopathies of the World Heart Federation to test the reproducibility of detection of viral genomes, such as enteroviruses, adenovirus, cytomegalovirus and HCV in formalin-fixed tissues. In this study, autopsy and biopsy materials were analysed blindly. We found HCV genomes in 2 out of 11 hearts (18%) of patients with dilated cardiomyopathy and myocarditis from Italy, and in 4 out of 11 hearts (36%) from the United States, two of which were from patients with myocarditis, and the other two from patients with arrhythmogenic right ventricular cardiomyopathy. The results suggest that HCV may cause arrhythmogenic right ventricular cardiomyopathy as well as myocarditis, dilated cardiomyopathy and hypertrophic cardiomyopathy. As the detection of HCV genomes in formalin-fixed sections seems less sensitive than in frozen sections, HCV infection may actually be a more prevalent cause of myocardial injury.

In a collaborative research project with the National Cardiovascular Center and Juntendo University, we have tried detecting HCV genomes in paraffin sections of autopsied hearts. Among 106 hearts examined, β-actin gene was amplified in 61 (52.6%). Among these, HCV RNA was detected in 13 (21.3%), and negative strands in 4 hearts (6.6%). HCV RNA was found in 6 hearts (26.0%) with hypertrophic cardiomyopathy, 3 hearts (11.5%) with dilated cardiomyopathy and 4 hearts (33.3%) with myocarditis (Table 6). These HCV RNA-positive samples were obtained between 1979 and 1990, indicating that HCV RNA can be amplified from paraffin-embedded hearts preserved for many years (Matsumori et al. 2000).

We also analysed autopsied hearts with dilated cardiomyopathy from the University of Utah as a collaborative research project, and found HCV RNA in 8 of 23 hearts (35%) with positive actin genes (Table 7). The sequences of HCV genomes recovered from these hearts were highly homologous to the standard strain of HCV (Fig. 3). These observations lend support to the previous findings of an important role played by the HCV in the pathogenesis of hypertrophic cardiomyopathy and dilated cardiomyopathy. However, there were wide variations in the frequency of detection of HCV genomes in cardiomyopathy among different cities.

Table 6. Detection of hepatitis C virus genomes from the autopsied hearts of patients with myocarditis, and dilated and hypertrophic cardiomyopathies with positive-actin gene

Diagnosis	Positive, *n*	Frequency, %	*p* (Fisher)
Myocarditis	4/12	33.3%	0.0008
DCM	3/26	11.5%	0.034
HCM	6/23	26.0%	0.0005
Myocarditis+DCM+HCM	13/61	21.3%	0.0002
Controls*	0/52	0%	–

HCM, hypertrophic cardiomyopathy; IHD, ischemic heart disease *IHD $n = 32$, Non-cardiac death $n = 20$

(Reproduced from Matsumori et al. 2000, with permission of Nature Publishing Group)

Table 7. Detection of HCV genomes from the formalin-fixed paraffin sections of autopsied DCM hearts

	HCV genomes (+)	Total (%) (*n*)	HCV genomes (+)	Positive (%) Actin (*n*)
University of Utah	18/72	(25%)	8/23	(34.8%)
LDS Hospital, Utah	0/31	(0%)	0/12	(0%)
St. Paul's Hospital, Vancouver	0/24	(0%)	0/11	(0%)
Japan	5/50	(10%)	3/26	(11.5%)

HCV genomes were detected in none of 24 hearts from St. Paul's Hospital in Vancouver, Canada (Matsumori 2005). These results suggest that the frequency of cardiomyopathy caused by HCV infection may be different in different regions or in different populations. Some European investigators have reported negative associations between HCV infection and dilated cardiomyopathy, though the disparity in results may be due to inappropriate controls, incomplete clinical investigation or other factors such as regional or racial difference.

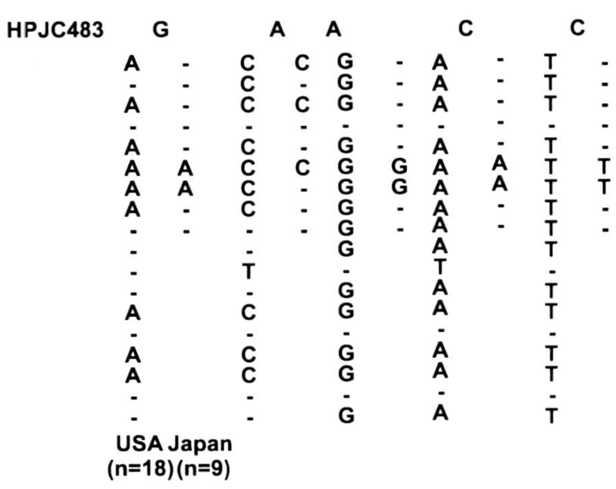

Fig. 3. Differences in the nucleotide sequences of HCV recovered from the hearts of cardiomyopathies obtained from Japan and the USA

7.3.5 Association of the Genes of the Major Histocompatibility Complex Class II with Cardiomyopathies Due to HCV Infection

The human major histocompatibility complex (MHC) is located on the short arm of chromosome 6 and encodes for several protein products involved in immune function, including complement, tumour necrosis factor (TNF)-α and the human leukocyte antigen (HLA) complex, whose polymorphisms are often proposed as candidate of susceptibility of various diseases. Associations of MHC class II antigens have been reported with patients with hypertrophic cardiomyopathies (Matsumori et al. 1981). More recently, MHC class II genes have also been analysed at the DNA level, though the results were inconsistent. In a Japanese study, the frequencies of DRB1*1401, DQB1*0503 and DRB1*1101 were increased in patients with dilated cardiomyopathy (Nishi et al. 1995). However, the development of dilated cardiomyopathy cannot be solely explained by the presence or absence of a single MHC class II allele. Since the aetiology of dilated cardiomyopathy is heterogeneous

(Matsumori 1997), different disease entities may be linked to different MHC class II genes.

Genetic studies to date have examined three aspects of HCV infection: (1) clearance, (2) progression (cirrhosis) and (3) susceptibility to infection. Recent studies on HCV hepatitis showed that DQB1*0301 was associated with clearance of the virus (Cramp et al. 1998; Alric et al. 1997). DRB1*04 and DQA1*03 were identified as protective alleles (Cramp et al. 1998), which are in strong linkage disequilibrium with DQB1*0301. DRB1*1101, which is also in linkage disequilibrium with DQB1*0301, was associated with clearance (Alric et al. 1997), and DRB1*11 was associated in other study. Several other studies have considered the association of MHC alleles with progression of liver disease. Two Japanese studies compared HCV carriers with normal liver function tests and normal histology with patients with abnormal liver function tests and cirrhosis, respectively (Kuzushita et al. 1998; Aikawa et al. 1996). In both studies, DQB1*0401, DRB1*0405 and a two-locus haplotype consisting of these alleles were more frequent in those who developed chronic liver disease.

We have recently studied association analyses of the distribution of alleles using phenotype frequencies in patients with hypertrophic or dilated cardiomyopathy and healthy controls. The frequency of HLA-DQB1*0303 was the most significantly increased in patients with hypertrophic cardiomyopathy (Table 8). HLA-DRB1*0901 was also significantly increased in patients with hypertrophic cardiomyopathy. However, there was no increase in either allele in patients with dilated cardiomyopathy. HLA-DRB1*1201 was slightly increased in patients with dilated cardiomyopathy, but not in patients with hypertrophic cardiomyopathy (Table 9, Fig. 4; Matsumori et al. 2003). MHC class II genes may play a role in the clearance of HCV and the susceptibility to HCV infection, and may influence the development of different phenotypes of cardiomyopathy.

7.3.6 Treatment of HCV Cardiomyopathies

In patients with HCV hepatitis, the success of treatment can be measured by the biochemical (normalization of alanine aminotransferase levels) and virological (undetectability of serum HCV RNA) responses.

Table 8. HLA alleles in 34 patients with hypertrophic cardiomyopathy associated with hepatitis C infection versus 136 control individuals

HLA allele	Patients		Control		RR	χ^2	*p*
	n	%	*n*	%			
DRB1*0101						0.10	
*0401	1	2.9	2	1.5	2.03	0.34	0.56
*0403	2	5.9	12	8.8	0.65	0.31	0.58
*0405	7	20.6	44	32.4	0.54	1.79	0.18
*0406	1	2.9	4	2.9	1.00	0.00	1
*0407	2	5.9	2	1.5	4.19	2.03	0.13
*0410	1	2.9	3	2.2	1.34	0.06	0.80
*0802	2	5.9	13	9.6	0.59	0.46	0.50
*0803	8	23.5	19	14.0	1.89	1.86	0.17
*0901	16	47.1	36	26.5	2.47	5.43	0.020
*1101	1	2.9	8	5.9	0.48	0.47	0.49
*1201	1	2.9	6	4.4	0.66	0.15	0.70
*1302	6	17.6	22	16.2	1.11	0.04	0.84
*1403	0	0	1	0.7	0.00	0.25	0.62
*1407	1	2.9	0	0	–	4.02	0.045
*1501	2	5.9	20	14.7	0.36	1.88	0.17
*1502	5	14.7	23	16.9	0.85	0.10	0.76
*1602	0	0	1	0.7	0.00	0.25	0.62
DQA1*0101	5	14.7	44	32.4	0.36	4.13	0.045
*0102	9	26.5	36	26.5	1.00	0.00	1
*0103	13	38.2	39	28.7	1.54	1.17	0.28
*0104	0	0	17	12.5	0.00	4.72	0.030
*0301	23	67.6	88	64.7	1.14	0.10	0.75
*0401	1	2.9	13	9.6	0.29	1.58	0.21
*0501	2	5.9	16	11.8	0.47	0.99	0.32
*0601	1	2.9	4	2.9	1.00	0.00	1
DQB1*0301	3	8.8	25	18.4	0.43	1.81	0.18
*0302	5	14.7	24	17.6	0.80	0.17	0.68
*0303	17	50.0	36	26.5	2.78	7.02	0.0081
*0401	7	20.6	43	31.6	0.56	1.59	0.21
*0402	2	5.9	8	5.9	1.00	0.00	1
*0501	5	14.7	27	19.9	0.70	0.47	0.49
*0502	2	5.9	10	7.4	0.79	0.09	0.76
*0601	13	38.2	38	27.9	1.60	1.37	0.24
*0602	1	2.9	19	14.0	0.19	3.19	0.074
*0604	6	17.6	22	16.2	1.11	0.04	0.84

RR, relative risk

Table 9. HLA alleles in 19 patients with dilated cardiomyopathy associated with hepatitis C virus infection versus 136 control individuals

HLA allele	Patients		Control		RR	χ^2	p
	n	%	n	%			
DRB1*0101	5	26.3	23	16.9	1.75	0.10	0.32
*0401	0	0	2	1.5	0.00	0.28	0.59
*0403	0	0	12	8.8	0.00	1.82	0.18
*0405	6	31.6	44	32.4	0.97	0.005	0.95
*0406	0	0	4	2.9	0.00	0.57	0.45
*0407	0	0	2	1.5	0.00	0.28	0.59
*0410	0	0	3	2.2	0.00	0.43	0.51
*0802	1	5.3	13	9.6	0.53	0.37	0.54
*0803	3	15.8	19	14.0	1.15	0.05	0.83
*0901	5	26.3	36	26.5	0.99	0.0002	0.99
*1101	2	10.5	8	5.9	1.88	0.60	0.44
*1201	3	15.8	6	4.4	4.06	3.95	0.047
*1302	0	0	22	16.2	0.00	3.58	0.058
*1403	1	5.3	1	0.7	7.50	2.68	0.10
*1407	0	0	0	0	–	–	–
*1501	3	15.8	20	14.7	1.09	0.02	0.90
*1502	4	21.1	23	16.9	1.31	0.20	0.66
*1602	1	5.3	1	0.7	7.50	2.68	0.10
DQA1*0101	5	26.3	44	32.4	0.75	0.28	0.60
*0102	6	31.6	36	26.5	1.28	0.22	0.64
*0103	7	36.8	39	28.7	1.45	0.53	0.47
*0104	0	0	17	12.5	0.00	2.67	0.10
*0301	12	63.2	88	64.7	0.94	0.02	0.89
*0401	1	5.3	13	9.6	0.53	0.37	0.54
*0501	3	15.8	16	11.8	1.41	0.25	0.62
*0601	0	0	4	2.9	0.00	0.57	0.45
DQB1*0301	3	15.8	25	18.4	0.83	0.08	0.78
*0302	0	0	24	17.6	0.00	3.97	0.046
*0303	6	31.6	36	26.5	1.28	0.22	0.64
*0401	6	31.6	43	31.6	1.00	0.00001	1.00
*0402	1	5.3	8	5.9	0.89	0.01	0.91
*0501	5	26.3	27	19.9	1.44	0.43	0.51
*0502	1	5.3	10	7.4	0.70	0.11	0.74
*0601	7	36.8	38	27.9	1.50	0.64	0.42
*0602	3	15.8	19	14.0	1.15	0.05	0.83
*0604	2	10.5	22	16.2	0.61	0.41	0.52

RR, relative risk

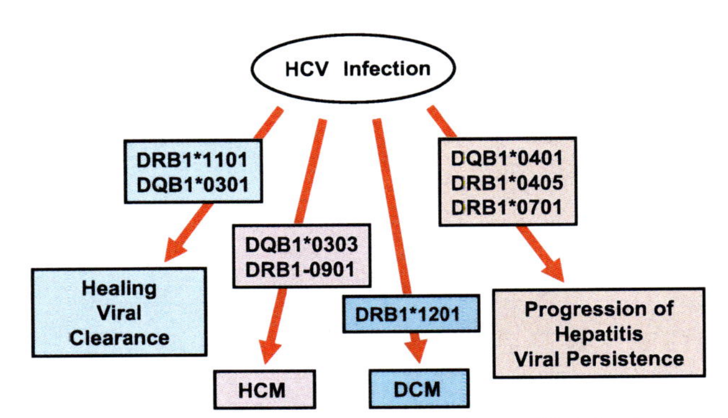

Fig. 4. Major histocompatibility complex genes and HCV infection

However, therapeutic markers to follow HCV cardiomyopathies have not been established in clinical practice. We have examined the effect of interferon on myocardial injury associated with active HCV hepatitis in collaboration with Shimane University. Since TL-201-SPECT is a more sensitive method than electrocardiography or echocardiography to detect myocardial injury induced by HCV, we used T1-SPECT scores to measure the effects of interferon on myocardial injury. SPECT scores improved in 8 out of 15 patients (53%) whose interferon treatment was completed. Circulating HCV disappeared after interferon therapy in all 11 patients, with either a decrease or no change in SPECT scores, but HCV genomes persisted in the blood in 2 patients whose clinical status worsened (Ooyake et al. 1999). This preliminary study suggests that interferon is a promising treatment for myocardial diseases caused by HCV infection.

We have recently reported that patients with dilated cardiomyopathy whose prognosis is poor have abnormally high serum concentrations of cardiac troponin T in the absence of an increase in serum creatine kinase concentrations, and that, in that population, cardiac troponin T is a prognostic marker (Sato et al. 2001). Serial measurements of serum cardiac troponin T concentrations seem to be a reliable indicator of myocyte injury, and we have hypothesized that, in patients with cardiomyopathy, therapeutic interventions for heart failure which ultimately improve the prognosis, should be associated with a fall in cardiac troponin T.

Therefore, in patients with HCV cardiomyopathies, monitoring of HCV RNA and cardiac troponin T appears appropriate.

We have reported the treatment with interferon for a patient with dilated cardiomyopathy and striated myopathy associated with HCV infection, guided by serial measurements of serum HCV RNA and cardiac troponin T (Sato et al. 2000). In that patient, serum concentrations of cardiac troponin T remained abnormally high over a 3-year period despite treatment of heart failure with angiotensin-converting enzyme inhibitors, β-adrenergic blockers, calcium antagonists, dopamine and dobutamine. Clinical manifestations of heart failure progressed, while echocardiographic left ventricular ejection fraction decreased from 49% to 29%, and left ventricular end-diastolic dimension increased from 60 mm to 69 mm. HCV RNA in heart tissue was positive by PCR. Interferon therapy was introduced with monitoring of cardiac troponin T concentration, which fell in parallel with a decline in serum HCV RNA during treatment. It is also noteworthy that, after cessation of interferon therapy, serum concentrations of cardiac troponin T and serum HCV RNA returned toward their baseline values (Sato et al. 2000; Fig. 5, left). These observations strongly suggest that the myocyte injury documented in

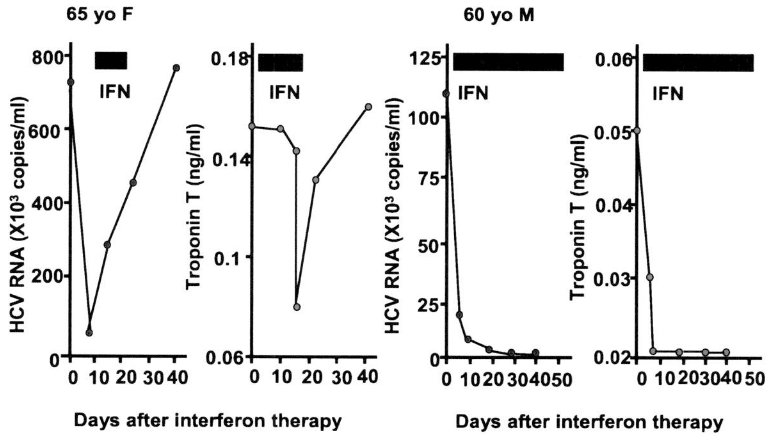

Fig. 5. Interferon therapy in patients with HCV cardiomyopathy guided by serum HCV RNA and cardiac troponin T

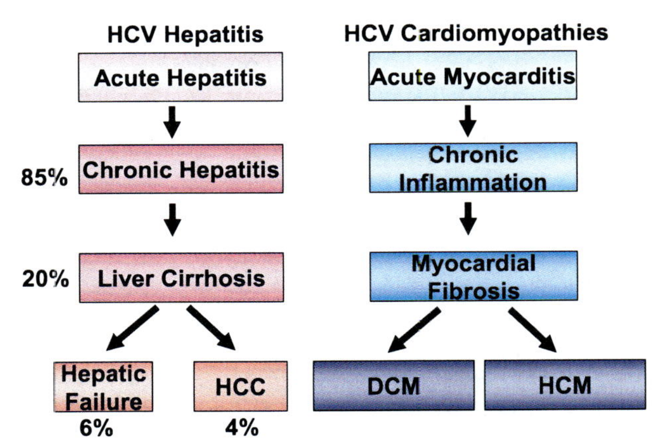

Fig. 6. Comparison of the pathogenesis of HCV hepatitis and cardiomyopathies

this patient was related to HCV infection. We have now treated another patient with HCV cardiomyopathy with interferon, and HCV RNA and cardiac troponin T both fell concomitantly during treatment (Fig. 5, right).

Pathogenesis of HCV hepatitis and cardiomyopathies is compared in Fig. 6. In HCV hepatitis, most patients develop chronic hepatitis, and years later liver cirrhosis, hepatic failure and hepatocellular carcinoma. In HCV heart diseases, most patients may develop chronic inflammation in the heart, and later dilated cardiomyopathy due to necrosis and loss of myocytes. However, myocytes in adult heart may not replicate. Proliferative stimuli induced by HCV infection may develop into myocyte hypertrophy and hypertrophic cardiomyopathy.

We are proposing collaborative work on a global network on myocarditis/cardiomyopathies to clarify the prevalence of cardiac involvement in HCV infection, and to perform treatment trials.

Acknowledgements. This work was supported in part by a research grant from the Japanese Ministry of Health, Labour and Welfare and a Grant for Scientific Research from the Japanese Ministry of Education, Culture, Sports, Science and Technology.

We would like to thank Ms. M. Ozone for preparing the manuscript.

References

Abelmann WH, Lorell BH (1989) The Challenge of cardiomyopathy. J Am Coll Cardiol 13:1219–1239

Aikawa T, Kojima M, Onishi H, Tamura R, Fukuda S, Suzuki T, Tsuda F, Okamoto H, Miyakawa Y, Mayumi M (1996) HLA DRB1 and DQB1 alleles and haplotypes influencing the progression of hepatitis C. J Med Virol 49:274–278

Alric L, Fort M, Izopet J, Vinel JP, Charlet JP, Selves J, Puel J, Pascal JP, Duffaut M, Abbal M (1997) Genes of the major histocompatibility complex class II influence the outcome of hepatitis C virus infection. Gastroenterology 113:1675–1681

Beld M, Habibuw MR, Rebers S PH, Boom R, Reesink H W (2000) Evaluation of automated RNA-extraction technology and a qualitative HCV assay for sensitivity and detection of HCV RNA in pool-screening system. Transfusion 40:575–579

Caforio A LP, Stewart JT, McKenna W J (1990) Idiopathic dilated cardiomyopathy. Br Med J 300:890–891

Cohen J (1999) The scientific challenge of hepatitis C. Science 285:26–30

Cramp ME, Carucci P, Underhill J, Naoumov NV, Williams R, Donaldson P T (1998) Association between HLA class II genotype and spontaneous clearance of hepatitis C viraemia. J Hepatol 29:207–213

Feldman AM, McNamara D (2000) Myocarditis. N Engl J Med 343:1388–1398

Higuchi M, Tanaka E, Kiyosawa K (2002) Epidemiology and clinical aspects on hepatitis C. Jpn J Infect Dis 55:69–77

Ishizaka N, Ishizaka Y, Takahashi E, Tooda E, Hashimoto H, Nagai R, Yamakoda M (2002) Association between hepatitis C virus seropositivity, carotid-artery plaque, and intima-media thickening. Lancet 359:133–135

Johnson RA, Palacios I (1982) Dilated cardiomyopathies of the adult. N Engl J Med 307:1119–1126

Kawai C, Matsumori A, Fujiwara H (1987) Myocarditis and dilated cardiomyopathy. Annu Rev Med 38:221–239

Kawano H, Kawai S, Nishijo T, Shirai T, Inagaki Y, Okada R (1994) An autopsy case of hypertrophic cardiomyopathy with pathological findings suggesting chronic myocarditis. Jpn Heart J 35:95–105

Kuzushita N, Hayashi N, Moribe T, Katayama K, Kanto T, Nakatani S, Kaneshige T, Tatsumi T, Ito A, Mochizuki K, Sasaki Y, Kasahara A, Hori M (1998) Influence of HLA haplotypes on the clinical coursed of individuals infected with hepatitis C virus. Hepatology 27:240–244

Liu PP, Mason JW (2001) Advances in the understanding of myocarditis. Circulation 140:1076–1082

Matsumori A (1993) Animal models: pathological findings and therapeutic considerations. In: Banatvala JE, Edward A (eds) Viral infection of the heart. Edward Arnold, London, pp 110–137
Matsumori A (1997) Molecular and immune mechanisms in the pathogenesis of cardiomyopathy. Jpn Circ J 61:275–291
Matsumori A (2001) Myocardial diseases, nephritis, and vasculitis associated with hepatitis virus. Intern Med 40:78–79
Matsumori A (2003) Introductory chapter. In: Matsumori A (ed) Cardiomyopathies and heart failure. Biomolecular, infectious and immune mechanisms. Kluwer Academic Publishers, Boston, pp 1–15
Matsumori A (2005) Hepatitis C virus infection and cardiomypathies. Circ Res 96:144–147
Matsumori A, Kawai C (1982a) An animal model of congestive (dilated) cardiomyopathy: dilation and hypertrophy of the heart in the chronic stage in DBA/2 mice with myocarditis caused by encephalomyocarditis virus. Circulation 66:377–380
Matsumori A, Kawai C (1982b) An experimental model for congestive heart failure after encephalomyocarditis virus myocarditis in mice. Circulation 65:1230–1235
Matsumori A, Kawai C, Wakabayashi A, Terasaki PI, Park MS, Sakurami T, Ueno Y (1981) HLA-DRW4 antigen linkage in patients with hypertrophic obstructive cardiomyopathy. Am Heart J 101:14–16
Matsumori A, Matoba Y, Sasayama S (1995) Dilated cardiomyopathy associated with hepatitis C virus infection. Circulation 92:2519–2525
Matsumori A, Matoba Y, Nishio R, Ono K, Sasayama S (1996) Detection of hepatitis C virus RNA from the heart of patients with hypertrophic cardiomyopathy. Biochem Biophys Res Commun 222:678–682
Matsumori A, Ohashi N, Hasegawa K, Sasayama S, Eto T, Imaizumi T, Izumi T, Kawamura K, Kawana M, Kimura A, Kitabatake A, Matsuzaki M, Nagai R, Tanaka H, Horie M, Hori M, Inoko H, Seko Y, Sekiguchi M, Shimotohno K, Sugishita Y, Takeda N, Takihara T, Tanaka M, Tokuhisa T, Toyo-oka T, Yokoyama M (1998a) Hepatitis C virus infection and heart disease. A multicenter study in Japan. Jpn Circ J 62:389–391
Matsumori A, Ohashi N, Sasayama S (1998b) Hepatitis C virus infection and hypertrophic cardiomyopathy. Ann Intern Med 129:749–750
Matsumori A, Ohashi N, Nishio R, Kakio T, Hara M, Furukawa Y, Ono K, Shioi T, Sasayama S (1999) Apical hypertrophic cardiomyopathy and hepatitis C virus infection. Jpn Circ J 63:433–438
Matsumori A, Yutani C, Ikeda Y, Kawai S, Sasayama S (2000) Hepatitic C virus from the hearts of patients with myocarditis and cardiomyopathy. Lab Invest 80:1137–1142

Matsumori A, Furukawa Y, Hasegawa K, Sato K, Nakagawa H, Morikawa Y, Miura K, Ohno Y, Tamakoshi A, Inaba Y, Sasayama S (2002) Epidemiologic and clinical characteristics of cardiomyopathis in Japan – results from nationwide surveys. Circ J 66:323–336

Matsumori A, Ohashi N, Ito H, Furukawa Y, Hasegawa K, Sasayama S, Naruse T, et al. (2003) Genes of the major histocompatibility complex class II influence the phenotype of cardiomyopathies associated with hepatitis C virus infection. In: Matsumori A (ed) Cardiomyopathies and heart failure. Kluwer Academic Publishers, Boston, pp 515–521

Miura K, Nakagawa H, Morikawa Y, Sasayama S, Matsumori A, Hasegawa K, Ohno Y, Tamakoshi A, Kawamura T, Inaba Y (2002) Epidemiology of idiopathic cardiomyopathy in Japan: results from a nationwide survey. Heart 87:126–130

Nakamura K, Matsumori A, Kusano K, Banba K, Taniyama M, Nakamura Y, Morita H, Matsubara H, Yamanari H, Ohe T (2000) Hepatitis C virus infection in a patient with dermatomyositis and left ventricular dysfunction. Jpn Circ J 64:617–618

Nishi H, Koga Y, Koyanagi T, Harada H, Imaizumi T, Toshima H, Sasazuki T, Kimura A (1995) DNA Typing of HLA class II genes in Japanese patients with dilated cardiomyopathy. J Mol Cell Cardiol 27:2385–2392

Okabe M, Fukuda K, Arakawa K, Kikuchi M (1997) Chronic variant of myocarditis associated with hepatitis C virus infection. Circulation 96:22–24

Olinde KD, O'Connell JB (1994) Inflammatory heart disease: pathogenesis, clinical manifestation, and treatment of myocarditis. Annu Rev Med 45:481–490

Ooyake N, Kuzuo H, Hirano Y, Shimada T, Matsumori A (1999) Myocardial injury induced by hepatitis C virus and interferon therapy (abstract). Presented at the 96th Annual Scientific Meeting of the Japanese Society of Internal Medicine, Tokyo

Richardson P, Mckenna W, Bristow M, Maisch B, Mautner B, O'Connell J, Olsen E, Thiene G, Goodwin J, Gyarfas I, Martin I, Nordet P (1996) Report of the 1995 World Health Organization/International Society and Federation of Cardiology Task Force on the definition and classification of cardiomyopathies. Circulation 93:841–842

Sato Y, Takatsu Y, Yamada T, Kataoka K, Taniguchi R, Mimura R, Sasayama S, Matsumori A (2000) Interferon treatment for dilated cardiomyopathy and striated myopathy associated with hepatitis C virus infection based on serial measurements of serum concentration of cardiac troponin T. Jpn Circ J 64:321–324

Sato Y, Yamada T, Taniguchi R, Nagai K, Makiyama T, Okada H, Kataoka K, Ito H, Matsumori A, Sasayama S, Takatsu Y (2001) Persistently increased serum concentrations of cardiac troponin T in patients with idiopathic dilated cardiomyopathy are predictive of adverse outcomes. Circulation 103:369–374

Takeda A, Sakata A, Takeda N (1999) Detection of hepatitis C virus RNA in the hearts of patients with hepatogenic cardiomyopathy. Mol Cell Biochem 195:257–261

Teragaki M, Nishiguchi S, Takeuchi K, Yoshiyama M, Akioka K, Yoshikawa J (2003) Prevalence of hepatitis C virus infection among patients with hypertrophic cardiomyopathy. Heart Vessels 18:167–170

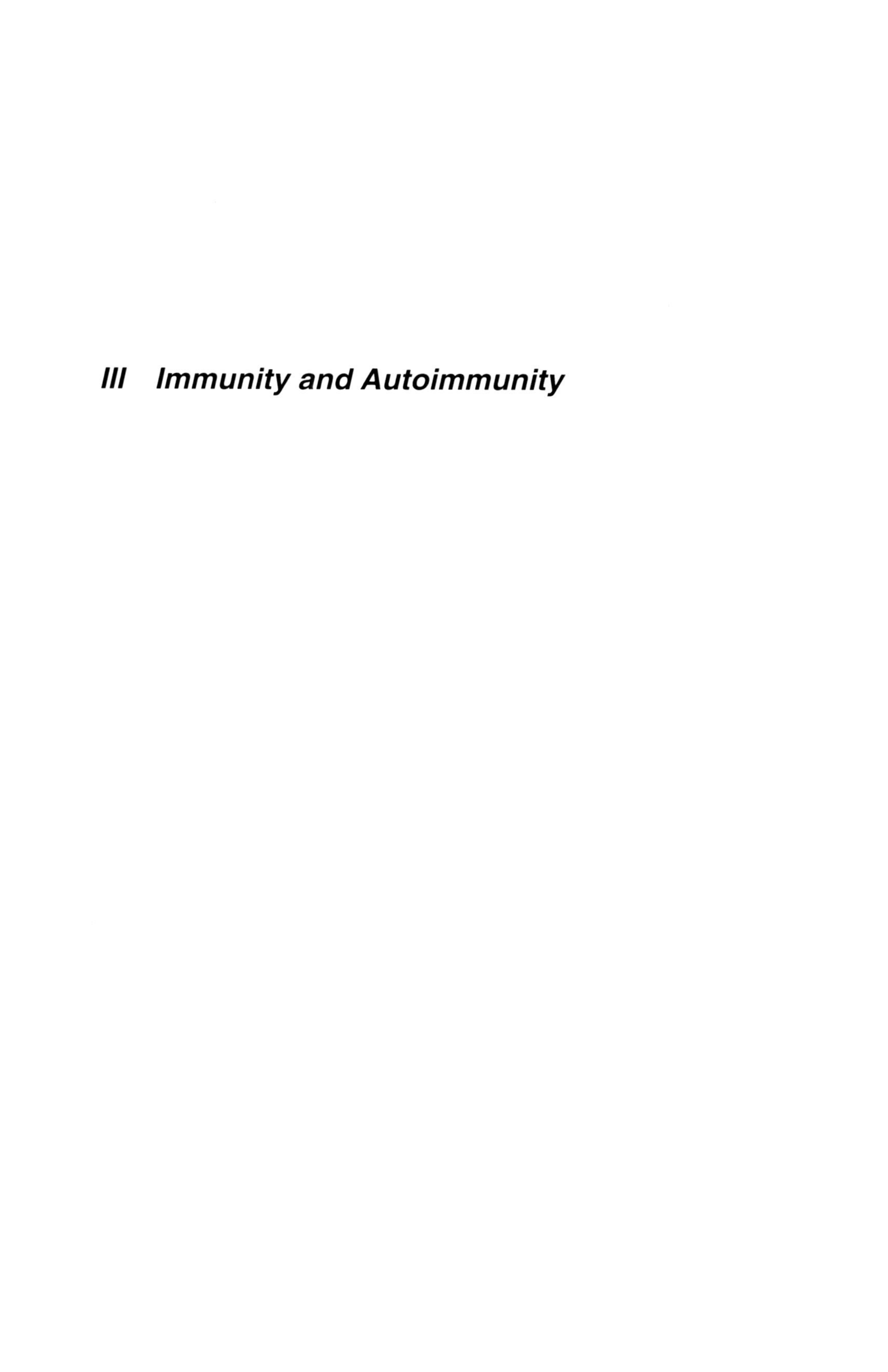

III Immunity and Autoimmunity

8 Recent Insights into the Role of Host Innate and Acquired Immunity Responses

P. Liu, K. Fuse, G. Chu, Y. Liu, A. Opavsky

Abstract. Viral myocarditis can present as dramatic heart failure in the young, and chronic indolent cardiomyopathy in the older adult. The outcome of the disease is still poor, associated with high mortality during long-term follow-up. Enteroviral myocarditis serves as an excellent model to understand virus and host interactions. The virus enters the target cells via collaborating receptors, and this process triggers an inflammatory response in the host. The immune reaction is a two-edged sword, with appropriate activation of the immune system capable of clearing the virus, but excessive activation leads to a chronic inflammatory process that triggers the remodeling of the heart and consequent clinical heart failure. Through genetic dissection strategies, we have identified that the acquired immune system is activated through the T cell receptor and signaling amplification systems, such as the tyrosine kinase p56lck, phosphatase CD45 and downstream ERK1/2, and the family of cytokines. This signaling system not only promotes inflammatory cell clonal expansion but paradoxically also promotes viral proliferation. The innate immune system is now recognized as playing an ever-expanding role in coordinating the host immune response through the Toll-like receptors, triggering downstream signaling adaptors such as MyD88, IRAK, and TRIF/IRFs. These lead to activation of cytokines or interferons, depending on the balance of the signal contributions. The ongoing research in this area should help us to understand the immune response of the heart to viral infection, while identifying potential targets for therapy.

8.1 Continuum of Viral Myocarditis and Dilated Cardiomyopathy

Myocarditis is defined as inflammation of the heart muscle, which most commonly presents as dramatic heart failure in the young, and chronic heart failure and dilated cardiomyopathy in the older adult. The incidence of myocarditis is estimated at 1% to 5% of all deaths (Liu and Mason 2001) and about 1/3 of cases of dilated cardiomyopathy. Enteroviral infection remains one of the common causes of myocarditis, representing about 25% prevalence by polymerase chain reaction (PCR) (Jin et al. 1990; Pauschinger et al. 1999). The enteroviruses most commonly linked with myocarditis are coxsackievirus group B (CVB), especially the B3 serotype, adenovirus, and more controversially parvovirus and hepatitis C virus. The outcome of myocarditis and dilated cardiomyopathy is still poor. In the NIH myocarditis trial, patients with positive diagnosis based on the Dallas biopsy criteria had a mortality of 20% at 1 year, and mortality of 56% at 4.3 years, with many cases of chronic heart failure, despite optimal medical management (Mason et al. 1995).

Myocarditis is a classic model in which to examine the triggers and consequences of cardiac inflammation. Myocarditis can result from a diverse repertoire of etiology, but infection and subsequent immune activation are believed to be the fundamental factors leading to the development of heart failure. We have previously conceptually divided the myocarditis process into an initial viral phase, followed by an immune phase and finally a cardiomyopathy phase incorporating cardiac remodeling and dilatation (Fig. 1; Liu and Mason 2001).

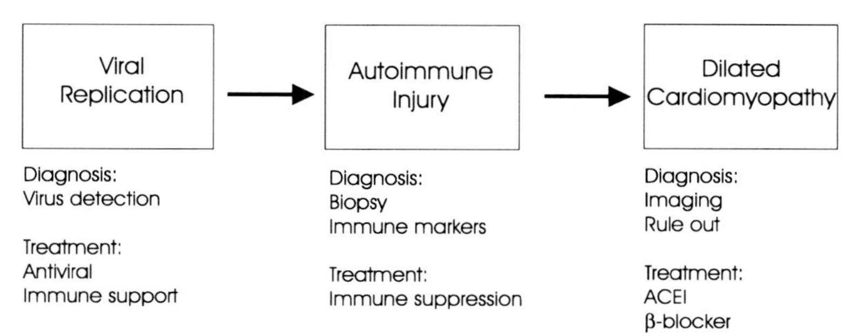

Fig. 1. The conceptual 3 stages of viral myocarditis leading to dilated cardiomyopathy. The initial phase is viral, where the virus infects a susceptible host, and is able to engage the receptor and mediate viral proliferation. The appropriate host immune response will clear the virus and infected cells. However, in the setting of inappropriate host immune controls, the inflammatory process will be unable to effectively clear the virus, leading to continued inflammation. The presence of additional viral proteins and damaged host tissue will expose the immune system to additional sources of antigens, and stimulate continued proliferation of inflammatory cells of the acquired immunity (T and B cells), as well as innate immunity. Finally, the inflammatory process including cytokine and matrix metalloproteases will alter the structure of the myocardium to such an extent that dilated cardiomyopathy eventually results. This is one of the known causes of heart failure in adults

8.2 The Virus and Viral Receptors

The current understanding of the pathogenesis of viral myocarditis has recently been reviewed by our laboratory (Ayach et al. 2003; Liu and Mason 2001), and summarized as follows. The disease represents a delicate interaction between the virus and the host. The disease is initiated by the introduction of a virus of pathogenic strain (e.g., enterovirus such as coxsackievirus CVB3) that invades the susceptible host through a portal of entry via virus internalizing receptor, and ultimately reaches the myocardium through hematogenous or lymphangitic spread. The virus is initially processed in the spleen after complement C3 activation, where the virus will proliferate in both T and B lymphocytes and macrophages. It is through the immune activation that the viruses reach the target organs (heart and pancreas in the cases of CVB3). Once the virus reaches the susceptible myocyte, it will land on its specific receptor or receptor complex that includes the coxsackie–adenoviral receptor or CAR (Bergelson et al. 1997; Martino et al. 2000; Noutsias et al. 2001), and the attachment- and virulence-determining co-receptor known as decay-accelerating factor (DAF) or CD55 (Fig. 2; Liu and Opavsky 2000; Martino et al. 1998; Shafren et al. 1995).

The main internalizing receptor, CAR (hence the high epidemiological frequency of these viruses) – which has been demonstrated to be important for all coxsackieviruses to gain tissue entry – is a member of the immunoglobulin superfamily with adhesion molecule properties (Liu and Opavsky 2000; Martino et al. 2000). The viral internalization is significantly facilitated by a co-receptor, DAF, which is important in deflecting complement through steric interactions. There is a variety of splice variants of CAR molecules that correlates with host susceptibility (Martino et al. 1998, 2000).

Through the activation of this receptor complex, the negative strand RNA of the virus will enter the cell and is reverse transcribed into a pos-

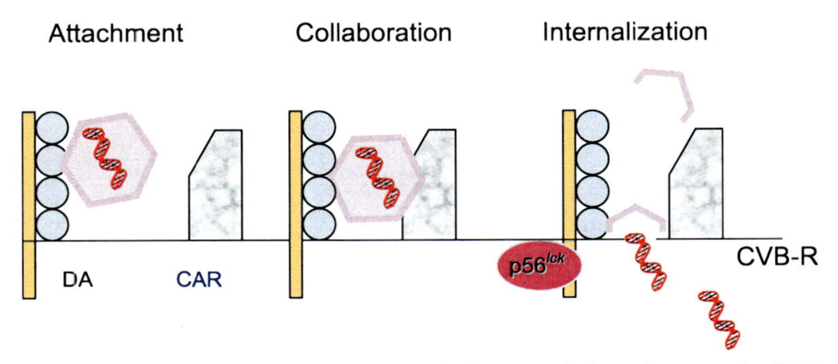

Fig. 2. The enteroviruses, such as coxsackieviruses and adenovirus, use the CAR (coxsackie–adenoviral receptor) for viral internalization. CAR is a member of the immunoglobulin superfamily, and is an adhesion molecule likely under immune regulation. The viral attachment and entry, however, is markedly enhanced in the presence of a co-receptor, DAF (decay-accelerating factor, or CD55). The proximity of the collaborating receptors in the coated pits on the cell membrane surface ensures successful engagement by the virus. The DAF also uses the T cell receptor tyrosine kinase, p56lck, as its key internal signaling mediator

itive strand to act as a template for subsequent viral RNA duplication. The polycistronic RNA encodes a large polyprotein that contains its own cleavage enzyme and important viral capsid subunits VP1–VP4. Exuberant viral replication in a susceptible host lacking suitable immunity defenses can cause acute myocardial damage and early death of the host.

8.3 Contribution of Immune Activation to Myocarditis

Viral myocarditis has always been associated with prominent features of inflammation, both systemically but particularly histologically in the myocardium. Indeed the presence of virus was not firmly established until molecular techniques became available in the last decade to document the presence of the viral genome. Regrettably, the current definition of the so-called Dallas criteria (simultaneous presence of both inflammatory cell infiltrate and myocyte necrosis) focuses only on the consequences of immune activation (Aretz et al. 1987) and makes no reference at all to the etiological viral infective agent.

Pathological examination of the myocardium of patients diagnosed with myocarditis reveals the presence of inflammatory cells such as macrophages and T lymphocytes, and an increased concentration of cytokines (Chow et al. 1989; Dec et al. 1990; Liu et al. 1993; Matsumori et al. 1994). When we carefully dissect out the contributions of the various components of immunity, we appreciate the duality of the immune response in the setting of viral myocarditis (Godeny and Gauntt 1987; Kishimoto et al. 1987, 1988; Kishimoto and Abelmann 1990; Lange and Schreiner 1994). It appears that the innate immunity responses, such as interferon regulatory factor (IRF)-1, inducible nitric oxide synthase (iNOS), and tumor necrosis factor (TNF), are intrinsically protective. However, acquired immunity, such as selective T cell activation amplification pathways, is detrimental to the host (Liu et al. 1996; Liu and Mason 2001).

8.4 Role of Acquired Immunity in Viral Myocarditis

The presence of inflammatory cell infiltration, including macrophages and T cells, indicates that acquired immunity also plays an important role in the pathogenesis of myocarditis. The viral peptide fragments are processed in the Golgi apparatus of the host cell, and presented to the cell surface in an MHC-restricted manner. This immune activation is teleologically protective initially, as the T cells attempt to seek out the infected cells and destroy them using mechanisms such as cytokine-mediated signaling (Liu et al. 2000; Opavsky et al. 2002; Wada et al. 2001) or perforin mediated cell deaths (Seko et al. 1991, 1993). Similarly, macrophages with their antiviral and immunotactic potential are also initially protective (Cook et al. 1995; Sin et al. 1997). However, the continuous activation of these cells is ultimately detrimental for the host, as both cytokine-mediated and T cell-directed damage to the myocytes subsequently reduces the number of contractile units in a terminally differentiated organ (Henke et al. 1995; Kishimoto and Abelmann 1990; Opavsky et al. 1999; Zoller et al. 1994). Similarly, macrophages contribute to the ongoing disease pathogenesis, as mice genetically deficient in macrophage inflammatory protein-1α (MIP-1α) (Cook et al. 1995) or depleted of Mac1+ macrophages, have much-improved outcome.

To understand why certain individuals develop overwhelming myocarditis after exposure to the virus and rapidly die from the disease, while others do not even show inflammation, we have been methodically mapping the major determinants of the host immune system using molecular targeting strategies in knockout mice. Earlier work has identified that components of innate immunity is critical for host survival, yet T cell signaling and activation is injurious to the host. Through $CD4^-/CD8^-$ knockout mice, we have established that both CD4 and CD8 T cells contribute to the host autoimmune inflammatory disease, accompanied by a shift of cytokine profile from T helper (Th)1 to Th2 response (Opavsky et al. 1999). Furthermore, we have identified that the T cell co-stimulatory tyrosine kinase, p56lck, is critical for both virus proliferation in the heart, as well activation of the T cells that target the heart (Fig. 3; Liu et al. 2000). Indeed, p56lck$^{-/-}$ homozygous knockout animals do not develop any myocarditis, despite exposure to large doses of the coxsackievirus. More recently, we have identified that

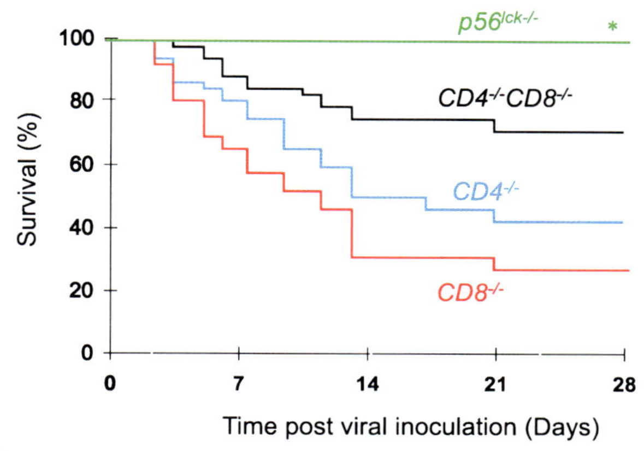

Fig. 3. Comparison of survival curves of A/J mice harboring specific targeted gene knockout, when exposed to coxsackievirus B3 (Gaunt variety). Double knockout of $CD4^-/CD8^-$ leads to improved survival, with shift of cytokines from Th2 to Th1 response. However, the most complete protection of the host is accomplished by the targeted removal of p56lck, or CD45. Targeted removal of the former led to decreased cytokine production, T cell proliferation, and decreased viral proliferation. Removal of the latter led to increased JAK/STAT (Janus kinase/signal transducer and activator of transcription) signaling and increased production of type I interferons

p56lck triggers downstream extracellular signal regulated kinase (ERK) activation in the host target cell, and appears to be critical for the determinant of host susceptibility (Opavsky et al. 2002). To validate these observations, we have also investigated the function of the associated tyrosine phosphatase CD45 linked in function to the p56lck kinase, and confirmed that $CD45^{-/-}$ animals are also resistant to viral myocarditis (Irie-Sasaki et al. 2001). After careful dissection, it was apparent that CD45 is an important src as well as JAK/STAT phosphatase, and virally triggered CD45 activation shuts down interferon production. Interferon levels are markedly increased once CD45 is removed, and the host is indeed rescued.

Most recently, we have shown that direct activation of co-receptor signaling pathways involving p56lck may lead to activation of signal pathways and downstream cytokine modulation. This recently has been shown to be a function of the mitogen-activated protein (MAP) kinases ERK1 and ERK2, which has dual function of modulating viral proliferation as well as myocardial phenotype. This begins to unravel the link between virus receptor signaling pathways of CAR, its co-receptor DAF, and the T cell signaling regulators of p56lck and CD45, to intracellular targets such as ERK1/2 (Fig. 4). This reaffirms the opportunistic nature of viruses in taking advantages of potential weaknesses of the host immune system. In effect, the virus uses a host receptor/signaling system to its own advantage in terms of the ability to internalize and proliferate. At the same time, the activation of the immune system brings in more potential substrates that further regulate the viral receptors, leading to a positive feedback system for continued disease progression in the host.

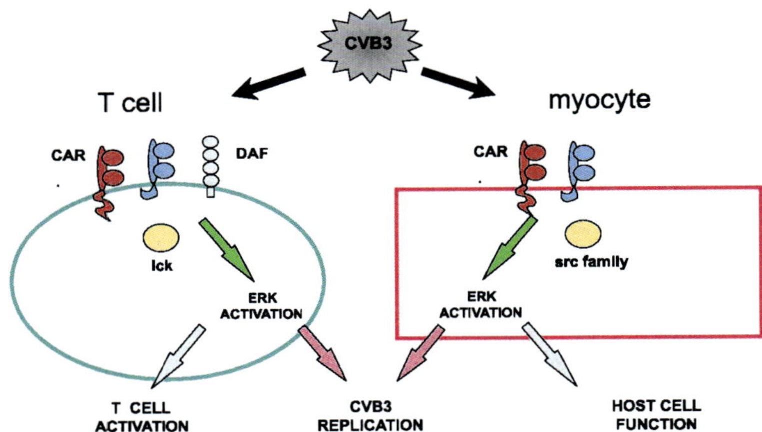

Fig. 4. A current model of host immune response in viral proliferation and host cell remodeling. The presence of coxsackievirus B3 (*CVB3*) can engage the *CAR* and *DAF* receptors. This triggers downstream signaling cascades including lck and in turn *ERK*1/2 phosphorylation. The latter is critical for increase in calcium transients that triggers T cell clonal expansion. However, ERK1/2 activation is also critical for viral proliferation directly in situ. The myocytes also are exposed to the same pathway, supporting viral proliferation and production of cytokines. However, ERK activation in the myocyte also leads to cellular remodeling, including hypertrophy and, when excessive, cell deaths

8.5 Components of Innate Immune Response

Innate immunity can be conceptually divided into two broad categories: intrinsic and extrinsic (Ayach et al. 2003). Extrinsic innate immunity involves cellular systems that respond to the presence of an external pathogen or foreign DNA by mobilization without the requirement of specific individualized antigenic recognition, such as a T cell receptor. These cells may include macrophages, granulocytes, or natural killer cells that are derived from hematopoietic stem cells and can provide a general defense against external pathogens (Dai et al. 1997). Intrinsic innate immunity, on the other hand, involves the local production of protective mediators such as interferons and defensins by any host cells faced with pathogen, as a result of signaling through cell surface Toll-like receptors (TLRs). These TLRs recognize general molecular patterns often associated with invading pathogens or foreign genetic material, and trigger a specific network of signaling pathways that leads to production of cytokines and interferons (Hemmi et al. 2000). Recent cloning and identification of entire families of TLRs and their adaptors indicated that each TLR member recognizes a unique general molecular pattern (Fig. 5), but share a number of signaling modules. The activation of innate immunity is now thought to also provide the first initial critical step for the subsequent development of acquired immunity through specific antigen recognition via T or B cell receptors and ultimate production of T killer cells or production of antigen-specific antibodies.

Binding of these patterns to TLR leads to a cascade of signaling pathways that is conserved from plants to insects to mammals and humans, and ultimately activates NF-κB and interferon production (Dangl 1998). The TLRs are type 1 transmembrane receptors that have extracellular leucine-rich repeat domain and cytoplasmic domain homologous to interleukin-1 receptor (IL-1R). Upon molecular pattern recognition, the TLR recruits the IL-1R-associated kinase (IRAK) via the adapter molecule MyD88 (Suzuki et al. 2002), or its family member Mal (MyD88 adaptor like), which in turn activates NF-κB via TRAF6 and other intermediates (Yeh and Chen 2003). Differential utilization of these adaptor members ultimately leads to distinct TLR responses.

In mammalian systems, TLR also leads to the induction of type I interferons (IFN-α/β) leading to an antiviral and heightened defense

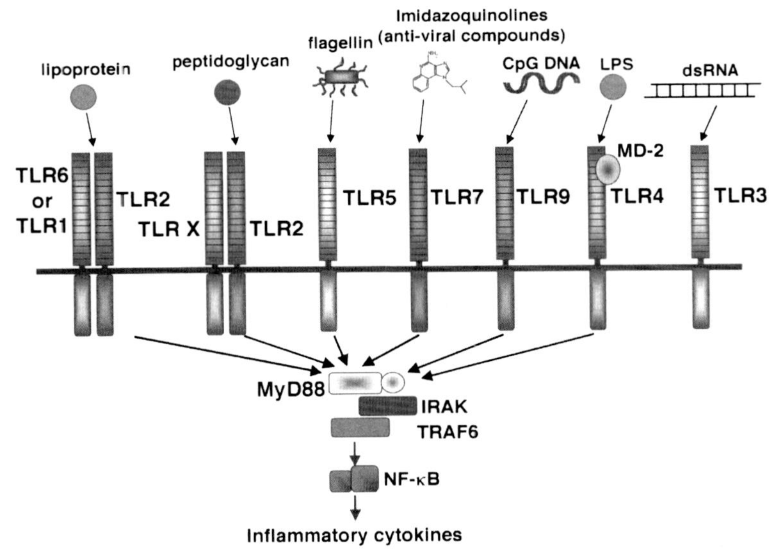

Fig. 5. An expanding repertoire of Toll-like receptor (TLR) mediating signaling in innate immunity. The TLR recognizes generalized molecular patterns, and *TLR3* recognizes double-stranded RNA, while *TLR4* recognizes lipopolysaccharide (*LPS*) and other bacterial antigens. *TLR9* typically recognizes *CpG* motifs, while *TLR7* may also participate in viral recognition. The downstream signaling includes the important adaptor *MyD88*, followed by *IRAK*s and *TRAF6*, before activating nuclear factor (NF)-κB and the cytokine repertoire

state in host cells. This effect appears to be mediated by select members of the TLR family that activate interferon regulator factor (IRF)-3, and IRF-9, possibly by recruitment of TRIF (TIR domain-containing adapter inducing IFNs) to the receptor as an adaptor (Yeh and Chen 2003). This leads to IFN production, but can also activate cytokine such as TNF-α and IL-1, and anti-inflammatory cytokines such as IL-10 and IL-12. IL-10 also can act as a bridge between the innate immune response to acquired immune response. Data to date suggest that innate immunity is required for the proper induction of acquired immunity involving T cell activation, using amplification systems such as p56lck tyrosine kinase.

8.6 Evolving Role of Innate Immunity in Viral Myocarditis

Even though recognition of the innate immunity system is relatively recent, there is already a significant collection of data to suggest it is an important component of the pathogenesis of viral myocarditis. Analysis of human cardiomyopathy hearts confirms the presence of family of Toll-like receptors including TLR-2 and TLR-4s (Satoh et al. 2003). In animal models of myocarditis, we have demonstrated the upregulation of many components of innate immunity, including MyD88 and IRAK4. It has been known for some time that excessive NF-κB activation is potentially harmful and NF-κB decoys are protective (Liu et al. 2001). Conversely, the administration of exogenous interferon is protective for the host in terms of viral attenuation and improved survival (Matsumori et al. 1987, 1988).

To investigate the potential contributions of interferon regulation in this disease condition, we have recently examined the role of IRF-1 (interferon regulatory factor, the key transcription factor for interferon production) by inoculating CVB3 in IRF-$1^{-/-}$ mice. In a manuscript submitted for publication, we have identified overwhelming mortality in IRF-1 deficient mice (100% vs 30% in wild-type), with orders of magnitude increase in viral titers in the myocardium. However, IRF-1 in mice not only regulates type I/II interferon production, but also inducible nitric oxide synthase (Kimura et al. 1994; Matsuyama et al. 1993). Indeed, previously, we have demonstrated along with others the importance of iNOS induction in attenuating viral proliferation – likely also an important component of innate immunity (Bachmaier et al. 1997; Irie-Sasaki et al. 2001; Wei et al. 1995). We now realize that the more direct inducer of interferon following viral exposure involves the IRF-3/9 and downstream IRF-7 systems, which are more specific for interferon, and IRF-3 is a potentially downstream target for TRIF (TIR domain-containing adapter inducing IFNs), potentially a TLR adaptor in balance with MyD88.

However, to more directly examine interferon's role in myocarditis, we have collaborated with Dr. Eleanor Fish, a member of the Department of Immunology here at the University of Toronto, through the evaluation of interferon-β receptor knockout mice (Deonarain et al. 2004). Following inoculation with CVB3, we have found that the IFN-$\beta^{-/-}$

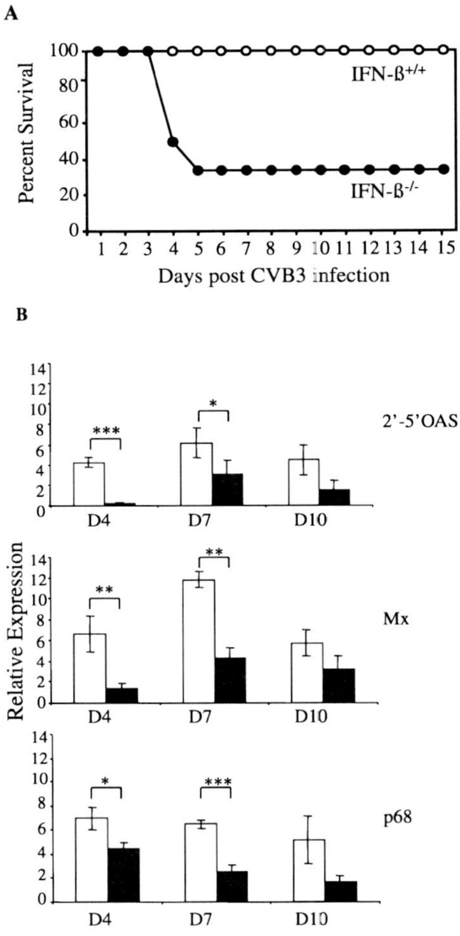

Fig. 6a,b. The importance of type I interferon in innate immunity is illustrated in this diagram. Interferon (*IFN*)-$\beta^{-/-}$-null animals are exquisitely sensitive to CVB3 infection with significantly increased mortality. The downstream intracellular effectors of interferon in attenuating viral proliferation, including 2′-5′ oligoadenylate synthetase (*OAS*), the small protein GTPase *Mx*, and serine/threonine protein kinase *p68* are all downregulated at the same time (*,**,***: $p < 0.01$)

mice had again increased viral titers, slower clearance of virus from the heart and spleen, and more aggressive immunological damage to the myocardium accompanied by a worse mortality (Fig. 6). This was accompanied by a downregulation of IFN-stimulated gene targets, including 2′-5′ oligoadenylate synthase, serine/threonine protein kinases, and the GTPase Mx, and evidence of cardiomyocyte disruption in the IFN-$\beta^{-/-}$ mice. This identifies the importance of interferon-β in viral myocarditis, and raises the intriguing observation that interferon activation (part of the innate immunity system) may have implications for subsequent T cell activation.

More directly examining the TLR system, Dr. Noel Rose's group at Johns Hopkins has evaluated CVB3 infection in TLR-$4^{-/-}$ mice (Fairweather et al. 2003). They have identified a worse outcome in that viral titers were higher in the knockout mice, yet the late survival is improved possibly through a more favorable shift in the Th1/Th2 proinflammatory cytokine balance. However, the contributions of the converging systems of signaling pathways, including MyD88, IRAK4, TRIF, and IRF-3, are currently being investigated for viral myocarditis. The results of these investigations will help significantly in our understanding of the initial immune triggers in viral myocarditis.

8.7 Therapeutic Opportunities

As we develop additional insights into the field, we realize that the delicate balance between the virus and host response, which in turn includes the balance between innate and acquired immunity, leads to either successful attenuation of the virus or exuberant inflammatory response that paradoxically aids the virus infection. Modulating the harmful effects of acquired immunity and rebalancing the adverse components of innate immunity will be potentially most efficacious.

Potential targets that may be useful for modulation include the T cell receptor tyrosine kinase p56lck or its counterpart CD45. However, the presence of novel targets, such as the MyD88-IRAK4 pathway, may also be promising. For now, most consistently, the upregulation of interferon/IRF pathways or the exogenous provision of interferons will likely be the most effective treatment for patients with acute infection or persistent infection. Interferon therapy may not only eliminate the viral infective agent, but also provide salutary modulation of the host inflammatory response to provide the most optimal outcome.

References

Aretz HT, Billingham ME, Edwards WD, Factor SM, Fallon JT, Fenoglio JJ, Olsen EG, Schoen FJ (1987) Myocarditis, a histopathologic definition and classification. Am J Cardiovasc Pathol 1:3–14

Ayach B, Fuse K, Martino T, Liu P (2003) Dissecting mechanisms of innate and acquired immunity in myocarditis. Curr Opin Cardiol 18:175–181

Bachmaier K, Neu N, Pummerer C, Dunca GS, Mak TW, Matsuyama T, Penninger JM (1997) iNOS expression and nitrotyrosine formation in response to inflammation is controlled by the interferon regulatory transcription factor 1. Circulation 96:585–591

Bergelson JM, Cunningham JA, Droguett G, Kurt-Jones EA, Krithivas A, Hong JS, Horwitz MS, Crowell RL, Finberg RW (1997) Isolation of a common receptor for Coxsackie B viruses and adenoviruses 2 and 5. Science 275:1320–1323

Chow LH, Radio SJ, Sears TD, McManus BM (1989) Insensitivity of right ventricular endomyocardial biopsy in the diagnosis of myocarditis. J Am Coll Cardiol 14:915–920

Cook DN, Beck MA, Coffman TM, Kirby SL, Sheridan JF, Pragnell IB, Smithies O (1995) Requirement of MIP-1 alpha for an inflammatory response to viral infection. Science 269:1583–1585

Dai WJ, Bartens W, Kohler G, Hufnagel M, Kopf M, Brombacher F (1997) Impaired macrophage listericidal and cytokine activities are responsible for the rapid death of Listeria monocytogenes-infected IFN-gamma receptor-deficient mice. J Immunol 158:5297–5304

Dangl J (1998) Innate immunity. Plants just say NO to pathogens. Nature 394:525

Dec GW, Fallon JT, Southern JF, Palacios I (1990) Borderline myocarditis: an indication for repeat endomyocardial biopsy. J Am Coll Cardiol 15:283–289

Deonarain R, Cerullo D, Fuse K, Liu PP, Fish EN (2004) Protective role for interferon-beta in coxsackievirus B3 infection. Circulation 110:3540–3543

Fairweather D, Yusung S, Frisancho S, Barrett M, Gatewood S, Steele R, Rose NR (2003) IL-12 receptor beta 1 and Toll-like receptor 4 increase IL-1 beta- and IL-18-associated myocarditis and coxsackievirus replication. J Immunol 170:4731–4737

Godeny EK, Gauntt CJ (1987) In situ immune autoradiographic identification of cells in heart tissues of mice with Coxsackievirus B3-induced myocarditis. Am J Pathol 129:267–276

Hemmi H, Takeuchi O, Kawai T, Kaisho T, Sato S, Sanjo H, Matsumoto M, Hoshino K, Wagner H, Takeda K, Akira S (2000) A Toll-like receptor recognizes bacterial DNA. Nature 408:740–745

Henke A, Huber S, Stelzner A, Whitton JL (1995) The role of CD8+ T lymphocytes in coxsackievirus B3-induced myocarditis. J Virol 69:6720–6728

Irie-Sasaki J, Sasaki T, Matsumoto W, Opavsky A, Cheng M, Welstead G, Griffith E, Krawczyk C, Richardson CD, Aitken K, Iscove N, Koretzky G, Johnson P, Liu P, Rothstein DM, Penninger JM (2001) CD45 is a JAK phosphatase and negatively regulates cytokine receptor signaling. Nature 409:349–354

Jin O, Sole MJ, Butany JW, Chia WK, McLaughlin PR, Liu P, Liew CC (1990) The detection of enterovirus RNA in myocardial biopsies from patients with myocarditis and cardiomyopathy using gene amplification by the polymerase chain reaction. Circulation 82:8–16

Kimura T, Nakayama K, Penninger J, Kitagawa M, Harada H, Matsuyama T, Tanaka N, Kamijo R, Vilcek J, Mak TW (1994) Involvement of the IRF-1 transcription factor in antiviral responses to interferons. Science 264:1921–1924

Kishimoto C, Abelmann WH (1990) In vivo significance of T cells in the development of coxsackievirus B3 myocarditis in mice. Immature but antigen-specific T cells aggravate cardiac injury. Circ Res 67:589–598

Kishimoto C, Kuribayashi K, Fukuma K, Masuda T, Tomioka N, Abelmann WH, Kawai C (1987) Immunologic identification of lymphocyte subsets in experimental murine myocarditis with encephalomyocarditis virus. Different kinetics of lymphocyte subsets between the heart and the peripheral blood, and significance of Thy 1.2+ (pan T) and Lyt 1+, 23+ (immature T) subsets in the development of myocarditis. Circ Res 61:715–725

Kishimoto C, Misaki T, Crumpacker CS, Abelmann WH (1988) Serial immunologic identification of lymphocyte subsets in murine coxsackievirus B3 myocarditis: different kinetics and significance of lymphocyte subsets in the heart and in peripheral blood. Circulation 77:645–653

Lange LG, Schreiner GF (1994) Immune mechanisms of cardiac disease. N Engl J Med 330:1129–1135

Liu P, Mason J (2001) Advances in the understanding of myocarditis. Circulation 104:1076–1082

Liu P, Wang E, Sole M (1993) Pathology and pathogenesis of myocarditis: A 1990's perspective. Cardiovasc Pathol 2:247–257

Liu P, Martino T, Opavsky A, Penninger J (1996) Myocarditis-balance between viral infection and immune response. Can J Cardiol 12:935–943

Liu P, Aitken K, Kong YY, Martino T, Dawood F, Wen WH, Opavsky MA, Kozieradzki I, Bachmaier K, Straus D, Mak T, Penninger J (2000) Essential role for the tyrosine kinase p56lck in coxsackievirus B3 mediated heart disease. Nat Med 6:429–434

Liu PP, Opavsky MA (2000) Viral myocarditis: Receptors that bridge that immune with the cardiovascular systems. Circ Res 86:253–254

Liu PP, Le J, Nian M (2001) NF-kB Decoy: targeting the heart of inflammatory heart disease. Circ Res 89:850–852

Martino T, Petric M, Weingartl H, Bergelson JM, Opavsky MA, Richardson CD, Modlin JF, Finberg RW, Kain K, Willis N, Gauntt CJ, Liu PP (2000) The coxsackie-adenovirus receptor (CAR) is used by reference strains and clinical isolates representing all 6 serotypes of coxsackievirus group B, and by swine vesicular disease virus. J Virol 271:99–108

Martino TA, Petric M, Brown M, Aitken K, Gauntt CJ, Richardson CD, Chow LH, Liu PP (1998) Cardiovirulent coxsackieviruses and the decay-accelerating factor (CD55) receptor. Virology 244:302–314

Mason JW, O'Connell JB, Herskowitz A, Rose NR, McManus BM, Billingham ME, Moon TE, Investigators MTT (1995) A clinical trial of immunosuppressive therapy for myocarditis. N Engl J Med 333:269–275

Matsumori A, Crumpacker CS, Abelmann WH (1987) Prevention of viral myocarditis with recombinant human leukocyte interferon alpha A/D in a murine model. J Am Coll Cardiol 9:1320–1325

Matsumori A, Tomioka N, Kawai C (1988) Protective effect of recombinant alpha interferon on coxsackievirus B3 myocarditis in mice. Am Heart J 115:1229–1232

Matsumori A, Yamada T, Suzuki H, Matoba Y, Sasayama S (1994) Increased circulating cytokines in patients with myocarditis and cardiomyopathy. Br Heart J 72:561–566

Matsuyama T, Kimura T, Kitagawa M, Pfeffer K, Kawakami T, Watanabe N, Kundig TM, Amakawa R, Kishihara K, Wakeham A, Potter J, Furlonger CL, Narendran A, Suzuki H, Ohashi PS, Paige CJ, Taniguchi T, Mak TW (1993) Targeted disruption of IRF-1 or IRF-2 results in abnormal type I IFN gene induction and aberrant lymphocyte development. Cell 75:83–97

Noutsias M, Fechner H, de Jonge H, Wang X, Dekkers D, Houtsmuller AB, Pauschinger M, Bergelson J, Warraich R, Yacoub M, Hetzer R, Lamers J, Schultheiss HP, Poller W (2001) Human coxsackie-adenovirus receptor is colocalized with integrins alpha(v)beta(3) and alpha(v)beta(5) on the cardiomyocyte sarcolemma and upregulated in dilated cardiomyopathy: implications for cardiotropic viral infections. Circulation 104:275–280

Opavsky MA, Penninger J, Aitken K, Wen WH, Dawood F, Mak T, Liu P (1999) Susceptibility to myocarditis is dependent on the response of ab T lymphocytes to coxsackieviral infection. Circ Res 85:551–558

Opavsky MA, Martino T, Rabinovitch M, Penninger J, Richardson CD, Petric M, Trinidad C, Butcher L, Chan J, Liu PP (2002) Enhanced ERK-1/2 activation in mice susceptible to coxsackievirus-induced myocarditis. J Clin Invest 109:1561–1569
Pauschinger M, Bowles NE, Fuentes-Garcia FJ, Pham V, Kuhl U, Schwimmbeck PL, Schultheiss HP, Towbin JA (1999) Detection of adenoviral genome in the myocardium of adult patients with idiopathic left ventricular dysfunction. Circulation 99:1348–1354
Satoh M, Nakamura M, Akatsu T, Iwasaka J, Shimoda Y, Segawa I, Hiramori K (2003) Expression of Toll-like receptor 4 is associated with enteroviral replication in human myocarditis. Clin Sci 104:577–584
Seko Y, Shinkai Y, Kawasaki A, Yagita H, Okumura K, Takaku F, Yazaki Y (1991) Expression of perforin in infiltrating cells in murine hearts with acute myocarditis caused by coxsackievirus B3. Circulation 84:788–795
Seko Y, Matsuda H, Kato K, Hashimoto Y, Yagita H, Okumura K, Yazaki Y (1993) Expression of intercellular adhesion molecule-1 in murine hearts with acute myocarditis caused by coxsackievirus B3. J Clin Invest 91:1327–1336
Shafren DR, Bates RC, Agrez MV, Herd RL, Burns GF, Barry RD (1995) Coxsackievirus B1, B3 and B5 use decay accelerating factor as a receptor for cell attachment. J Virol 69:3873–3877
Sin JI, Sung JH, Suh YS, Lee AH, Chung JH, Sung YC (1997) Protective immunity against heterologous challenge with encephalomyocarditis virus by VP1 DNA vaccination: effect of coinjection with a granulocyte-macrophage colony stimulating factor gene. Vaccine 15:1827–1833
Suzuki N, Suzuki S, Duncan GS, Millar DG, Wada T, Mirtsos C, Takada H, Wakeham A, Itie A, Li S, Penninger JM, Wesche H, Ohashi PS, Mak TW, Yeh WC (2002) Severe impairment of interleukin-1 and Toll-like receptor signalling in mice lacking IRAK-4. Nature 416:750–756
Wada H, Saito K, Kanda T, Kobayashi I, Fujii H, Fujigaki S, Maekawa N, Takatsu H, Fujiwara H, Sekikawa K, Seishima M (2001) Tumor necrosis factor-alpha (TNF-alpha) plays a protective role in acute viral myocarditis in mice: a study using mice lacking TNF-alpha. Circulation 103:743–749
Wei XQ, Charles IG, Smith A, Ure J, Feng GJ, Huang FP, Xu D, Muller W, Moncada S, Liew FY (1995) Altered immune responses in mice lacking inducible nitric oxide synthase. Nature 375:408–411
Yeh WC, Chen NJ (2003) Immunology: another toll road. Nature 424:743–748
Zoller J, Partridge T, Olsen I (1994) Interactions between cardiomyocytes and lymphocytes in tissue culture: an in vitro model of inflammatory heart disease. J Mol Cell Cardiol 26:627–638

9 The Significance of Autoimmunity in Myocarditis

N.R. Rose

Abstract. A growing body of evidence supports the view that some forms of human myocarditis and dilated cardiomyopathy result from a pathogenic autoimmune response. The evidence is based first on the presence of heart-specific antibodies in many patients with these diseases, including antibodies with demonstrated functional effects. These antibodies may be present before the onset of dilated cardiomyopathy and may be predictive of the course of disease in terms of deterioration of cardiac function. Depletion of the heart-specific antibodies by extracorporeal immunoadsorption may result in amelioration of disease in some patients, often continuing for long periods of time. Clinical investigations show that a subpopulation of patients with dilated cardiomyopathy benefit from immunosuppressive treatment. In one report, this subpopulation was identified as autoantibody-positive and virus-negative. Finally, animal experiments have shown that autoimmune myocarditis can be induced by viral infection and that this autoimmune response can be duplicated by immunization with a well-characterized antigen, cardiac myosin. Based on this evidence, we propose that some forms of dilated cardiomyopathy and myocarditis result from pathogenic autoimmune responses that represent the final common pathogenetic pathway of various infectious and even non-infectious injuries.

9.1 Introduction

Myocarditis, defined by the Dallas criteria as "the presence of an inflammatory infiltrate in the myocardium with necrosis and/or degeneration of adjacent myocytes" remains an etiologic dilemma and a therapeutic problem (Aretz et al. 1987). Different agents can cause the same pathologic picture and, consequently, different approaches to treatment can be taken (Feldman and McNamara 2000). A number of infectious microorganisms have been cited as possible causes, including enteroviruses, adenoviruses, cytomegaloviruses, parvoviruses, human immunodeficiency virus, measles virus, mumps virus, hepatitis A and C viruses, and herpes simplex virus (Towbin et al. 1999). Any of these viruses can induce a similar response in the heart. Historically, the infectious agent most frequently associated with human cases of myocarditis has been coxsackievirus B3, one of the enteroviruses, but the predominant causative agent varies with time and place (Grist and Bell 1969). In a large proportion of cases, no infectious agent can be found and drugs and toxins have been implicated in the disease. These etiologic uncertainties have contributed to the difficulty of developing rational therapeutic strategies. Therefore, myocarditis continues to represent a major cause of heart failure, especially in young adults (Drory et al. 1991). Moreover, chronic myocarditis sometimes evolves to dilated cardiomyopathy, a disease often requiring cardiac transplantation.

A remarkable finding in most cases of human myocarditis is the presence of autoantibodies specific for heart tissue. This observation, coupled with multiple etiologies, has led to the widely held proposition that myocarditis and dilated cardiomyopathy are the consequences of an autoimmune response to heart that represent a final common pathogenetic pathway. The present article summarizes the points raised to support a significant contribution of autoimmunity to the pathogenesis of human myocarditis. The evidence for autoimmunity has been summarized under five major headings:

1. Autoantibodies are present in many humans with autoimmune myocarditis.
2. Autoantibodies are relevant to the development and severity of disease.
3. Reduction of antibody benefits patients with the disease.
4. Immunosuppressive therapy can be used selectively to reduce disease.
5. In animal models, autoimmunity can definitively be shown to induce myocarditis and dilated cardiomyopathy.

9.2 Human Studies

The first step in assigning an autoimmune etiology to a human disease is to demonstrate relevant autoantibodies. Autoantibodies may be themselves pathogenic, or may be markers of a concurrent, cell- mediated autoimmune response.

9.3 Autoantibodies Are Present in Myocarditis

The first test used to demonstrate autoantibodies in the serum of patients with myocarditis was immunofluorescence on sections of frozen human heart tissue. Serum from individuals with no history of heart disease usually is negative or reacts only in low titer. Moreover, the antibody found in some normal humans is not specific for heart muscle but reacts with skeletal muscle as well. In contrast, we have found high titers of heart-specific autoantibodies in 59% of samples from patients with myocarditis and 20% of samples from patients with dilated cardiomyopathy (Neumann et al. 1990). In a more extensive study, Caforio found that heart-specific autoantibodies in high titer were present in 26% of the patients with dilated cardiomyopathy, compared with only 1% of patients with other cardiac diseases (Caforio et al. 2002). At least three immunofluorescent patterns were observed, fibrillary, sarcolemmal, and cytoplasmic. The presence of anti-fibrillary antibodies suggested that cardiac myosin might be a prominent antigen in human myocarditis. This premise was confirmed directly in western immunoblots (Neumann et al. 1990).

Based on studies from immunofluorescence as well as on investigations using western immunostaining, further tests were carried out measuring antibodies to purified myosin in patients with myocarditis or dilated cardiomyopathy. In our initial experience, we found that patients recovering from many viral infections have antibodies to myosin. Generally, however, these antibodies are not specific for the cardiac form of myosin, are transient, and are often of the IgM class. The antibodies could be removed by absorption with either skeletal or cardiac muscle myosin. In contrast, patients with myocarditis characteristically develop an organ-specific autoantibody that is inhibited by cardiac myosin but not by skeletal myosin, and is usually of the IgG class. Konstadoulakis et al. (1993) reported that antibody to cardiac myosin was found in 66% of patients with dilated cardiomyopathy, compared with none in healthy controls. Caforio et al. (1992) also reported that 86% of dilated cardiomyopathy patients reacted with the alpha (atrial-specific) and beta (ventricular and slow skeletal) myosin heavy chain isoforms. Going one step further, Warraich et al. (1999) found that autoantibodies against the cardiac isoforms of myosin heavy chains were characteristic of patients with dilated cardiomyopathy, compared with patients with ischemic heart disease.

Based on these observations, some centers are using antibodies to cardiac myosin as an aid in the assessment of myocarditis and dilated cardiomyopathy. Antibodies rise in the course of disease. Lauer and colleagues (2000), for example, found that 52% of myocarditis patients had antibodies at the time of initial assessment. Six months later, antimyosin antibodies were found in 76% of these individuals. As additional evidence for the diagnostic or prognostic value of these antibodies, Warraich et al. (2000) reported that patients with antibodies to cardiac myosin rejected transplanted hearts earlier than antibody-negative patients.

Although myosin is usually considered to be an intracellular antigen, it may be expressed on the surface of myocardiocytes following cardiac damage. In fact, monoclonal antibodies to myosin have actually been used to assess cardiac damage in patients (Kühl et al. 1998). The question remains open whether autoantibodies to cardiac myosin can act upon normal heart cells and are directly responsible for cardiac damage or even whether they contribute secondarily to the damage in vivo.

Antibodies to cardiac beta-1 adrenoceptor are likely to have direct clinical importance, since the receptors are present on the surface of cardiac cells. They are, however, not entirely heart-specific. Antibodies to beta-1 adrenoceptor have been studied in detail, particularly by Limas and colleagues (1990). They found such antibodies in about half of their patients with dilated cardiomyopathy. Moreover, these antibodies are strongly associated with particular HLA haplotypes HLA DR4, and HLA DR1 (Limas and Limas 1991). The antibodies are relatively specific for B1 adrenergic receptors with little or no response to the B2 receptor (Chiale et al. 1995). Moreover, they have been reported to react specifically with the second extra-cellular loop of the beta-1 adrenoceptor (Magnusson et al. 1990; Wallukat et al. 2003).

A number of other antigens have been incriminated in cases of myocarditis and dilated cardiomyopathy. For example, antibodies to the adenine nucleotide (ADP/ATP) translocator (ANT) have been described in detail by Schulze and colleagues (1990). ANT is located on the surface of mitochondrial membrane, and antibody directed to it might be expected to impair energy metabolism and calcium flux. Indeed, ex vivo studies have suggested that these antibodies may have such effects on cardiomyocytes. Whether they are active in patients is still a subject requiring investigation. Other mitochondrial antibodies found in patients with dilated cardiomyopathy are directed to the branched chain alpha-ketoacid dehydrogenase (BCKD) (Ansari et al. 1994). Although these antigens are not unique for the heart, they may well be more accessible in heart tissue than in other sites.

A great deal of effort has been directed recently to studies of antibodies to troponin as another important antigen in dilated cardiomyopathy. Okazaki and colleagues (2003) found that mice deficient in the T cell receptor costimulatory molecule PD1 spontaneously developed autoimmune dilated cardiomyopathy accompanied by the production of high titers of antibodies to the cardiac marker troponin. Passive transfer of monoclonal antibodies to cardiac troponin are capable of inducing dilated cardiomyopathy in mice. In parallel, elevated titers of autoantibodies to troponin I have been described in patients with dilated cardiomyopathy.

9.4 Autoantibodies Are Relevant to the Course of Disease

There is growing evidence that autoantibodies precede the development of disease.

Caforio et al. (2001) found that cardiac antibodies detected by immunofluorescence in symptom-free relatives of patients with dilated cardiomyopathy were associated with echocardiographic changes suggestive of early disease. Autoantibody can serve as a predictive marker of disease susceptibility in healthy subjects at risk of myocarditis or dilated cardiomyopathy, such as first-degree relatives of patients (Mahon et al. 2002).

Autoantibodies can also predict the course of disease. Lauer et al. (2000) measured cardiac myosin antibodies in 33 patients with biopsy-proven chronic myocarditis, and found that functional studies correlated with the presence or absence of autoantibody. Antibodies to myosin were associated with a poorer left ventricular systolic function and diastolic stiffness in patients with chronic myocarditis.

Peripartum cardiomyopathy is a unique form of the disease that is prevalent in certain areas (Heider et al. 1999; Sundstrom et al. 2002). Women with this disease develop high titers of antibodies to normal human cardiac tissue in the course of their pregnancy, suggesting that the disease is related to a mounting pathogenic autoimmune process.

9.5 Antibody Depletion Benefits Disease

Recent studies have suggested that immunoadsorption designed to remove circulating autoantibody produces clinical benefit (Felix et al. 2001). In these experiments, patients with idiopathic dilated cardiomyopathy were treated by extracorporeal adsorption of immunoglobulin with an anti-IgG column. In a preliminary study, antibody titers in patients who benefited from the treatment remained low for long periods of time. In contrast patients whose cardiac function deteriorated had rising antibody titers (Dorffel et al. 2004). Immunoadsorbent columns with protein A rather than anti-IgG failed to remove IgG3 subclass antibody and had little therapeutic value, suggesting that autoantibodies in this subclass are pathogenic (Staudt et al. 2002). Although these investigations show that depletion of immunoglobulin improves cardiac function, it cannot be with certainty stated which antibody has been removed by the treatment.

9.6 Immunosuppressive Therapy Can Sometimes Reduce Disease

A number of clinical studies have been carried out in an effort to determine whether methods of immunosuppression will benefit patients with myocarditis or dilated cardiomyopathy. In general, the outcome of these studies has been disappointing. In the largest investigation reported to date, Mason et al. (1995) found, on a statistical basis, no consistent improvement in patients with myocarditis diagnosed on the basis of the Dallas criteria. Closer examination of the data, however, clearly showed that some patients did improve significantly from immunosuppressive treatment, whereas others actually deteriorated. This observation has stimulated efforts to distinguish patients who are likely to benefit from immunosuppression from those who will not. An investigation carried out by Wojnicz (2001) used up-regulation of HLA in cardiac biopsies to distinguish patients with an "immune-mediated" form of myocarditis. He found improvement of left ventricular function in the cohort with HLA up-regulation. Frustaci et al. (2003) showed that immunosuppression in patients with active myocarditis is most likely to improve cardiac function when there are circulating cardiac autoantibodies and no viral genome in the myocardium. In his protocol, patients with a histologic diagnosis of active myocarditis who failed to respond to conventional supportive therapy were treated with prednisone and azathioprine for 6 months. They were carefully monitored for 1 year. Of 41 patients, 21 showed prompt improvement in left ventricular ejection fraction and showed evidence of healed myocarditis when they underwent biopsy. Of the other 20 patients who did not respond to immunosuppressive therapy, 12 remained the same, 3 required cardiac transplantation, and 5 died. Polymerase chain reaction studies of frozen heart tissue showed evidence of viral genome in 17 of the 20 poor responders. Among the viruses identified were enteroviruses, Epstein-Barr virus, adenovirus, influenza A virus and parvovirus B19. These results strongly suggest that the presence of circulating cardiac autoantibodies, as evaluated by indirect immunofluorescence, and the absence of viral genome can serve as predictors of a favorable response to immunosuppression.

9.7 Animal Studies

Although autoantibodies are present in most patients with myocarditis, a cause-and-effect relationship between a human disease and autoimmunity is difficult to establish. Animal models can be employed to investigate plausible mechanisms.

9.8 Animal Models Show that Autoimmunity Can Induce Myocarditis and Dilated Cardiomyopathy

Although viruses have long been implicated in instigating autoimmune disease, there are very few examples where the mechanisms have been studied in detail. A number of years ago, we undertook a systematic investigation in mice of myocarditis induced by coxsackievirus B3, using a strain with known cardiotropic properties (Rose et al. 1987). We found that all strains of mice infected with this virus developed an acute, focal myocarditis that reached a peak in approximately 7–9 days, and then gradually subsided. By 21 days, little or no evidence of the preceding acute viral myocarditis could be found. Infectious virus could not be isolated from the hearts of these mice after day 9. In a few strains of mice, notably A/J or strains sharing the A background, the myocarditis failed to resolve but gradually developed into a chronic form of myocarditis. Viral genome persisted in the heart whether or not the disease continued. Histologically, this chronic myocarditis differed from the original acute viral disease. It was characterized by a generalized mononuclear infiltration of the myocardium. Accompanying this histologically distinct form of myocarditis was the appearance of heart-specific antibodies. Further studies demonstrated these antibodies reacted specifically with cardiac myosin and were predominantly of the IgG1 subclass (Rose and Baughman 1998; Fairweather et al. 2001).

With these experiments as a point of departure, we prepared cardiac myosin from mouse hearts. Using skeletal myosin as control, a number of strains of mice were injected with these preparations accompanied by a powerful adjuvant (Neu et al. 1987). The strains of mice susceptible to the virus-induced form of chronic myocarditis developed the typical mononuclear infiltration. The response was seen with the cardiac, but not skeletal, myosin injections, and was accompanied by the production of cardiac-specific autoantibodies especially of the IgG1 subclass. This disease could be adoptively transferred with the CD4 T cells, establishing its autoimmune etiology (Smith and Allen 1991). Based on the parallel between the virus-induced chronic myocarditis and the chronic form of myocarditis induced by immunization with purified cardiac myosin, we proposed that the late phase of myocarditis that develops in genetically susceptible mice is an autoimmune response induced by the virus. Moreover, we concluded that the virus itself is not required for the autoimmune disease, since purified cardiac myosin (or a synthetic peptide derived from myosin) induces myocarditis and dilated cardiomyopathy in the absence of virus.

We then asked if it is possible to distinguish the mice that are susceptible to the late-phase of autoimmune myocarditis from those that resolved the acute myocarditis with no lasting consequences. Experiments indicated that the early steps of the innate immune response were distinctive in the susceptible strains. Blocking either IL-1 beta or TNF-alpha during the initiation of the immune response prevented the development of late-phase autoimmune myocarditis, even in the most susceptible mice (Lane et al. 1992, 1993; Smith and Allen 1992). Equally important, ordinarily non-susceptible strains of mice could be induced to develop the disease, following either viral infection or myosin immunization if co-treated with recombinant IL-1 or TNF-alpha. Lipopolysaccharide, a potent inducer of innate immunity, was equally effective in converting "non-susceptible" to "responder" strains of mice. Recent studies have confirmed that differences in IL-1 beta levels were detectable between "responder" and "non-responder" strains as early as 6 h following coxsackievirus infection (Fairweather et al. 2004).

Further investigations indicated that mice infected with coxsackievirus B3 sometimes progressed to a form of dilated cardiomyopathy (Afanasyeva et al. 2004). Using the myosin-induced model of myocarditis where no virus was present, we succeeded in demonstrating that the transition from chronic myocarditis to dilated cardiomyopathy was enhanced if interferon (IFN)-gamma was depleted. Depletion could be accomplished either by administering a monoclonal antibody to IFN-gamma or using IFN-gamma knockout mice. The severe, often fatal, form of dilated cardiomyopathy in these mice was characterized by the development of typical cardiac dysfunction, similar to that seen in human patients with dilated cardiomyopathy. This step-wise progression to dilated cardiomyopathy could be prevented by administering recombinant IFN-gamma to the mice. Although the role of IFN-gamma in regulating late stages of this immunopathic response have not been fully elucidated, the cytokine limits autoimmune responses by diminishing antigen-induced cell death, a major regulatory mechanism that prevents autoimmune disease in normal mice (Afanasyeva et al. 2005).

9.9 Conclusions

The precise etiology of human myocarditis remains obscure, but circumstantial evidence indicates viral infection acts as an initiating cause in many cases. Immunologic studies show that most patients with myocarditis associated with coxsackievirus infection produce heart-reactive antibodies. Sometimes antibodies are directly pathogenic; in other instances, antibodies may serve as markers of a concurrent cell-mediated autoimmune response. Depletion of circulating antibody by immunoadsorption benefits some patients and immunosuppressive therapy has a favorable effect in patients with heart-reactive autoantibodies but no evidence of persistent viral infection.

Using experimental models, it has been established that coxsackievirus infection can cause chronic myocarditis and dilated cardiomyopathy in genetically susceptible strains. These later phases of disease can be reproduced in the same susceptible strains of mice by immunization with purified cardiac myosin in the absence of virus. Based on these observations, we propose that some cases of human cardiomyopathy are due to an autoimmune response to cardiac myosin that has been mobilized from cardiomyocytes during preceding viral infection. Autoimmunity may represent a final common pathogenetic pathway of multiple infections and even non-infectious etiologies.

Acknowledgements. The author's research was supported by NIH research grants HL67290, HL70729 and AI51835. He thanks Dr. Daniela Cihakova for careful review of the manuscript and Hermine Bongers for editorial assistance.

References

Afanasyeva M, Georgakopoulos D, Rose NR (2004) Autoimmune myocarditis: cellular mediators of cardiac dysfunction. Autoimmun Rev 3:476–486

Afanasyeva M, Georgakopoulos D, Belardi DF, Bedja D, Fairweather D, Warig Y, Kaya Z, Gabrielson KL, Rodriguez ER, Caturegli P, Kass DA, Rose NR (2005) Impaired up-regulation of CD25 on CD4+ T cells in IFN-gamma knockout mice is associated with progression of myocarditis to heart failure. Proc Natl Acad Sci U S A 102:180–185

Ansari AA, Neckelmann N, Villinger F, Leung P, Danner DJ, Brar SS, Zhao S, Gravanis MB, Mayne A, Gershwin ME, et al (1994) Epitope mapping of the branched chain alpha-ketoacid dehydrogenase dihydrolipoyl transacylase (BCKD-E2) protein that reacts with sera from patients with idiopathic dilated cardiomyopathy. J Immunol 153:4754–4765

Aretz HT, Billingham ME, Edwards WD, Factor SM, Fallon JT, Fenoglio JJ Jr, Olsen EG, Schoen FJ (1987) Myocarditis: a histopathological definition and classification. Am J Cardiovasc Pathol 1:3–14

Caforio AL, Grazzini M, Mann JM, Keeling PJ, Bottazzo GF, McKenna WJ, Schiaffino S (1992) Identification of alpha- and beta-cardiac myosin heavy chain isoforms as major autoantigens in dilated cardiomyopathy. Circulation 85:1734–1742

Caforio AL, Mahon NJ, Mckenna WJ (2001) Cardiac autoantibodies to myosin and other heart-specific autoantigens in myocarditis and dilated cardiomyopathy. Autoimmunity 34:199–204

Caforio AL, Mahon NJ, Tona F, McKenna WJ (2002) Circulating cardiac autoantibodies in dilated cardiomyopathy and myocarditis: pathogenic and clinical significance. Eur J Heart Fail 4:411–417

Chiale PA, Rosenbaum MB, Elizari MV, Hjalmarson A, Magnussen Y, Wallukat G, Hoebeke J (1995) High prevalence of antibodies against beta 1- and beta 2-adrenoceptors in patients with primary electrical cardiac abnormalities. J Am Coll Cardiol 26:864–869

Dorffel WV, Wallukat G, Dorffel Y, Felix SB, Baumann G (2004) Immunoadsorption in idiopathic dilated cardiomyopathy, a 3-year follow-up. Int J Cardiol 97:529–534

Drory Y, Turetz Y, Hiss Y, Lev B, Fisman EZ, Pines A, Kramer MR (1991) Sudden unexpected death in persons less than 40 years of age. Am J Cardiol 68:1388–1392

Fairweather D, Kaya Z, Shellam GR, Lawson CM, Rose NR (2001) From infection to autoimmunity. J Autoimmun 16:175–186
Fairweather D, Frisancho-Kiss S, Rose NR (2004) Viruses as adjuvants for autoimmunity: evidence from Coxsackievirus-induced myocarditis. Rev Med Virol 15:17–27
Feldman AM, McNamara D (2000) Myocarditis. N Engl J Med 343:1388–1398
Felix SB, Staudt A, Friedrich GB (2001) Improvement of cardiac function after immunoadsorption in patients with dilated cardiomyopathy. Autoimmunity 34:211–215
Frustaci A, Chimenti C, Calabrese F, Pieroni M, Thiene G, Maseri A (2003) Immunosuppressive therapy for active lymphocytic myocarditis: virological and immunologic profile of responders versus nonresponders. Circulation 107:857–863
Grist NR, Bell EJ (1969) Coxsackie viruses and the heart. Am Heart J 77:295–300
Heider AL, Kuller JA, Strauss RA, Wells SR (1999) Peripartum cardiomyopathy: a review of the literature. Obstet Gynecol Surv 54:526–531
Konstadoulakis MM, Kroumbouzou H, Tsiamis E, Trikas A, Toutouzas P (1993) Clinical significance of antibodies against tropomyosin, actin and myosin in patients with dilated cardiomyopathy. J Clin Lab Immunol 40:61–67
Kühl U, Lauer B, Souvatzoglu M, Vosberg H, Schultheiss HP (1998) Antimyosin scintigraphy and immunohistologic analysis of endomyocardial biopsy in patients with clinically suspected myocarditis C evidence of myocardial cell damage and inflammation in the absence of histologic signs of myocarditis. J Am Coll Cardiol 32:1371–1376
Lane JR, Neumann DA, Lafond-Walker A, Herskowitz A, Rose NR (1992) Interleukin 1 or tumor necrosis factor can promote Coxsackie B3-induced myocarditis in resistant B10.A mice. J Exp Med 175:1123–1129
Lane JR, Neumann DA, Lafond-Walker A, Herskowitz A, Rose NR (1993) Role of IL-1 and tumor necrosis factor in coxsackie virus-induced autoimmune myocarditis. J Immunol 151:1682–1690
Lauer B, Schanwell M, Kühl U, Strauer BE, Schultheiss HP (2000) Antimyosin autoantibodies are associated with deterioration of systolic and diastolic left ventricular function in patients with chronic myocarditis. J Am Coll Cardiol 35:11–18
Limas CJ, Limas C (1991) Beta-adrenoceptor antibodies and genetics in dilated cardiomyopathy C an overview and review. Eur Heart J 12(Suppl D):175–177
Limas CJ, Goldenberg IF, Limas C (1990) Influence of anti-beta receptor antibodies on cardiac adenylate cyclase in patients with idiopathic dilated cardiomyopathy. Am Heart J 119:1322–1328

Magnusson Y, Marullo S, Hoyer S, Waagstein F, Andersson B, Vahlne A, Guillet JG, Strossberg AD, Hjalmarson A, Hoebeke J (1990) Mapping of a functional autoimmune epitope on the beta 1-adrenergic receptor in patients with idiopathic dilated cardiomyopathy. J Clin Invest 86:1658–1663

Mahon NG, Madden BP, Caforio ALP, Elliott PM, Haven AJ, Keogh BE, Davies MJ, McKenna WJ (2002) Immunohistologic evidence of myocardial disease in apparently healthy relatives of patients with dilated cardiomyopathy. J Am Coll Cardiol 39:455–562

Mason JW, O'Connell JB, Herskowitz A, Rose NR, McManus BM, Billingham ME, Moon TE (1995) A clinical trial of immunosuppressive therapy for myocarditis. The Myocarditis Treatment Trial Investigators. N Engl J Med 333:269–275

Neu N, Rose NR, Beisel KW, Herskowitz A, Gurri-Glass G, Craig SW (1987) Cardiac myosin induces myocarditis in genetically predisposed mice. J Immunol 139:3630–3636

Neumann DA, Burek CL, Baughman KL, ROSE NR, Herskowitz A (1990) Circulating heart-reactive antibodies in patients with myocarditis or cardiomyopathy. J Am Coll Cardiol 16:839–846

Okazaki T, Tanaka Y, Nishio R, Mitsuiye T, Mizoguchi A, Wang J, Ishida M, Hiai H, Matsumori A, Minato N, Honjho T (2003) Autoantibodies against cardiac troponin I are responsible for dilated cardiomyopathy in PD-1 deficient mice. Nat Med 9:1477–1483

Rose NR, Baughman K (1998) Immune-mediated cardiovascular disease. In: Rose NR, Mackay IR (eds) Autoimmune diseases, 3rd edn. Academic Press, San Diego, pp 623–636

Rose NR, Beisel KW, Herskowitz A, Neu N, Wolfgram LJ, Alvarez FL, Traystman MD, Craig SW (1987) Cardiac myosin and autoimmune myocarditis. In: Evered D, Whelan J (eds) Autoimmunity and autoimmune disease. Ciba Foundation Symposium 129. John Wiley and Sons, Chichester, pp 3–24

Schulze K, Becker BF, Schauer R, Schultheiss P (1990) Antibodies to ADP-ATP carrier – an autoantigen in myocarditis and dilated cardiomyopathy – impair cardiac function. Circulation 81:959–969

Smith SC, Allen PM (1991) Myosin-induced acute myocarditis is a T cell-mediated disease. J Immunol 147:2141–2147

Smith SC, Allen PM (1992) Neutralization of endogenous tumor necrosis factor ameliorates the severity of myosin-induced myocarditis. Circ Res 70:856–863

Staudt A, Böhm M, Knebel F, Grosse Y, Bischoff C, Hummel A, Dahm JB, Borges A, Jochmann N, Wernecke K, Wallukat G, Baumann G, Felix SB (2002) Potential role of autoantibodies belonging to the immunoglobulin

G-3 subclass in cardiac dysfunction among patients with dilated cardiomyopathy. Circulation 106:2448–2453

Sundstrom JB, Fett JD, Carraway RD, Ansari AA (2002) Is peripartum cardiomyopathy an organ-specific autoimmune disease? Autoimmun Rev 1:73–77

Towbin JA, Bowles KR, Bowles NE (1999) Etiologies of cardiomyopathies and heart failure. Nat Med 5:266–267

Wallukat G, Podlowski S, Nissen E, Morwinski R, Csonka C, Tosaki A, Blasig IE (2003) Functional and structural characterization of anti-beta1-adrenoceptor autoantibodies of spontaneously hypertensive rats. Mol Cell Biochem 251:67–75

Warraich RS, Dunn MJ, Yacoub MH (1999) Subclass specificity of autoantibodies against myosin in patients with idiopathic dilated cardiomyopathy: pro-inflammatory antibodies in DCM patients. Biochem Biophys Res Commun 259:255–261

Warraich RS, Pomerance A, Stanley A, Banner NR, Dunn MJ, Yacoub MH (2000) Cardiac myosin autoantibodies and acute rejection after heart transplantation in patients with dilated cardiomyopathy. Transplantation 69:1609–1617

10 The Roles of Immunity and Autoimmunity in Chronic Heart Failure

S. von Haehling, W. Doehner, S.D. Anker

Abstract. Chronic heart failure (CHF) represents a major public health burden in developed countries. The introduction of new treatments has helped to improve its prognosis in recent years. However, it is still not possible to directly target the immunological aspects of the disease. In fact, chronic immune activation with the up-regulation of pro-inflammatory substances in the plasma remains an important feature of the disease, independently of its aetiology. Autoimmune mechanisms play a significant role in a subgroup of patients with dilated cardiomyopathy. The interplay between the two systems has not been established so far. This review briefly summarizes immune and autoimmune mechanisms in CHF.

10.1 Introduction

Chronic heart failure (CHF) is a multisystem disorder that affects various bodily systems and not merely the cardiovascular system. Indeed, convincing evidence has accumulated over the last years to suggest that CHF represents a state of chronic inflammation (Anker and von Haehling 2004). The original discovery of elevated levels of tumour necrosis factor-α (TNF-α), a pro-inflammatory cytokine, in advanced stages of the disease (Levine et al. 1990) triggered an avalanche of research 15 years ago that has lasted to the present day. Thus, attention has mostly focussed on the role of TNF-α and other pro-inflammatory substances. However, the origin of pro-inflammatory activation remains enigmatic. Several hypotheses have been suggested to explain this phenomenon; there is only indirect evidence available to support these theories. However, it is clear that pro-inflammatory activation largely contributes to the progression of CHF and that it triggers the deterioration of the clinical status of the patients.

Pro-inflammatory cytokine activation occurs independently of CHF aetiology. A subgroup of approximately 25% of CHF patients present with dilated cardiomyopathy (DCM), a disorder associated with progressive dilatation of the (predominantly left) ventricle and loss of cardiac function in the absence of known causes. DCM represents the most important cause for severe CHF in younger adults in developed countries (Centers for Disease Control and Prevention 1998). A genetic background (mutations in genes encoding for myocyte structural proteins) is suspected in about 30% of these patients (Graham and Owens 1999; Seidman and Seidman 2001). In many other cases, myocarditis appears to be the underlying disorder that eventually yields DCM; however, in some patients the aetiology remains unclear. Autoantibodies also appear to play an important role. Current concepts regarding exogenous causes of DCM, therefore, comprise chronic myocarditis and primary abnormalities of the immune system (Kühl et al. 1996; Luppi et al. 1998). This, for example, is the case in giant-cell myocarditis, a rare disease associated with autoimmune disorders (Eriksson and Penninger 2005). The article will briefly summarize the roles of both immune and autoimmune mechanisms in CHF.

10.2 Immune Activation in CHF

The cytokine TNF-α belongs to a large family of pharmacologically active proteins called cytokines. Inflammatory cytokines are produced by monocytes/macrophages and different types of lymphocytes. Interestingly, it has been demonstrated that cardiomyocytes of CHF patients are also capable of its production (Mann 2001a). After transcription, which is controlled by various physiological mechanisms, TNF-α is inserted into the cell membrane of the respective cell (Kriegler et al. 1988). Proteolytic cleavage by TNF-α-converting enzyme (TACE) yields its soluble form. TNF-α signals via two distinct surface receptors, of which TNFR-1 (p55) is more abundantly expressed. Indeed, TNFR-1 appears to be the main signalling receptor, which mostly mediates deleterious and cytotoxic effects (Beutler and van Huffel 1994; von Haehling et al. 2004a). The other TNF-α receptor, TNFR-2 (p75), seems to play a more protective role in the heart, although it may induce cytotoxic effects as well (Heller et al. 1992).

The "cytokine hypothesis" holds that CHF progresses, at least in part, as a consequence of the toxic effects exerted by endogenous cytokine cascades on the heart and the peripheral circulation (Seta et al. 1996). Indeed, TNF-α confers a number of detrimental effects: the development of left ventricular dysfunction, left ventricular remodelling (Bozkurt et al. 1998; Pagani et al. 1992), increased cardiac myocyte apoptosis (Krown et al. 1996), the development of anorexia and cachexia (Anker et al. 1997a), reduced skeletal muscle blood flow and endothelial dysfunction (Anker et al. 1998), severity of insulin resistance (Leyva et al. 1998), activation of the inducible form of nitric oxide synthase, β-receptor uncoupling from adenylate cyclase (Anker and von Haehling 2004; Mann 2002) and other effects. The discovery of a close association of TNF-α plasma levels with both a poor short-term (Ferrari et al. 1995) and long-term (Rauchhaus et al. 2000; Deswal et al. 2001) survival of CHF patients was therefore not surprising. However, cytokines do not *cause* CHF per se. In fact, these substances "only" lead to disease progression (Seta et al. 1996; Mann 2002).

10.2.1 Competing Theories

Several hypotheses have been suggested to explain the origin of immune activation in CHF. These hypotheses attribute the origin of pro-inflammatory mediators to either mononuclear cells in the bloodstream or to peripheral tissues, or to the failing myocardium itself. In essence, TNF-α might be produced in response to myocardial injury (Matsumori et al. 1994), under-perfusion of peripheral tissues (Tsutamoto et al. 1998) or as a consequence of bacterial translocation through the oedematous gut wall (Anker et al. 1997b). Endotoxin (lipopolysaccharide, LPS), a cell-wall component of gram-negative bacteria, is one of the strongest inducers of TNF-α and other pro-inflammatory mediators (Genth-Zotz et al. 2002; Niebauer et al. 1999). Only indirect evidence is available to support the aforementioned hypotheses. It is also well possible that both cardiac and peripheral tissues contribute to elevated levels of pro-inflammatory substances in CHF.

10.2.2 Therapeutic Approaches

It is currently not possible to directly target immune activation as a means to stop CHF from progressing. A number of small studies showed that drugs like pentoxifylline (Skudicky et al. 2001; Sliwa et al. 1998; Bahrmann et al. 2004) and thalidomide (Gullestad et al. 2002; Agoston et al. 2002) are able to reduce TNF-α plasma levels in CHF patients. The results of these studies have recently been discussed elsewhere (von Haehling and Anker 2005). The most promising approach, however, directly targeting the TNF-α signalling pathway, has ultimately failed. Indeed, a large trial programme to study the effects of 24 weeks of subcutaneous application of etanercept, a TNFR-2 fusion protein, or placebo in a total of 2,048 CHF patients was terminated early because of missing clinical benefit (Mann et al. 2004). The combined analysis of the two study arms RENAISSANCE (Randomized Etanercept North AmerIcan Strategy to Study ANtagonism of CytokinEs) and RECOVER (Research into Etanercept: CytOkine Antagonism in VEntriculaR Dysfunction), showed that the number of patients classified "improved", "unchanged" or "worsened" was similar for patients on placebo or any dose of etanercept. These results are difficult to explain. One of the explanations might be that the patients were not screened for TNF-α plasma levels at study entry and that patients without an immunological dysregulation were studied. It is also possible that targeting a single cytokine is not sufficient.

A number of other therapeutic approaches are currently underway. Statins, also known as 3-hydroxy-3-methylglutaryl-coenzyme A reductase inhibitors, may prove beneficial in CHF treatment (von Haehling et al. 2004b). Another promising method is the so-called immune modulation therapy. A small double-blind, placebo-controlled study in 75 patients with moderate to severe CHF showed that this therapy significantly reduced the risk of death ($p = 0.022$) and of hospitalization ($p = 0.008$) after 6 months (Torre-Amione et al. 2004). Plasma levels of TNF-α, interleukin (IL)-6, interferon-γ, IL-10 and C-reactive protein (CRP) were left unaffected. A large-scale study is currently underway (von Haehling and Anker 2005). The procedure requires 10 ml of venous blood that is exposed to ultraviolet light and ozone gas in a special blood-treatment unit. Thereafter, the blood is re-injected into the respective patients. The result is an up-regulation of the anti-inflammatory cytokine IL-10 and of transforming growth factor-β (TGF-β; Fadok et al. 1998; Voll et al. 1997) and the induction of a specific T helper cell type 1 response (Gallucci et al. 1999; Jonuleit et al. 2000).

10.3 Autoimmunity in CHF

10.3.1 Autoantibodies

Autoimmunity is characterized by defined self-antigens, organ-specificity, autoreactive T-cells and/or autoantibodies that can transfer disease (Eriksson and Penninger 2005). Studies of autoimmunity in CHF have largely focussed on the role of different autoantibodies in DCM patients. The question of whether or not autoimmunity plays a significant role in CHF first arose almost three decades ago, when Sterin-Borda and colleagues found that an antibody from patients with Chagas' disease, a common cause of DCM in South America, interacts with the plasma membrane of cells from isolated rat atrial preparations (Sterin-Borda et al. 1976). Later studies confirmed that sera from some DCM patients contain different types of autoantibodies.

Autoantibodies against the second extracellular loop of the cardiac β_1-adrenergic receptor (β_1-EC$_{II}$) have been studied in greatest detail (Jahns et al. 2004). The β_1-EC$_{II}$ epitope represents a potent autoantigen (Magnusson et al. 1989), and antibodies against β_1-EC$_{II}$ activate the β_1-adrenergic receptor signalling cascade in vitro (Magnusson et al. 1994). These autoantibodies may yield continuous activation of the adrenergic system and inhibition of receptor desensitization (Hoebeke 2001). In vivo data support this view, as the presence of such antibodies is associated with significantly poorer left ventricular function [left ventricular ejection fraction (LVEF)] in DCM with vs without autoantibodies against β_1-adrenergic receptor epitopes (30.9 ± 2.2 vs $38.5 \pm 1.8\%$, $p < 0.05$; Jahns et al. 1999), and a higher prevalence of ventricular arrhythmias (Chiale et al. 2001). Moreover, the presence of autoantibodies against β-adrenergic receptors in sera of 104 patients with idiopathic DCM independently predicted sudden cardiac death ($p = 0.03$; Iwata et al. 2001). It appears that the autoantibodies exert their agonist-like effect via the β-adrenoceptor–adenylate cyclase–protein kinase A cascade (Wallukat et al. 1999). The effects are nullified by both β_1-selective and non-selective adrenergic antagonists. Indeed, the addition of bisoprolol, a β_1-selective antagonist, removed the antibodies from their binding sites (Magnusson et al. 1994). This is noteworthy, because it could contribute significantly to the beneficial effects of β-blocker treatment in DCM (Waagstein et al. 1993).

A number of other antibodies has also been identified in DCM that are directed against muscarinic receptors, mitochondrial antigens, adenosine diphosphate, adenosine triphosphate carrier proteins and cardiac myosin (Mann 2001b). Indeed, monoclonal antibodies directed against coxsackievirus group B, a virus known to cause myocarditis in some cases, cross-react with epitopes on murine cardiac myosin (Weller et al. 1989). However, the extent of genetic homology appears to be low (Rose and Hill 1996). A study in 26 DCM patients found 14 patients to be positive for cardiac autoantibodies (Caforio et al. 1992). Of the 14 DCM sera-containing heart-specific antibodies, 12 reacted with both the α- and β-myosin heavy chain isoforms (Caforio et al. 1992). These studies are in line with the finding that immunization of mice with α-myosin heavy chain can result in the development of a dilated cardiac phenotype (Liao et al. 1993).

An antibody specific to adenosine diphosphate (ADP)/adenosine triphosphate (ATP) translocator protein of the mitochondria was shown to cross-react with a cell surface molecule, most probably the calcium channel, on rat cardiac myocytes (Schultheiss et al. 1988). Its binding enhanced the transmembrane calcium current, and produced calcium-dependent cell-lysis in the absence of complement. Later studies revealed that this autoantibody might impair cardiac function (Schulze et al. 1990). Antibody-binding to its antigen usually yields complement activation and subsequent lysis of the antigen-coated cell (Mobini et al. 2004). Binding of C5b-9 complement, the terminal membrane attack complex, to cardiac myocytes itself has recently been shown to induce TNF-α synthesis and release from these cells (Zwaka et al. 2002). TNF-α release and C5b-9 correlated with immunoglobulin (Ig) G deposition in the myocardium. In vitro, a C5b-9 attack on cardiac myocytes induces nuclear factor (NF)-κB activation as well as transcription, synthesis and secretion of TNF-α. These authors observed abundant deposition of IgG in the myocardium of patients with DCM and assumed that IgG is a triggering factor that leads to complement activation in damaged tissue and induction of mitogenesis (Zwaka et al. 2002).

However, it has to be kept in mind that tissue injury (secondary to ischaemia or infection) can lead also to the production of autoantibodies (Mann 2001b). If this is the case, autoantibody-production is the result, rather than the cause of the tissue injury. If, on the other hand, autoantibodies play a role in DCM initiation and/or progression, autoantibody removal would be expected to have beneficial effects. Indeed, immunoadsorption, which was first used in the treatment of familial heterozygous hypercholesterolaemia, was effectively used to remove circulating antibodies against the β_1-adrenoceptor (Wallukat et al. 1996). This technique was also able to decrease markers of oxidative stress, such as thiobarbituric acid-reactive substances, lipid peroxides and anti-oxidized low-density lipoprotein-autoantibodies (all $p < 0.05$) within five consecutive days of treatment (Schimke et al. 2001).

10.3.2 T Cell-Mediated Mechanisms

An important role of T cells in the pathogenesis of coxsackievirus B infection was proposed over two decades ago. Indeed, this fact is underscored by the finding that T cell-depleted mice infected with coxsackievirus B3 exhibit a marked reduction in myocardial disease as compared to immunologically intact mice (Woodruff and Woodruff 1974). Moreover, inflammation of the heart muscle can be transferred to non-immunized mice by purified T cells from mice with active myocarditis (Pummerer et al. 1996). In this study, T cell-mediated autoimmune heart disease was induced by immunization with cardiac myosin. Shortly before the onset of disease, the authors recognized a significant increase in the number of cardiac interstitial cells expressing major histocompatibility complex (MHC) class II molecules (Pummerer et al. 1996). The authors therefore concluded that in cardiac myosin-induced myocarditis, expression of interstitial MHC II molecules (and possibly endothelial intracellular adhesion molecule-1) is a prerequisite for target organ recognition by autoreactive T cells. An abnormal expression of MHC class II molecules on endothelial cells is another common finding (Eriksson and Penninger 2005). Interestingly, the idea of autoimmune mechanisms in CHF is supported by mouse models, in which an association of the disease with certain specific haplotypes has been reported (Fairweather et al. 2001). It was demonstrated that transgenic mice expressing the human leucocyte antigen (HLA) DQ8 molecule only, develop spontaneous cardiomyopathy (Elliott et al. 2003; Taylor et al. 2004). However, the precise mechanisms that contribute to T cell autoimmune mechanisms are still not entirely understood.

10.4 Conclusion

The pathophysiology of chronic heart failure CHF – regardless of its disease aetiology – shares common immunological features. Particularly prominent is the presence of inflammatory immune activation with elevated levels of pro-inflammatory substances. Cardiac autoantibodies are a frequent finding in heart failure patients suffering from DCM, but rarely in CHF cases of other aetiologies. Therapeutic approaches may vary with different CHF backgrounds. Anti-cytokine strategies are most likely to be of benefit in patients with an inflammatory "problem", and it is necessary to identify the right patients for the right treatment. DCM patients may additionally benefit from immuno-adsorption of autoantibodies and treatments against oxidative stress. This also requires identification of the right patients before a treatment is initiated. We believe research in this area will provide the opportunity to improve the outcome for these patients – but this remains to be proved.

References

Agoston I, Dibbs ZI, Wang F, Muller G, Zeldis JB, Mann DL, Bozkurt B (2002) Preclinical and clinical assessment of the safety and potential efficacy of thalidomide in heart failure. J Card Fail 8:306–314

Anker SD, von Haehling S (2004) Inflammatory mediators in chronic heart failure: an overview. Heart 90:464–470

Anker SD, Chua TP, Ponikowski P, Harrington D, Swan JW, Kox WJ, Poole-Wilson PA, Coats AJ (1997a) Hormonal changes and catabolic/anabolic imbalance in chronic heart failure and their importance for cardiac cachexia. Circulation 96:526–534

Anker SD, Egerer KR, Volk HD, Kox WJ, Poole-Wilson PA, Coats AJ (1997b) Elevated soluble CD14 receptors and altered cytokines in chronic heart failure. Am J Cardiol 79:1426–1430

Anker SD, Volterrani M, Egerer KR, Felton CV, Kox WJ, Poole-Wilson PA, Coats AJ (1998) Tumour necrosis factor alpha as a predictor of impaired peak leg blood flow in patients with chronic heart failure. QJM 91:199–203

Bahrmann P, Hengst UM, Richartz BM, Figulla HR (2004) Pentoxifylline in ischemic, hypertensive and idiopathic-dilated cardiomyopathy: effects on left-ventricular function, inflammatory cytokines and symptoms. Eur J Heart Fail 6:195–201

Beutler B, van Huffel C (1994) Unraveling function in the TNF ligand and receptor families. Science 264:667–668

Bozkurt B, Kribbs SB, Clubb FJ Jr, Michael LH, Didenko VV, Hornsby PJ, Seta Y, Oral H, Spinale FG, Mann DL (1998) Pathophysiologically relevant concentrations of tumor necrosis factor-alpha promote progressive left ventricular dysfunction and remodeling in rats. Circulation 97:1382–1391

Caforio AL, Grazzini M, Mann JM, Keeling PJ, Bottazzo GF, McKenna WJ, Schiaffino S (1992) Identification of alpha- and beta-cardiac myosin heavy chain isoforms as major autoantigens in dilated cardiomyopathy. Circulation 85:1734–1742

Centers for Disease Control and Prevention (1998) Changes in mortality from heart failure, United States. J Am Med Assoc 280:874–875

Chiale PA, Ferrari I, Mahler E, Vallazza MA, Elizari MV, Rosenbaum MB, Levin MJ (2001) Differential profile and biochemical effects of antiautonomic membrane receptor antibodies in ventricular arrhythmias and sinus node dysfunction. Circulation 103:1765–1771

Deswal A, Petersen NJ, Feldman AM, Young JB, White BG, Mann DL (2001) Cytokines and cytokine receptors in advanced heart failure: an analysis of the cytokine database from the Vesnarinone trial (VEST). Circulation 103:2055–2059

Elliott JF, Liu J, Yuan ZN, Bautista-Lopez N, Wallbank SL, Suzuki K, Rayner D, Nation P, Robertson MA, Liu G, Kavanagh KM (2003) Autoimmune cardiomyopathy and heart block develop spontaneously in HLA-DQ8 transgenic IAbeta knockout NOD mice. Proc Natl Acad Sci USA 100:13447–13452

Eriksson U, Penninger JM (2005) Autoimmune heart failure: new understandings of pathogenesis. Int J Biochem Cell Biol 37:27–32

Fadok VA, Bratton DL, Konowal A, Freed PW, Westcott JY, Henson PM (1998) Macrophages that have ingested apoptotic cells in vitro inhibit proinflammatory cytokine production through autocrine/paracrine mechanisms involving TGF-beta, PGE2, and PAF. J Clin Invest 101:890–898

Fairweather D, Kaya Z, Shellam GR, Lawson CM, Rose NR (2001) From infection to autoimmunity. J Autoimmun 16:175–186

Ferrari R, Bachetti T, Confortini R, Opasich C, Febo O, Corti A, Cassani G, Visioli O (1995) Tumor necrosis factor soluble receptors in patients with various degrees of congestive heart failure. Circulation 92:1479–1486

Gallucci S, Lolkema M, Matzinger P (1999) Natural adjuvants: endogenous activators of dendritic cells. Nat Med 5:1249–1255

Genth-Zotz S, von Haehling S, Bolger AP, Kalra PR, Wensel R, Coats AJ, Anker SD (2002) Pathophysiologic quantities of endotoxin-induced tumor necrosis factor-alpha release in whole blood from patients with chronic heart failure. Am J Cardiol 90:1226–1230

Graham RM, Owens WA (1999) Pathogenesis of inherited forms of dilated cardiomyopathy. N Engl J Med 341:1759–1762

Gullestad L, Semb AG, Holt E, Skardal R, Ueland T, Yndestad A, Froland SS, Aukrust P (2002) Effect of thalidomide in patients with chronic heart failure. Am Heart J 144:847–850

Heller RA, Song K, Fan N, Chang DJ (1992) The p70 tumor necrosis factor receptor mediates cytotoxicity. Cell 70:47–56

Hoebeke J (2001) Molecular mechanisms of anti-G-protein-coupled receptor autoantibodies. Autoimmunity 34:161–164

Iwata M, Yoshikawa T, Baba A, Anzai T, Mitamura H, Ogawa S (2001) Autoantibodies against the second extracellular loop of beta1-adrenergic receptors predict ventricular tachycardia and sudden death in patients with idiopathic dilated cardiomyopathy. J Am Coll Cardiol 37:418–424

Jahns R, Boivin V, Siegmund C, Inselmann G, Lohse MJ, Boege F (1999) Autoantibodies activating human beta1-adrenergic receptors are associated with reduced cardiac function in chronic heart failure. Circulation 99:649–654

Jahns R, Boivin V, Hein L, Triebel S, Angermann CE, Ertl G, Lohse MJ (2004) Direct evidence for a beta 1-adrenergic receptor-directed autoimmune attack as a cause of idiopathic dilated cardiomyopathy. J Clin Invest 113:1419–1429

Jonuleit H, Schmitt E, Schuler G, Knop J, Enk AH (2000) Induction of interleukin 10-producing, nonproliferating CD4(+) T cells with regulatory properties by repetitive stimulation with allogeneic immature human dendritic cells. J Exp Med 192:1213–1322

Kriegler M, Perez C, DeFay K, Albert I, Lu SD (1988) A novel form of TNF/cachectin is a cell surface cytotoxic transmembrane protein: ramifications for the complex physiology of TNF. Cell 53:45–53

Krown KA, Page MT, Nguyen C, Zechner D, Gutierrez V, Comstock KL, Glembotski CC, Quintana PJ, Sabbadini RA (1996) Tumor necrosis factor alpha-induced apoptosis in cardiac myocytes. Involvement of the sphingolipid signaling cascade in cardiac cell death. J Clin Invest 98:2854–2865

Kühl U, Noutsias M, Seeberg B, Schultheiss HP (1996) Immunohistological evidence for a chronic intramyocardial inflammatory process in dilated cardiomyopathy. Heart 75:295–300

Levine B, Kalman J, Mayer L, Fillit HM, Packer M (1990) Elevated circulating levels of tumor necrosis factor in severe chronic heart failure. N Engl J Med 323:236–241

Leyva F, Anker SD, Godsland IF, Teixeira M, Hellewell PG, Kox WJ, Poole-Wilson PA, Coats AJ (1998) Uric acid in chronic heart failure: a marker of chronic inflammation. Eur Heart J 19:1814–1822

Liao L, Sindhwani R, Leinwand L, Diamond B, Factor S (1993) Cardiac alpha-myosin heavy chains differ in their induction of myocarditis. Identification of pathogenic epitopes. J Clin Invest 92:2877–2882

Luppi P, Rudert WA, Zanone MM, Stassi G, Trucco G, Finegold D, Boyle GJ, Del Nido P, McGowan FX Jr, Trucco M (1998) Idiopathic dilated cardiomyopathy: a superantigen-driven autoimmune disease. Circulation 98:777–785

Magnusson Y, Hoyer S, Lengagne R, Chapot MP, Guillet JG, Hjalmarson A, Strosberg AD, Hoebeke J (1989) Antigenic analysis of the second extracellular loop of the human beta-adrenergic receptors. Clin Exp Immunol 78:42–48

Magnusson Y, Wallukat G, Waagstein F, Hjalmarson A, Hoebeke J (1994) Autoimmunity in idiopathic dilated cardiomyopathy. Characterization of antibodies against the beta 1-adrenoceptor with positive chronotropic effect. Circulation 89:2760–2767

Mann DL (2001a) Recent insights into the role of tumor necrosis factor in the failing heart. Heart Fail Rev 6:71–80

Mann DL (2001b) Autoimmunity, immunoglobulin adsorption and dilated cardiomyopathy: has the time come for randomized clinical trials? J Am Coll Cardiol 38:184–186

Mann DL (2002) Inflammatory mediators and the failing heart: past, present, and the foreseeable future. Circ Res 91:988–998

Mann DL, McMurray JJ, Packer M, Swedberg K, Borer JS, Colucci WS, Djian J, Drexler H, Feldman A, Kober L, Krum H, Liu P, Nieminen M, Tavazzi L, van Veldhuisen DJ, Waldenstrom A, Warren M, Westheim A, Zannad F, Fleming T (2004) Targeted anticytokine therapy in patients with chronic heart failure: results of the Randomized Etanercept Worldwide Evaluation (RENEWAL). Circulation 109:1594–1602

Matsumori A, Yamada T, Suzuki H, Matoba Y, Sasayama S (1994) Increased circulating cytokines in patients with myocarditis and cardiomyopathy. Br Heart J 72:561–566

Mobini R, Maschke H, Waagstein F (2004) New insights into the pathogenesis of dilated cardiomyopathy: possible underlying autoimmune mechanisms and therapy. Autoimmun Rev 3:277–284

Niebauer J, Volk HD, Kemp M, Dominguez M, Schumann RR, Rauchhaus M, Poole-Wilson PA, Coats AJ, Anker SD (1999) Endotoxin and immune activation in chronic heart failure: a prospective cohort study. Lancet 353:1838–1842

Pagani FD, Baker LS, Hsi C, Knox M, Fink MP, Visner MS (1992) Left ventricular systolic and diastolic dysfunction after infusion of tumor necrosis factor-alpha in conscious dogs. J Clin Invest 90:389–398

Pummerer CL, Grassl G, Sailer M, Bachmaier KW, Penninger JM, Neu N (1996) Cardiac myosin-induced myocarditis: target recognition by autoreactive T cells requires prior activation of cardiac interstitial cells. Lab Invest 74:845–852

Rauchhaus M, Doehner W, Francis DP, Davos C, Kemp M, Liebenthal C, Niebauer J, Hooper J, Volk HD, Coats AJ, Anker SD (2000) Plasma cytokine parameters and mortality in patients with chronic heart failure. Circulation 19;102:3060–3067

Rose NR, Hill SL (1996) The pathogenesis of postinfectious myocarditis. Clin Immunol Immunopathol 80:S92–99

Schimke I, Muller J, Priem F, Kruse I, Schon B, Stein J, Kunze R, Wallukat G, Hetzer R (2001) Decreased oxidative stress in patients with idiopathic dilated cardiomyopathy one year after immunoglobulin adsorption. J Am Coll Cardiol 38:178–183
Schultheiss HP, Kühl U, Janda I, Melzner B, Ulrich G, Morad M (1988) Antibody-mediated enhancement of calcium permeability in cardiac myocytes. J Exp Med 168:2105–2119
Schulze K, Becker BF, Schauer R, Schultheiss HP (1990) Antibodies to ADP-ATP carrier – an autoantigen in myocarditis and dilated cardiomyopathy – impair cardiac function. Circulation 81:959–969
Seidman JG, Seidman C (2001) The genetic basis for cardiomyopathy: from mutation identification to mechanistic paradigms. Cell 104:557–567
Seta Y, Shan K, Bozkurt B, Oral H, Mann DL (1996) Basic mechanisms in heart failure: the cytokine hypothesis. J Card Fail 2:243–249
Skudicky D, Bergemann A, Sliwa K, Candy G, Sareli P (2001) Beneficial effects of pentoxifylline in patients with idiopathic dilated cardiomyopathy treated with angiotensin-converting enzyme inhibitors and carvedilol: results of a randomized study. Circulation 103:1083–1088
Sliwa K, Skudicky D, Candy G, Wisenbaugh T, Sareli P (1998) Randomised investigation of effects of pentoxifylline on left-ventricular performance in idiopathic dilated cardiomyopathy. Lancet 351:1091–1093
Sterin-Borda L, Cossio PM, Gimeno MF, Gimeno AL, Diez C, Laguens RP, Meckert PC, Arana RM (1976) Effect of chagasic sera on the rat isolated atrial preparation: immunological, morphological and function aspects. Cardiovasc Res 10:613–622
Torre-Amione G, Sestier F, Radovancevic B, Young J (2004) Effects of a novel immune modulation therapy in patients with advanced chronic heart failure: results of a randomized, controlled, phase II trial. J Am Coll Cardiol 44:1181–1186
Tsutamoto T, Hisanaga T, Wada A, Maeda K, Ohnishi M, Fukai D, Mabuchi N, Sawaki M, Kinoshita M (1998) Interleukin-6 spillover in the peripheral circulation increases with the severity of heart failure, and the high plasma level of interleukin-6 is an important prognostic predictor in patients with congestive heart failure. J Am Coll Cardiol 31:391–398
Voll RE, Herrmann M, Roth EA, Stach C, Kalden JR, Girkontaite I (1997) Immunosuppressive effects of apoptotic cells. Nature 390:350–351
von Haehling S, Anker SD (2005) Future prospects of anti-cytokine therapy in chronic heart failure. Expert Opin Investig Drugs 14:163–176
von Haehling S, Jankowska EA, Anker SD (2004a) Tumour necrosis factor-alpha and the failing heart: pathophysiology and therapeutic implications. Basic Res Cardiol 99:18–28

von Haehling S, Okonko DO, Anker SD (2004b) Statins: a treatment option for chronic heart failure? Heart Fail Monit 4:90–97

Waagstein F, Bristow MR, Swedberg K, Camerini F, Fowler MB, Silver MA, Gilbert EM, Johnson MR, Goss FG, Hjalmarson A (1993) Beneficial effects of metoprolol in idiopathic dilated cardiomyopathy. Metoprolol in Dilated Cardiomyopathy (MDC) Trial Study Group. Lancet 342:1441–1446

Wallukat G, Reinke P, Dorffel WV, Luther HP, Bestvater K, Felix SB, Baumann G (1996) Removal of autoantibodies in dilated cardiomyopathy by immunoadsorption. Int J Cardiol 54:191–195

Wallukat G, Muller J, Podlowski S, Nissen E, Morwinski R, Hetzer R (1999) Agonist-like beta-adrenoceptor antibodies in heart failure. Am J Cardiol 83:75H-79H

Weller AH, Simpson K, Herzum M, Van Houten N, Huber SA (1989) Coxsackievirus-B3-induced myocarditis: virus receptor antibodies modulate myocarditis. J Immunol 143:1843–1850

Woodruff JF, Woodruff JJ (1974) Involvement of T lymphocytes in the pathogenesis of coxsackie virus B3 heart disease. J Immunol 113:1726–1734

Zwaka TP, Manolov D, Ozdemir C, Marx N, Kaya Z, Kochs M, Hoher M, Hombach V, Torzewski J (2002) Complement and dilated cardiomyopathy: a role of sublytic terminal complement complex-induced tumor necrosis factor-alpha synthesis in cardiac myocytes. Am J Pathol 161:449–457

11 Clinical Implications of Anti-cardiac Immunity in Dilated Cardiomyopathy

A.L.P. Caforio, N.G. Mahon, W.J. McKenna

Abstract. Criteria of organ-specific autoimmunity are fulfilled in a subset of patients with myocarditis/dilated cardiomyopathy (DCM). In particular, circulating heart-reactive autoantibodies are found in such patients and symptom-free relatives. These autoantibodies are directed against multiple antigens, some of which are expressed in the heart (organ-specific), others in heart and some skeletal muscle fibres (partially heart-specific) or in heart and skeletal muscle (muscle-specific). Distinct autoantibodies have different frequency in disease and normal controls. Different techniques detect one or more antibodies, thus they cannot be used interchangeably for screening. It is unknown whether the same patients produce more antibodies or different patient groups develop autoimmunity to distinct antigens. IgG antibodies, shown to be cardiac- and disease-specific for myocarditis/DCM, can be used as autoimmune markers for relatives at risk as well as for identifying patients in whom immunosuppression may be beneficial. Some autoantibodies may also have a functional role, but further work is needed.

11.1 Introduction

Autoimmune disease takes place due to the loss of tolerance to self-antigens that is maintained under physiological conditions. To be classified as autoimmune, a disease must fulfill at least two of the major criteria proposed by Witebsky and later updated by Rose (Witebsky et al. 1957; Rose and Bona 1993). There are also minor criteria, some of which are common to all autoimmune conditions. These criteria are shown in Table 1.

Circulating autoantibodies, a feature of autoimmune disease, are not always pathogenic but represent markers of ongoing tissue damage. In non-organ specific autoimmune conditions the autoantibodies are against ubiquitous autoantigens (e.g. nuclear antigens in systemic lupus erythematosus) and organ damage is generalized. In organ-specific autoimmune disease, immunopathology is restricted to one tissue or apparatus, and the antibody and/or cell-mediated immune damage is directed against autoantigens that are unique to the affected organ (e.g. thyroid peroxidase in Hashimoto's thyroiditis). An early histological finding of organ-specific autoimmunity is a mononuclear cell infiltrate in the affected target tissue, e.g. insulitis in type 1 insulin-dependent diabetes mellitus (IDDM), with inappropriate or increased expression of HLA class II and of adhesion molecules. At a later stage, inflammatory cells are reduced and fibrotic changes occur, leading to tissue atrophy and organ dysfunction (such as in Hashimoto's thyroiditis). However, in other instances, organ-specific autoimmunity may be associated with enhanced target organ function (e.g. Basedow's disease).

Organ-specific autoimmune diseases occur as a result of genetic and environmental factors. The genetic predisposition explains why different autoimmune conditions may be associated in patients or in their relatives, as well as for the well-known feature that single autoimmune diseases often have familial aggregation. The inheritance of susceptibility is usually polygenic. Organ-specific autoimmune diseases are commonly associated with specific HLA class II antigens, but the mechanisms by which

Table 1. Criteria of autoimmune disease

Major

- Mononuclear cell infiltration and abnormal HLA expression in the target organ (organ-specific disease) or in various organs (non organ-specific disease)
- Circulating autoantibodies and/or autoreactive lymphocytes in patients and in unaffected family members
- Autoantibody and/or autoreactive lymphocytes in situ within the affected tissue
- Identification and isolation of autoantigen (s) involved
- Disease induced in animals by immunisation with relevant autoantigen, and/or passive transfer of serum, purified autoantibody and/or lymphocytes
- Efficacy of immunosuppressive therapy

Minor

(a) Common to all autoimmune disorders
- Middle-aged women most frequently affected
- Familial aggregation
- HLA association
- Hypogammaglobulinaemia
- Clinical course characterized by exacerbations and remissions
- Autoimmune diseases associated in the same patient or in family members

(b) Typical of organ-specific autoimmune disorders
- Autoantigens at low concentration
- Autoantibodies directly against organ-specific autoantigens
- Immunopathology mediated by type II, IV, V, VI reactions

HLA is involved in disease predisposition are largely undefined. The majority of organ-specific autoimmune diseases are chronic and previously classified as "idiopathic". Organ- and disease-specific antibodies are detected in the affected patients. These antibodies are also found in family members years before disease development and thus identify asymptomatic relatives at risk (Bottazzo et al. 1986). Organ-specific autoimmunity is involved in a proportion of the idiopathic forms of dilated cardiomyopathy and myocarditis (MacLellan and Lusis 2003; Okazaki et al. 2003; Eriksson et al. 2003).

11.2 DCM and Myocarditis: Classification and Clinical Features

Dilated cardiomyopathy (DCM), is an important cause of heart failure and a common indication for heart transplantation. It is characterized by dilatation and impaired contraction of the left or both ventricles; it may be idiopathic, familial/genetic, viral, and/or immune-based (Richardson et al. 1996). Clinical presentation may be with symptoms/signs of congestive heart failure, bradyarrhythmia/tachyarrhythmia or thromboembolism. The duration of the pre-clinical phase is often uncertain. Asymptomatic DCM may be diagnosed following the detection of a systolic murmur of mitral insufficiency, of a left bundle branch block on ECG, or of enlarged cardiac chambers with systolic dysfunction on echocardiography. DCM is familial in at least 30% of cases and has adverse prognosis. The diagnosis of DCM requires the exclusion of known, specific causes of heart failure, including coronary artery disease. On endomyocardial biopsy there is myocyte loss, compensatory hypertrophy, fibrous tissue and immunohistochemical findings consistent with chronic inflammation (myocarditis) in 30%–40% of cases.

Myocarditis is an inflammatory disease of the myocardium, and is diagnosed by endomyocardial biopsy using established histological, immunological and immunohistochemical criteria; it may be idiopathic, infectious or autoimmune and may heal or lead to DCM (Aretz et al. 1985; Richardson et al. 1996; Caforio and McKenna 1996). The clinical features of myocarditis are heterogeneous. Cardiac involvement may be preceded (1–2 weeks) by a systemic flu-like illness. Myocarditis may be subclinical, causing minor symptoms (palpitation, atypical chest pain), ECG abnormalities (atrioventricular conduction disturbance, bundle branch block, ST and T wave changes), or arrhythmias (paroxysmal atrial fibrillation or ventricular arrhythmias) with or without global or regional left and/or right ventricular dysfunction. Pericardial involvement commonly coexists with myocarditis. Other presentations of myocarditis include syncope, sudden death, acute right or left ventricular failure, cardiogenic shock, or DCM. A syndrome mimicking acute myocardial infarction, but with normal coronary arteries, may also occur.

Prognosis of myocarditis is thought to be good, with complete recovery in the majority of patients. However, in neonates and young children, the elderly and the debilitated, the disease is often severe, causing fulminant and fatal heart failure. Relapses may occur, and a proportion of patients will develop residual mild left ventricular dysfunction or DCM. Thus, in a patient subset, myocarditis and DCM represent the acute and chronic stages of an inflammatory disease of the myocardium, which can be viral, post-infectious immune or primarily organ-specific autoimmune (Caforio 1994; Richardson et al. 1996; Caforio and McKenna 1996).

11.3 Involvement of Autoimmunity in Myocarditis and DCM

Autoimmune features in human myocarditis/DCM include familial aggregation (Baig et al. 1998), a weak association with HLA-DR4 (Anderson et al. 1984), lymphocytic mononuclear cell infiltrates, abnormal expression of HLA class II and adhesion molecules on cardiac endothelium in affected patients and family members (Caforio et al. 1990b; Caforio and McKenna 1996; Mahon et al. 2002) and increased levels of circulating cytokines and cardiac autoantibodies in patients and family members (reviewed by Caforio et al. 2002). In addition, there are experimental models of both antibody-mediated and cell-mediated autoimmune myocarditis/DCM following immunization with relevant autoantigen(s) (MacLellan and Lusis 2003; Okazaki et al. 2003; Eriksson et al. 2003), the best characterized of which is cardiac myosin (Rose

2000; Kuan et al. 2000). In this chapter, we mainly focus on the clinical implications of circulating cardiac autoantibodies.

11.3.1 Anti-heart Autoantibodies by s-I IFL

Several researchers reported antibodies to distinct cardiac antigens in myocarditis and DCM, but the organ- and disease-specificity of these antibody types were not always evaluated (reviewed by Caforio et al. 2002). Earlier studies by standard indirect immunofluorescence (s-I IFL) described antibodies to sarcolemmal and myofibrillar antigens, but these were either cross-reactive or untested on skeletal muscle. In addition, it remained unclear whether these antibodies were disease-specific for myocarditis/DCM, because controls with other cardiac disease were not always tested. These autoantibodies were found in 12%–75% of DCM/myocarditis patients and in 4%–34% of normal control subjects (Fletcher and Wenger 1968; Camp et al. 1969; Kirsner et al. 1973; Maisch et al. 1983a,b; Table 2). The observed antibody patterns were similar to those described in rheumatic heart disease, Dressler's and post-pericardiotomy syndromes, the "diffuse" being more frequent than the "sarcolemmal-sarcoplasmic" staining pattern. A more recent study on rat heart tissue sections showed high titre (1:20) antibodies of IgG class in 59% of patients with myocarditis, 20% with DCM and in none of the normal control subjects; interestingly, these authors suggested that the three main antibody patterns ("diffuse"; "peripheral" or "sarcolemmal"; "fibrillary" or "striated") could co-exist (Neumann et al. 1990).

Using indirect s-I IFL on 4-μm-thick unfixed fresh frozen cryostat sections of blood group O normal human atrium, ventricle and skeletal muscle, and absorption with human heart and skeletal muscle and rat liver, organ-specific antibodies of IgG class were found in about one third of myocarditis/DCM patients and their symptom-free family members, 1% of patients with other cardiac disease, 3% of normal subjects and 17% of patients without cardiac disease but with autoimmune polyendocrinopathy (Caforio et al. 1990a, 1991, 1994, 1997a; Table 2). Cardiac antibodies of the cross-reactive 1 type, which exhibited partial organ-specificity for heart antigens by absorption, were also more frequently detected in DCM/myocarditis and autoimmune polyendocrinopathy than in controls. Conversely, cardiac antibodies of the cross-reactive 2 type,

Table 2. Anti-heart autoantibodies in myocarditis (M) and DCM

Antibody	Antibody positive (%) Technique	M	DCM	OCD	Normals	Reference(s)
Muscle-specific						
ASA	s-I IFL	47*	10	NT	25	Maisch et al. 1983a,b
AMLA	AMC	41*	9	NT	12	Maisch et al. 1983a,b
AFA	s-I IFL	28*	24*	NT	6	Maisch et al. 1983a,b
IFA	s-I IFL	32*	41*	NT	3	Maisch et al. 1983a,b
Heart-reactive	s-I IFL	59*	20*	NT	0	Neumann et al. 1990
	s-I IFL	NT	12–28	21–33	4	Fletcher and Wenger 1968; Camp et al. 1969; Kirsner et al. 1973
Anti-s.Na/K-ATPase	ELISA+Western blot	NT	26*	NT	2	Baba et al. 2002
Organ-specific cardiac	s-I IFL+abs	34*,**	26*,**	1	3	Caforio et al. 1990a, 1997a,b
Anti-Mitochondrial						
M7	ELISA	13*	31*	10	0	Klein et al. 1984
ANT	SPRIA	91*,**	57*,**	0	0	Schultheiss and Bolte 1985; Schultheiss et al. 1990
BCKD-E2	ELISA	100*,**	60*,**	4	0	Ansari et al. 1994
Anti-laminin	ELISA	73	78	25–35	6	Wolff et al. 1989

AFA, anti-fibrillary antibody; AMC, antibody-mediated cytotoxicity; AMLA, anti-myolemmal antibody; ASA, anti-sarcolemmal antibody; IFA, anti-interfibrillary; LBI, ligand-binding inhibition; NT, not tested; OCD, other cardiac disease (other abbreviations as in text) * $p < 0.05$ vs normals ** $p < 0.05$ vs OCD +abs, +absorption

Table 2. (continued)

Antibody	Antibody positive (%) Technique	M	DCM	OCD	Normals	Reference(s)
Anti-β_1 Receptor						
Inhibiting	LBI	NT	30–75*,**	37	18	Limas et al. 1989, 1991
	ELISA	NT	31*,**	0	12	Magnusson et al. 1990, 1994
Stimulating	Bioassay	96*,**	95*,**	8	0	Wallukat et al. 1991
	ELISA	NT	38**	6	19	Chiale et al. 1995
	ELISA	NT	26*,**	10	1	Jahns et al. 1999
Anti-M2 Receptor	ELISA	NT	39*	NT	7.5	Fu et al. 1993
Anti-α-and -β-MHC	Western blot	NT	46*,**	8	0	Caforio et al. 1992
Anti-MLC 1v	Western blot	NT	35	25	15	Caforio et al. 1992
Non myofibrillar	Western blot	NT	46*,**	17	0	Caforio et al. 1992
Anti-MHC	Western blot	NT	67**	42	NT	Latif et al. 1993
Anti-MLC 1	Western blot	NT	17**	0	NT	Latif et al. 1993
Anti-tropomyosin	Western blot	NT	55**	21	NT	Latif et al. 1993
Anti-actin	Western blot	NT	71**	21	NT	Latif et al. 1993
Anti-HSP-60	Western blot	NT	85**	42	NT	Latif et al. 1993
Anti-HSP-60, 70	Western blot	NT	10–14**	1–2	3	Portig et al. 1997
Anti-β-MHC	ELISA	37*,**	44*,**	16	2.5	Lauer et al. 1994
Anti-α-MHC	ELISA	17*,**	20*,**	4	2	Goldman et al. 1995; Caforio et al. 1997a,b

AFA, anti-fibrillary antibody; M, myocarditis; AMC, antibody-mediated cytotoxicity; AMLA, anti-myolemmal antibody; ASA, anti-sarcolemmal antibody; IFA, anti-interfibrillary; LBI, ligand-binding inhibition; NT, not tested; OCD, other cardiac disease (other abbreviations as in text) *$p < 0.05$ vs normals **$p < 0.05$ vs OCD +abs, +absorption which were entirely skeletal muscle cross-reactive by absorption, were found in similar proportions among groups. No patients with Dressler's or post-pericardiotomy syndromes, or active rheumatic heart disease, were included in these studies (Caforio et al. 1990a, 1991, 1997a). The s-I IFL patterns (figures in Caforio et al. 1990a ; Betterle et al. 1997) were as follows:

1. "*Organ-specific*". Sera were observed which gave diffuse cytoplasmic staining of both atrial and ventricular myocytes. The staining was stronger in atrial than in ventricular myocytes. These sera were negative on skeletal muscle. The titre range of these antibodies was 1/10 to 1/80 on atrial tissue and 1/10 to 1/20 on ventricular tissue. All positive sera contained antibodies of IgG class and 10% of IgM class. Organ-specific cardiac antibodies titres fell after absorption with heart homogenate, but were not affected by incubation with skeletal muscle or rat liver.
2. "*Cross-reactive 1*" or "*Partially Organ-specific*". Antibodies which gave a fine striational staining pattern on atrium and to a lesser extent on ventricle, but were negative or only weakly stained skeletal muscle sections, were classified as "cross-reactive 1" or "partially organ-specific". Their titres ranged from 1/20 to 1/80 on the atrial substrate and from 1/10 to 1/40 on ventricular tissue. All positive sera contained antibodies of IgG class and few also contained IgM antibodies. Antibody titres in the sera classified as "cross-reactive 1" or "partially organ-specific" were reduced to the same extent after absorption with heart and with skeletal muscle, and were not affected by absorption with rat liver.
3. "*Cross-reactive 2*". Antibodies which gave a broad striational pattern on longitudinal sections of heart and skeletal muscle were classified as "cross-reactive 2". This pattern had been previously found in 30%–40% of sera from myasthenia gravis patients without thymoma and in all myasthenic patients with thymoma (Zweiman and Arnason 1987). These striational antibodies have been shown to react with the A band of the myofibrils of striated muscle, and to cross-react with thymus myoepithelial cells. Cardiac antibodies in the sera classified as "cross-reactive 2" were absorbed by human skeletal muscle and to a lesser extent by heart tissue, but not by rat liver.

11.3.2 Anti-heart Autoantibodies by s-I IFL: Technical Aspects and Suggested Nomenclature

1. O blood group human heart and skeletal muscle are employed to avoid false-positive reactions due to heterophile or anti-ABO antibodies. Testing sera on skeletal muscle is necessary to differentiate heart-specific (organ-specific) from cross-reactive patterns (positive on heart and skeletal muscle), and on rat liver and kidney to detect nonorgan-specific mitochondrial or smooth muscle antibodies which give false positive "muscle reactive" IFL (Nicholson et al. 1977; Caforio et al. 1990a; Betterle et al. 1997). The pattern defined as "intermyofibrillary" is rare and might represent anti-mitochondrial antibodies (Nicholson et al. 1977). A pseudo-sarcolemmal or "endomysial" interstitial pattern can be observed on some tissue substrates (heart and muscle). It lacks species and tissue specificity, gives a "brush border" staining on proximal tubules of rat kidney and it represents a false-positive reaction (Nicholson et al. 1977; Betterle et al. 1997).
2. Several studies suggest that there is not a pure "sarcolemmal" or "peripheral" pattern; in fact, sera giving "striated" patterns seem to react more intensely with the periphery of the myofibre if the section does not include longitudinally cut fibres (Nicholson et al. 1977; Neumann et al. 1990; Caforio et al. 1990a; Betterle et al. 1997). It is important that the section includes longitudinally cut fibres, in order to identify "striated" patterns not visible on transverse sections.
3. It is important to use standard positive and negative control sera titrated up to end-dilution in every assay, to minimize inter-assay variability (Caforio et al. 1990a; Betterle et al. 1997).
4. New potentially organ-specific patterns should be confirmed as heart-specific by absorption (Caforio et al. 1990a; Betterle et al. 1997). If absorption is not performed, patterns of anti-heart autoantibodies by s-I IFL should be classified according to those already described and characterized (Nicholson 1977; Caforio et al. 1990a; Betterle et al. 1997), as follows:

   (a) "*Diffuse*" (also defined as "diffuse- sarcoplasmic" or "organ-specific")

(b) "*Striated, non-myasthenic*" (also defined as "cross-reactive type 1", or "partially organ-specific" or "anti-fibrillary")
(c) "*Striated, myasthenic*" (also defined as "cross-reactive type 2")
(d) "*Diffuse/striated, non-myasthenic*", if a "striated, non-myasthenic" pattern is superimposed on a diffuse sarcoplasmic stain, resulting in a combination of patterns a and b
(e) "*Anti-intercalated disks*", isolated or in combination with diffuse or striated stain

11.3.3 Autoantibodies to Myosin Heavy Chain and Other Autoantigens by Immunoblotting Techniques

Two of the autoantigens recognized by the cardiac autoantibodies detected by IFL were identified as α- and β-myosin heavy chain (MHC) isoforms, as well as ventricular light chain type 1 (MLC-1v), by Western blotting; several bands due to unknown antigens were also detected in DCM-positive sera (Caforio et al. 1992). These unknown antigens had an apparent molecular weight of 30–35 kDa in 50% of positive sera, 55–60 kDa in 21%, 100 kDa in 14% and 130–150 kDa in 14% (Caforio et al. 1992). The β-MHC is expressed in slow skeletal and ventricular myosin and is therefore only partially cardiac specific. The α-isoform is expressed solely within the atrial myocardium. Antibodies to this molecule represent organ-specific cardiac autoantibodies. The identification of α- and β-MHC as autoantigens in human DCM parallels what is seen in the experimental model of autoimmune myocarditis/DCM (Neu et al. 1987; Smith and Allen 1991) and in human myocarditis (Caforio et al. 1997a,b; Neumann et al. 1990; Lauer et al. 2000). The finding of anti-MHC and MLC-1v antibodies of IgG class in DCM patients has been confirmed by several groups using Western blotting (Latif et al. 1993) or enzyme-linked immunosorbent assay (ELISA) (Limas et al. 1995; Goldman et al. 1995; Michels et al. 1994). A recent study has suggested that the disease-specific anti-myosin antibodies in DCM sera are mainly of IgG3 subclass (Warraich et al. 1999). By Western blot, antibodies to heat shock protein-60 (HSP-60), tropomyosin and actin have also been found more frequently in DCM than in ischaemic heart disease controls, but normal sera were not tested (Latif et al. 1993; Table 2). Others (Portig et al. 1997) found antibodies to HSP-60 and -70 at a higher frequency in DCM sera than in ischaemic or normal control subjects (Table 2). Latif et al. developed a microcytotoxicity assay and showed complement-mediated cytotoxic activity of DCM sera containing anti-heart antibodies by Western blot (Latif et al. 1994). DCM, ischaemic and normal control sera were screened used W1, a transformed human fetal cardiac cell line, and also EA.hy 926, an endothelial cell line, and IRB3, a fibroblast cell line. In the presence of complement, sera from 28 (62%) DCM patients showed greater killing of the W1 cell line as compared to sera from 13 (30%) of ischaemic patients ($p < 0.005$) and 3 (15%) normal subjects. Only 1 DCM patient showed killing of EA.hy 926 cell line, and 1 ischaemic showed killing of the fibroblast cell line. These in vitro data suggest that a complement-dependent, antibody-mediated mechanism of damage to cardiac myocytes may contribute to the pathogenesis of DCM.

11.3.4 Autoantibodies to Sarcolemmal Na-K-ATPase

A recent study, using porcine cerebral cortex Na-K-ATPase as an antigen in ELISA and as a substrate in enzyme activity measurement, tested sera from 100 DCM patients and 100 healthy individuals and found anti-Na-K-ATPase autoantibodies in 26% of DCM and in 2% of normal subjects (Baba et al. 2002; Table 2). Western blots showed that the antibodies recognized the α-subunit, and 3H-ouabain bindings in the presence of patient IgG showed that the dissociation constant was higher in DCM patients with antibodies than in those without, suggesting biological activity for the antibody. By multiple regression analysis, the presence of anti-Na-K-ATPase autoantibodies was an independent predictor for the occurrence of ventricular tachycardia. Cardiac sudden death was independently predicted by the presence of antibodies, as well as poor systolic function. The authors speculated that these antibodies may lead to electrical instability, because of abnormal Ca^{2+} handling by reduced Na-K-ATPase activity, and delayed afterdepolarizations via reverse-mode operation of the Na^{+}/Ca^{2+} exchanger, resulting from increased intracellular Na^{+} concentrations. Although this represents an interesting hypothesis, no definitive conclusions on the functional role of these antibodies can be drawn at present. It remains to be seen whether these antibodies are disease-specific for DCM, since no controls with heart failure from other aetiologies were studied. It is worth noting that sarcolemmal Na-K-ATPase does not seem to fulfill the strict criteria of organ-specific cardiac autoantigens: the α-1 subunit isoform is expressed in most tissues, the α-2 is predominant in skeletal muscle and can be detected in brain and heart, the α-3 is found in excitable tissues and the α-4 in testis (Urayama et al. 1989; Muller-Ehmsen et al. 2001). Similarly, the β-1 subunit isoform is fairly ubiquitous, whereas the β-2 and β-3 subunit isoforms are mostly found in skeletal muscle, neural tissues, lung and liver. In human heart, only α-1β-1, α-2β-1 and α-3β-1 heterodimers are present, and are thought to be involved in the actions of cardiac glycosides (Schwinger et al. 1999).

11.3.5 Autoantibodies to Mitochondrial and to Extracellular Matrix Antigens

Using ELISA, autoantibodies against laminin, a large basement membrane glycoprotein, were found in 73%–78% of myocarditis/DCM patients and in 6% of normal subjects; the authors did not include ischaemic heart disease controls, but they reported unpublished data indicating 25%–35% prevalence in coronary artery disease (Wolff et al. 1989; Table 2). Antibodies against distinct mitochondrial antigens – the M7 (Klein et al. 1984; Otto et al. 1998), the adenine nucleotide translocator (ANT) (Schultheiss and Bolte 1985; Schultheiss et al. 1990) and the branched chain α-ketoacid dehydrogenase dihydrolipoyl transacylase (BCKD-E2) (Ansari et al. 1994) and other respiratory chain enzymes (Pohlner et al. 1997) have also been detected. The M7 antibodies, detected by ELISA, were of IgG class and were found in 31% of DCM patients, 13% of those with myocarditis, 33% of controls with hypertrophic cardiomyopathy, but not in control subjects with other cardiac disease, other immune-mediated disorders or in normal subjects (Klein et al. 1984; Table 2). The test antigen was represented by different sub-cellular and sub-mitochondrial beef heart preparations; sera were also tested on sub-mitochondrial particles from pig kidney and rat liver. Using a indirect micro solid-phase radio immunoassay (SPRIA) and ANT, a protein of the internal mitochondrial membrane, purified from beef heart, liver and kidney as antigen, anti-ANT antibodies were found in 57%–91% of myocarditis/DCM sera, and in no controls with ischaemic heart disease or in normal subjects (Schultheiss et al. 1985, 1990; Table 2). Mitochondrial antigens have generally been classified as nonorgan-specific (Bottazzo et al. 1986; Rose et al. 1993). However, the heart-specificity of the M7 antibodies was shown by absorption studies, whereas these were not performed with the ANT and the BCKD-E2 antibodies. Experimentally induced affinity-purified anti-ANT antibodies cross-reacted with calcium channel complex proteins of rat cardiac myocytes, induced enhancement of transmembrane calcium current and produced calcium-dependent cell lysis in the absence of complement (Schultheiss et al. 1988, 1990). The authors suggested that such an enhancing effect of the antibodies in patients might lead to impaired function of the ANT, imbalance of energy delivery and demand within the myocyte, and subsequent cell death in vivo. The presence of this mechanism of antibody-dependent cell lysis has not been shown using the antibodies present in patients' sera.

11.3.6 Autoantibodies to β-Adrenergic and M2-Muscarinic Receptors

Several groups have demonstrated antibodies against the β_1-adrenoceptor (Wallukat et al. 1991; Limas et al. 1989; Limas and Limas 1991; Magnusson et al. 1990, 1994). Using a binding inhibition assay (inhibition of marked [^{3}H]dihydroalprenolol binding to rat cardiac membranes), a significant inhibitory activity, attributed to anti-β_1-adrenoceptor antibodies of IgG class, was found in 30%–75% of DCM sera, 37% of ischaemic or valvular heart disease controls and 18% of sera from normal subjects (Limas et al. 1989; Limas and Limas 1991). Positive DCM sera were also found to immunoprecipitate β-adrenoceptors from solubilized cardiac membranes. Antibody positive sera induced sequestration and endocytosis of β_1-receptors predominantly dependent on the β-receptor kinase, and selectively inhibited isoproterenol-sensitive adenylate cyclase activity (Limas et al. 1989; Limas and Limas 1991). Magnusson et al. (1990), using as antigens synthetic peptides analogous to the sequences of the second extra cellular loop of β_1- and β_2-adrenergic receptors by ELISA, found antibodies in 31% of DCM patients, 12% of normal subjects and in none of the controls with other cardiac disease. The antibodies from

DCM sera exhibited inhibitory activity of isoproterenol binding to the β-adrenergic receptor.

Other studies showed that, when analysed in a functional test system of spontaneously beating neonatal rat myocytes, antibody-positive DCM sera (Wallukat et al. 1991; Jahns et al. 1999, 2000) or the affinity-purified β_1-receptor antibodies (Magnusson et al. 1994) increased the beating frequency of isolated myocytes in vitro. β_1-Blocking drugs (propranolol, bisoprolol and metoprolol) inhibited the effect of the antibodies. These workers reported that the stimulating anti-β_1-receptor antibodies were present in 96% of myocarditis and 26%–95% of DCM sera, in 8%–10% of controls with ischaemic heart disease and 0%–19% of normal subjects (Table 2). They also suggested that this antibody-mediated stimulation of the β_1-receptor, observed in vitro, could occur in vivo and account for the accelerated decline in ventricular systolic function in some myocarditis/DCM patients.

Fu et al. (1993), using as antigen a synthetic peptide analogous to the 169–193 sequence of the second extra cellular loop of human M2 muscarinic receptors and an ELISA method, showed anti-M2 antibodies in 39% of DCM sera and 7% of the normal subjects (Table 2). The presence of the anti-M2 antibodies correlated with that of anti-β-receptor antibodies. A limitation of work involving the anti-receptor antibodies is that few disease controls have been studied. These receptors are not organ-specific cardiac autoantigens; in fact, their distribution is not restricted to the heart, and there are no cardiac-specific isoforms (Elalouf et al. 1993; Eglen et al. 1994).

11.4 Cardiac-Specific Antibodies in Myocarditis/DCM: Clinical Implications and Potential Functional Role

The presence of organ- and disease-specific cardiac antibodies of IgG class against myosin and other antigens supports the involvement of autoimmunity in at least one third of myocarditis/DCM patients (Caforio et al. 1990a; Neumann et al. 1990; Latif et al. 1993; Michels et al. 1994). These antibodies were associated with shorter duration and minor severity of symptoms, as well as with greater exercise capacity at diagnosis (Caforio et al. 1990a, 1997b, 2001). In many patients who were antibody positive at diagnosis, these markers became undetectable at follow-up (Caforio et al. 1997b). These findings suggest that cardiac-specific autoantibodies are early markers. The absence of antibodies at diagnosis in some patients could indicate that cell-mediated mechanisms are predominant and/or that autoimmunity is not involved; since the pre-clinical stage in DCM may be prolonged, it might also relate to reduction of antibody levels with disease progression (Caforio et al. 1997b). These findings have been obtained using standard autoimmune serology techniques, in particular s-I IFL, ELISA and immunoblotting, and have been confirmed by several groups (Neumann et al. 1990; Latif et al. 1993; Michels et al. 1994; Limas et al. 1995). The low frequency of cardiac-specific antibodies in control patients with heart dysfunction not due to myocarditis/DCM (Caforio 1990; Caforio et al. 1994; Goldman et al. 1995) and the decrease in antibody titres in advanced DCM (Caforio et al. 1997b, 2001) suggest that these markers are not epiphenomena associated with tissue necrosis of various causes, but represent specific markers of immune pathogenesis. The role of inflammatory cytokines (e.g. the IL-2/sIL-2R system) as markers of T lymphocyte activation in immune-mediated myocarditis/DCM and its relation to cardiac autoantibodies is a controversial issue (Limas et al. 1995; Caforio et al. 2001). Limas (1995) found that high-titre anti-β_1-receptor antibodies were more common among DCM patients with abnormal sIL-2R serum levels. Others found no association between the autoantibodies found by IFL and the anti-α-myosin antibodies detected by ELISA and sIL-2R levels (Caforio et al. 2001). sIL-2R may be related with distinct autoantibody specificities, e.g. in Graves' disease high sIL-2R was associated with anti-TSH receptor autoantibodies, but was unrelated to the autoantibodies to intracellular antigens (anti-microsomal and anti-thyroglobulin; Balazs and Farid 1991). The same may apply to DCM, high sIL-2R being present in association with antibodies to extracellular, e.g. the anti-β_1-receptor, rather than intracellular antigens, e.g. α-myosin and the other antigens involved in the IFL reaction. The cardiac-specific autoantibodies found by IFL and the anti-α-myosin antibodies detected by ELISA were found in similar proportions of patients with DCM and with biopsy-proven myocarditis according to the Dallas criteria, included in the Myocarditis Treatment Trial (Mason et al. 1995), suggesting that conventional histology does not distinguish between patients with and without an ongoing immune-mediated process in myocarditis/DCM (Caforio et al. 1997a). The Myocarditis Treatment Trial failed to show an improvement in survival in biopsy-proven myocarditis with immunosuppressive therapy; however, no immunohistochemical or serological markers (e.g. increased HLA expression on myocardial biopsy and/or detection of cardiac-specific autoantibodies in the serum in the absence of viral genome in myocardial tissue) were used to identify those patients with immune-mediated pathogenesis in whom immunosuppression could have been beneficial (Mason et al. 1995). Interestingly, a recent randomized, placebo-controlled study in DCM patients with HLA up-regulation on endomyocardial biopsy showed long-term benefit with immunosuppressive treatment (Wojnicz et al. 2001). Myocarditis/DCM patients with cardiac-specific autoantibodies should also be included in future trials of immunosuppressive therapy.

Myosin fulfilled the expected criteria for organ-specific autoimmunity, in that immunization with cardiac but not skeletal myosin reproduced, in susceptible mouse strains, the human disease phenotype of DCM (Neu et al. 1987; Smith et al. 1991). In this respect, less experimental data are available with other autoantigens. However, autoimmune diseases are often polyclonal, with production of autoantibodies to different autoantigens. Some of these autoantigens are involved earlier in disease and are more closely related to primary pathogenetic events compared to those which play a role in secondary immunopathogenesis (Rose et al. 1993). Both experimental and clinical evidence, in particular the multiplicity of autoantibody specificities identified so far (Table 2), exists that this also applies to myocarditis/DCM. Myosin is an intracellular protein, thus there are two major hypotheses – which may be not mutually exclusive – to explain interruption of tolerance to this autoantigen. These include molecular mimicry, since cross-reactive epitopes between cardiac myosin and infectious agents have been found, and myocyte necrosis due to viral infection or other tissue insults (Horwitz et al. 2000; Galvin et al. 2000; Rose 2000). Both mechanisms would explain the association of viral infection with autoimmune myocarditis/DCM. Infection with coxsackie B3 (CB3) virus triggers antimyosin reactivity and autoimmune myocarditis in many mouse strains, and immunization with cardiac myosin induces disease in the same susceptible strains (Neu et al. 1987; Smith et al. 1991). In some strains, such as Balb/c mice, CB3 virus-induced or myosin-induced myocarditis is T cell-mediated (Smith et al. 1991), whereas in other strains, such as DBA/2 mice, it is an antibody-mediated disease (Kuan et al. 2000). The same may apply to humans, so that the antimyosin antibodies may be directly pathogenic in some (Lauer et al. 2000) but not all patients with myocarditis/DCM (Caforio et al. 1997b) according to different immunogenetic backgrounds, isotype (Kuan et al. 2000) and/or subclass specificity of these antibodies (Warraich et al. 1999).

In relation to the proposed functional role of antibodies not against myosin, e.g. the anti-b and M2 receptor and some of anti-mitochondrial antibodies (Table 2) in man, passive transfer of the myocarditis/DCM phenotype to genetically susceptible animals by antibody-positive patients' sera would provide conclusive evidence for antibody-mediated pathogenesis. Non-antigen-specific IgG adsorption has recently been used in DCM patients with high titre antibodies to the β_1-receptor, and it has been suggested that it has beneficial clinical effects, accompanied by undetectable antibody titres during follow-up (Muller et al. 2000). This does not indicate a direct pathogenic effect of the anti-β_1-receptor antibodies. The adsorption technique used was non-antigen specific; in addition, in antibody-mediated disorders the antibody titres rise again at the end of plasmapheresis. However, the authors have recently provided new evidence in favour of the possibility that the beneficial effect of immunoadsorption is related to removal of pathogenic cardio-depressant autoantibodies of IgG3 subclass, although no conclusion is possible on the potential pathogenic role of a specific autoantibody (e.g. β_1-receptor antibody); Schimke et al. 2001; Felix et al. 2002; Staudt et al. 2002). It may be that this technique has a favourable immunomodulatory/immunosuppressive effect; in addition, IgG substitution performed after immunoadsorption to avoid infective complications of unselective IgG depletion, may have contributed to the observed haemodynamic improvement (Mann 2001; Gullestad et al. 2001). Thus, randomized studies are warranted. This does not undermine the possible role of any of the described antibodies (Table 2) as predictive markers. Subjects classified as negative for one antibody may be positive for another and combined testing may be advantageous. To this end, standardization of nomenclature and protocols for antibody detection and exchange of sera among laboratories currently assessing the individual antibodies will be useful.

In conclusion, several groups have shown that a subset of patients with myocarditis/idiopathic DCM and of their symptom-free relatives has circulating heart-reactive autoantibodies. These autoantibodies are directed against multiple antigens, some of which are strictly expressed in the myocardium (e.g. organ-specific for the heart), others are expressed in heart and skeletal muscle (e.g. muscle-specific). Distinct autoantibodies also have different prevalence in disease and normal controls (e.g. by IFL the organ-specific antibodies are disease-specific for DCM, some of the muscle-specific antibodies are not). Different antibody techniques detect one (e.g. ELISA for myosin or for anti-receptor antibodies) or more antibody specificities (e.g. indirect IFL), thus they cannot be used interchangeably as screening tools. Antibody frequency in DCM vs controls is expected to be different using distinct techniques; at present it is unknown whether the same subset (30%–40%) of patients produce more than one antibody or different patient groups develop autoimmunity to different antigens. Antibodies of IgG class, which are shown to be cardiac and disease-specific for myocarditis/DCM, can be used as reliable markers of autoimmune pathogenesis for identifying (1) patients in whom immunosuppression and/or immunomodulation therapy may be beneficial and (2) their relatives at risk. Some of these autoantibodies may also have a functional role in patients, as suggested by in vitro data as well as by preliminary clinical observations, but further work is needed to clarify this important issue.

References

Anderson JL, Carlquist JF, Lutz JR, et al. (1984) HLA A, B, and DR typing in idiopathic dilated cardiomyopathy: a search for immune response factors. Am J Cardiol 53:1326–1330

Ansari AA, Neckelmann N, Villinger F, et al. (1994) Epitope mapping of the branched chain a-ketoacid dehydrogenase dihydrolipoyl transacylase (BCKD-E2) protein that reacts with sera from patients with idiopathic dilated cardiomyopathy. J Immunol 153:4754–4765

Aretz HT, Billingham ME, Edwards WE, et al. (1985) Myocarditis: a histopathologic definition and classification. Am J Cardiovasc Pathol 1:1–10

Baba A, Yoshikawa T, Ogawa S (2002) Autoantibodies against sarcolemmal Na-K-ATPase: possible upstream targets of arrhythmias and sudden death in patients with dilated cardiomyopathy. J Am Coll Cardiol 40:1153–1159
Baig MK, Goldman JH, Caforio ALP, et al. (1998) Familial dilated cardiomyopathy: cardiac abnormalities are common in asymptomatic relatives and may represent early disease. J Am Coll Cardiol 31:195–201
Balazs CZ, Farid NR (1991) Soluble IL-2 receptor in sera of patients with Graves' disease. J Autoimmun 4:681–688
Betterle C, Spadaccino AC, Pedini B (1997) Autoanticorpi nelle malattie autoimmuni del muscolo scheletrico, della giunzione neuro-muscolare e del cuore. In: Betterle C (ed) Gli autoanticorpi. Manuale-Atlante a colori di diagnostica. Piccin, Padova, pp 313–337
Bottazzo GF, Todd I, Mirakian R, et al. (1986) Organ-specific autoimmunity. A 1986 overview. Immunol Rev 94:137–169
Caforio ALP (1994) Role of autoimmunity in dilated cardiomyopathy. Br Heart J 72:S30–S34
Caforio ALP, McKenna WJ (1996) Recognition and optimum treatment of myocarditis. Drugs 52:515–525
Caforio ALP, Bonifacio E, Stewart JT, et al. (1990a) Novel organ-specific circulating cardiac autoantibodies in dilated cardiomyopathy. J Am Coll Cardiol 15:1527–1534
Caforio ALP, Stewart JT, Bonifacio E, et al. (1990b) Inappropriate major histocompatibility complex expression on cardiac tissue in dilated cardiomyopathy. Relevance for autoimmunity? J Autoimmun 3:187–200
Caforio ALP, Wagner R, Gill JS, et al. (1991) Organ-specific cardiac antibodies: serological markers for systemic hypertension in autoimmune polyendocrinopathy. Lancet 337:1111–1115
Caforio ALP, Grazzini M, Mann JM, et al. (1992) Identification of α and β cardiac myosin heavy chain isoforms as major autoantigens in dilated cardiomyopathy. Circulation 85:1734–1742
Caforio ALP, Keeling PJ, Zachara E, et al. (1994) Evidence from family studies for autoimmunity in dilated cardiomyopathy. Lancet 344:773–777
Caforio ALP, Goldman JH, Haven AJ, et al. (1997a) Circulating cardiac autoantibodies as markers of autoimmunity in clinical and biopsy-proven myocarditis. Eur Heart J 18:270–275
Caforio ALP, Goldman JH, Baig KM, et al. (1997b) Cardiac autoantibodies in dilated cardiomyopathy become undetectable with disease progression. Heart 77:62–67

Caforio ALP, Goldman JH, Baig KM, et al. (2001) Elevated serum levels of soluble interleukin-2 receptor, neopterin and β-2 microglobulin in idiopathic dilated cardiomyopathy: relation to disease severity and autoimmune pathogenesis. Eur J Heart Fail 3:155–163
Caforio ALP, Mahon NJ, Tona F, et al. (2002) Circulating cardiac autoantibodies in dilated cardiomyopathy and myocarditis: pathogenetic and clinical significance. Eur J Heart Fail 4:411–417
Camp FT, Hess EV, Conway G, et al. (1969) Immunologic findings in idiopathic cardiomyopathy. Am Heart J 77:610–615
Chiale PA, Rosembaum MB, Elizari MV, et al. (1995) High prevalence of antibodies against b1- and b2-adrenoceptors in patients with primary electrical cardiac abnormalities. J Am Coll Cardiol 26:864–869
Eglen RM, Reddy H, Watson N, et al. (1994) Muscarinic acetylcholine receptor subtypes in smooth muscle. Trends Pharmacol Sci 15:114–119
Elalouf JM, Buhler JM, Tessiot C, et al. (1993) Predominant expression of beta 1-adrenergic receptor in the thick ascending limb of rat kidney: absolute mRNA quantitation by reverse transcription and polymerase chain reaction. J Clin Invest 91:264–272
Eriksson U, Ricci R, Hunziker L, et al. (2003) Dendritic cell-induced autoimmune heart failure requires cooperation between adaptive and innate immunity. Nat Med 9:1484–1490
Felix SB, Staudt A, Landsberger M, et al. (2002) Removal of cardiodepressant antibodies in dilated cardiomyopathy by immunoadsorption. J Am Coll Cardiol 39:646–652
Fletcher GF, Wenger NK (1968) Autoimmune studies in patients with primary myocardial disease. Circulation 37:1032–1035
Fu LX, Magnusson Y, Bergh CH, et al. (1993) Localization of a functional autoimmune epitope on the muscarinic acetylcholine receptor-2 in patients with idiopathic dilated cardiomyopathy. J Clin Invest 91:1964–1968
Galvin JE, Hemric ME, Ward K, et al. (2000) Cytotoxic mAb from rheumatic carditis recognizes heart valves and laminin. J Clin Invest 106:217–224
Goldman JH, Keeling PJ, Warraich RS, et al. (1995) Autoimmunity to α-myosin in a subset of patients with idiopathic dilated cardiomyopathy. Br Heart J 74:598–603
Gullestad L, Aass H, Fjeld JG, et al. (2001) Immunomodulating therapy with intravenous immunoglobulin in patients with chronic heart failure. Circulation 103:220–225
Horwitz MS, La Cava A, Fine C, et al. (2000) Pancreatic expression of interferon-γ protects mice from lethal coxsackievirus B3 infection and subsequent myocarditis. Nat Med 6:693–707

Jahns R, Boivin V, Siegmund C, et al. (1999) Autoantibodies activating human β1-adrenergic receptors are associated with reduced cardiac function in chronic heart failure. Circulation 99:649–654

Jahns R, Boivin V, Krapf T, et al. (2000) Modulation of β1-adrenoceptor activity by domain-specific antibodies and heart-failure associated autoantibodies. J Am Coll Cardiol 36:1280–1287

Kirsner AB, Hess EV, Fowler NO (1973) Immunologic findings in idiopathic cardiomyopathy: a prospective serial study. Am Heart J 86:625–630

Klein R, Maisch B, Kochsiek K, et al. (1984) Demonstration of organ specific antibodies against heart mitochondria (anti-M7) in sera from patients with some forms of heart diseases. Clin Exp Immunol 58:283–292

Kuan AP, Zuckier L, Liao L, et al. (2000) Immunoglobulin isotype determines pathogenicity in antibody-mediated myocarditis in naïve mice. Circ Res 86:281–285

Latif N, Baker CS, Dunn MJ, et al. (1993) Frequency and specificity of antiheart antibodies in patients with dilated cardiomyopathy detected using SDS-PAGE and western blotting. J Am Coll Cardiol 22:1378–1384

Latif N, Smith J, Dunn MJ, et al. (1994) Complement-mediated cytotoxic activity of anti-heart antibodies present in the sera of patients with dilated cardiomyopathy. Autoimmunity 19:99–104

Lauer B, Padberg K, Schultheiss HP, Strauer BE (1994) Autoantibodies against human ventricular myosin in sera of patients with acute and chronic myocarditis. J Am Coll Cardiol 23:146–153

Lauer B, Schannwell M, Kuhl U, et al. (2000) Antimyosin autoantibodies are associated with deterioration of systolic and diastolic left ventricular function in patients with chronic myocarditis. J Am Coll Cardiol 35:11–18

Limas CJ, Limas C (1991) β-Receptor antibodies and genetics in dilated cardiomyopathy. Eur Heart J 12(Suppl D):175–177

Limas CJ, Goldenberg IF, Limas C (1989) Autoantibodies against β-adrenoceptors in idiopathic dilated cardiomyopathy. Circ Res 64:97–103

Limas CJ, Goldenberg IF, Limas C (1995) Soluble interleukin-2 receptor levels in patients with dilated cardiomyopathy. Correlation with disease severity and cardiac autoantibodies. Circulation 91:631–634

MacLellan RW, Lusis AJ (2003) Dilated cardiomyopathy: learning to live with yourself. Nat Med 9:1455–1456

Magnusson Y, Marullo S, Hoyer S, et al. (1990) Mapping of a functional autoimmune epitope on the β-adrenergic receptor in patients with idiopathic dilated cardiomyopathy. J Clin Invest 86:1658–1663

Magnusson Y, Wallukat G, Waagstein F, et al. (1994) Autoimmunity in idiopathic dilated cardiomyopathy: characterization of antibodies against the β1-adrenoceptor with positive chronotropic effect. Circulation 89:2760–2767

Mahon NG, Madden B, Caforio ALP, et al. (2002) Immunohistochemical evidence of myocardial disease in apparently healthy relatives of patients with dilated cardiomyopathy. J Am Coll Cardiol 39:455–462

Maisch B, Deeg P, Liebau G, et al. (1983a) Diagnostic relevance of humoral and cytotoxic immune reactions in primary and secondary dilated cardiomyopathy. Am J Cardiol 52:1071–1078

Maisch B, Deeg P, Liebau G, et al. (1983b) Diagnostic relevance of humoral and cell-mediated immune reactions in patients with viral myocarditis. Clin Exp Immunol 48:533–545

Mann DL (2001) Autoimmunity, immunoglobulin adsorption and dilated cardiomyopathy: has the time come for randomized clinical trials? J Am Coll Cardiol 38:184–186

Mason JW, O'Connell JB, Herskowitz A, et al. (1995) A clinical trial of immunosuppressive therapy for myocarditis. N Engl J Med 333:269–275

Michels VV, Moll PP, Rodeheffer RJ, et al. (1994) Circulating heart autoantibodies in familial as compared with nonfamilial idiopathic dilated cardiomyopathy. Mayo Clin Proc 69:24–27

Muller J, Wallukat G, Dandel M, et al. (2000) Immunoglobulin adsorption in patients with idiopathic dilated cardiomyopathy. Circulation 101:385–391

Muller-Ehmsen J, Juvvadi P, Thompson CB, et al (2001) Ouabain and substrate affinities of human Na-K-ATPase α-1β-1, α-2β-1, α-3β-1 when expressed separately in yeast cells. Am J Physiol Cell Physiol 281:C1355–1364

Neu N, Rose NR, Beisel KW, et al. (1987) Cardiac myosin induces myocarditis in genetically predisposed mice. J Immunol 139:3630–3636

Neumann DA, Burek CL, Baughman KL, et al. (1990) Circulating heart-reactive antibodies in patients with myocarditis or cardiomyopathy. J Am Coll Cardiol 16:839–846

Nicholson GC, Dawkins RL, McDonald BL, et al. (1977) A classification of anti-heart antibodies: differentiation between heart-specific and heterophile antibodies. Clin Immunol Immunopathol 7:349–363

Okazaki T, Tanaka Y, Nishio R, et al. (2003) Autoantibodies against cardiac troponin I are responsible for dilated cardiomyopathy in PD-1-deficient mice. Nat Med 9:1477–1483

Otto A, Stahle I, Klein R, et al. (1998) Anti-mitochondrial antibodies in patients with dilated cardiomyopathy (anti-M7) are directed against flavoenzymes with covalently bound FAD. Clin Exp Immunol 111:541–547

Pohlner K, Portig I, Pankuweit S, et al. (1997) Identification of mitochondrial antigens recognized by antibodies in sera of patients with idiopathic dilated cardiomyopathy by two-dimensional gel electrophoresis and protein sequencing. Am J Cardiol 80:1040–1045

Portig I, Pankuweit S, Maisch B (1997) Antibodies against stress proteins in sera of patients with dilated cardiomyopathy. J Mol Cell Cardiol 29:2245–2251

Richardson P, McKenna WJ, Bristow M, et al. (1996) Report of the 1995 World Health Organization/International Society and Federation of Cardiology Task Force on the definition and classification of cardiomyopathies. Circulation 93:841–842

Rose NR (2000) Viral damage or 'molecular mimicry' – placing the blame in myocarditis. Nat Med 6:631–632

Rose NR, Bona C (1993) Defining criteria for autoimmune diseases (Witebsky's postulates revisited). Immunol Today 14:426–428

Schimke I, Muller J, Priem F, et al. (2001) Decreased oxidative stress in patients with idiopathic dilated cardiomyopathy one year after immunoglobulin adsorption. J Am Coll Cardiol 38:178–183

Schultheiss HP, Bolte HD (1985) Immunological analysis of auto-antibodies against the adenine nucleotide translocator in dilated cardiomyopathy. J Mol Cell Cardiol 17:603–617

Schultheiss HP, Ulrich G, Janda I, et al. (1988) Antibody mediated enhancement of calcium permeability in cardiac myocytes. J Exp Med 168:2105–2119

Schultheiss HP, Kuhl U, Schwimmbeck P, et al. (1990) Biomolecular changes in dilated cardiomyopathy. In: Baroldi G, Camerini F, Goodwin JF (eds) Advances in cardiomyopathies. Springer-Verlag, Berlin, Heidelberg, New York, p 221–234

Schwinger RH, Wang J, Frank K, et al. (1999) Reduced sodium pump α1-, α3-, β1-isoform protein levels and Na,K-ATPase activity but unchanged Na/Ca exchanger protein levels in human heart failure. Circulation 99:2105–2112

Smith SC, Allen PM (1991) Myosin-induced myocarditis is a T cell-mediated disease. J Immunol 147:2141–2147

Staudt A, Bohm M, Knebel F, et al. (2002) Potential role of autoantibodies belonging to the immunoglobulin G-3 subclass in cardiac dysfunction among patients with dilated cardiomyopathy. Circulation 106:2448–2453

Urayama O, Shutt H, Sweadner J (1989) Identification of three isozyme proteins of the catalytic subunit of the Na,K-ATPase in rat brain. J Biol Chem 264:8271–8280

Wallukat G, Morwinski M, Kowal K, et al. (1991) Antibodies against the β-adrenergic receptor in human myocarditis and dilated cardiomyopathy: β-adrenergic agonism without desensitization. Eur Heart J 12(Suppl D):178–181

Warraich RS, Dunn MJ, Yacoub MH (1999) Subclass specificity of autoantibodies against myosin in patients with idiopathic dilated cardiomyopathy: proinflammatory antibodies in dilated cardiomyopathy patients. Biochem Biophys Res Commun 259:255–261

Witebsky E, Rose NR, Terplan K, et al. (1957) Chronic thyroiditis and autoimmunization. JAMA 164:1439–1447

Wojnicz R, Nowalany-Kozielska E, Wojciechowska C, et al. (2001) Randomized, placebo controlled study for immunosuppressive treatment of inflammatory dilated cardiomyopathy. Two-year follow-up results. Circulation 104:39–45

Wolff PG, Kuhl U, Schultheiss HP (1989) Laminin distribution and autoantibodies to laminin in dilated cardiomyopathy and myocarditis. Am Heart J 117:1303–1309

Zweiman B, Arnason BGW (1987) Immunologic aspects of neurological and neuromuscular diseases. JAMA 258:2970–2973

IV Cardiac Remodeling

12 Inflammation and Cardiac Remodeling During Viral Myocarditis

S. Heymans

Abstract. Acute viral myocarditis is the main cause of cardiac failure in young patients and accounts for up to 60% of "idiopathic" dilated cardiomyopathy. The clinical course of viral myocarditis is mostly insidious with limited cardiac inflammation and dysfunction. However, overwhelming inflammation may occur in a subset of patients, leading to fulminant cardiac injury, whereas others develop chronic heart failure due to autoimmune myocarditis. Today, little effective treatment exists for patients, apart from general supportive therapy and antifailure regimens. Urokinase-type plasminogen activator (u-PA) and matrix metalloproteinases (MMP) have been implicated in cardiac inflammation, matrix remodeling, and wound healing after cardiac injury. The present review will assess the mechanism by which these proteinases mediate cardiac dilatation, fibrosis, and dysfunction after cardiac stress or injury, in order to understand how inhibition of proteinases may provide a novel therapeutic tool to prevent cardiac dilatation and failure during viral myocarditis.

12.1 Introduction

Viral myocarditis is an important cause of cardiac failure in young, otherwise healthy patients, and accounts for up to 60% of "idiopathic" dilated cardiomyopathy (Towbin 2001; Noutsias et al. 2003; Pauschinger et al. 2004). The clinical course of viral myocarditis varies. In most of the cases, viral myocarditis is insidious with limited cardiac inflammation and dysfunction. However, overwhelming inflammation may occur in a subset of patients, leading to fulminant cardiac injury and acute heart failure (Mason 2003; Vallejo and Mann 2003). In a third group of patients, viral myocarditis may progress to a chronic autoimmune myocarditis (Fairweather et al. 2001; Hill and Rose 2001; Liu and Mason 2001), eventually leading to a dilated cardiomyopathy. These patients only present with clinical symptoms weeks to months after the initial cardiac infection, and are therefore often classified as "idiopathic" dilated cardiomyopathy, since the link with the viral myocarditis is unclear or missing (Pauschinger et al. 1999, 2004a; Angelini et al. 2002).

Direct infection of myocytes by the virus is believed to play a minimal role in myocardial damage. The immune reaction against the virus, however, appears to be decisive in cardiac dilatation and failure during the acute phase of myocarditis (Kearney et al. 2001; Mason 2003; Mahrholdt et al. 2004). Interestingly, most of the patients presenting with fulminant acute myocarditis completely recover with normalization of cardiac dimensions and function, suggesting that initial failure is not caused by extensive necrosis of cardiomyocytes. These cardiomyocytes are surrounded by a strong interstitial matrix that is crucial to maintain cardiac geometry and function, by preventing myocyte slippage and by coordinating myocyte contraction (Tyagi 1998; Cleutjens and Creemers 2002; Spinale et al. 2002). Degradation of this matrix by proteinases produced by the inflammatory cells may therefore explain the pronounced cardiac dilatation and dysfunction during fulminant myocarditis. Inflammatory cells contain a bulk of proteinases that cause degradation of the interstitial matrix (Knowlton 2000), including matrix metalloproteinases, serine proteinases, and cathepsins. Once the virus is cleared due to the fulminant inflammatory response, inflammatory cells will disappear and the cardiac matrix may recover, resulting in normalization of cardiac dimensions. However, when chronic autoimmune myocarditis develops, inflammation-induced myocardial injury occurs at a continuous base, resulting in reparative fibrosis and progressive left ventricular (LV) dilatation.

The present review will focus on the possible implication of inflammatory cell-mediated proteinases in cardiac dilatation, fibrosis, and dysfunction during acute fulminant myocarditis or chronic autoimmune myocarditis. First, available data on an implication of proteinases in other cardiac diseases will be reviewed. Subsequently, the possible implication of proteinases in cardiac dilatation during acute fulminant myocarditis, and in dilatation and fibrosis during chronic autoimmune myocarditis, will be discussed.

12.2 Role of Matrix Metalloproteinases in Myocardial Remodeling

Myocardial matrix remodeling appears to play a pivotal role in the development of ventricular dilatation and heart failure (Creemers et al. 2001; Spinale 2002; Fedak et al. 2003). Matrix metalloproteinases (MMPs), which are readily present in the myocardium and are capable of degrading all the matrix components in the heart, are the driving force behind myocardial matrix degradation during remodeling. Their activity is counterbalanced by their physiological tissue inhibitors of MMPs (TIMPs), and the interplay between MMPs and TIMPs therefore determines the progression of both ventricular dilatation and fibrosis in the diseased hearts.

12.2.1 MMP Family

MMPs are a family of zinc-containing endoproteinases that share structural domains but differ in substrate specificity, cellular sources, and inducibility. The list of MMPs has grown rapidly in the past several years, and by now more than 20 mammalian members have been cloned and identified (reviewed in Creemers et al. 2001; Visse and Nagase

2003). Based on their substrate specificity and primary structure, the MMP family can be subdivided into four major groups; the collagenases, the gelatinases, the stromelysins, and membrane-type MMPs. The first group, the collagenases, include MMP-1 (interstitial collagenase), MMP-8 (neutrophil collagenase), and MMP-13 (collagenase-3), which can all cleave fibrillar collagens (types I, II, and III). Since fibrillar collagen is tightly apposed and highly cross-linked, these collagens are extremely resistant to cleavage by proteinases. MMP-1, -8, and -13 cleave fibrillar collagens in the heart at unique sites in the triple helix at 3/4 from the N-terminal end, generating 3/4 and 1/4 collagen fragments. These fragments unfold their triple helix and fall apart into fragmented single α-chains, the so-called gelatins. The second group, the gelatinases (MMP-2 and MMP-9) further degrade these gelatins, and are also capable of degrading type IV collagens in basement membranes. Group 3, the stromelysins (MMP-3, -10, and -11), so named because they are active against a broad spectrum of extracellular matrix (ECM) components, including proteoglycans, laminins, fibronectin, vitronectin, and some type of collagens. Finally, group 4, containing the membrane-type MMPs (MT-MMPs), degrade several ECM components and also trans-activate other MMPs at the level of the cellular membrane.

This substrate-based subdivision, however, is often more gradual than really absolute. Whereas it was assumed for a long time that the cleavage of fibrillar collagens was limited to the action of collagenases, previous evidence demonstrates that gelatinases are able to cleave collagens as well (Aimes and Quigley 1995; Okada et al. 1995). Since gelatinases (MMP-2 and -9) are produced in large amounts after cardiac injury (Heymans et al. 1999; Wilson et al. 2003; Lindsey 2004; Pauschinger et al. 2004b; Tao et al. 2004) and may cleave interstitial collagens as well as denatured collagens (gelatins), these gelatinases may play a crucial role in the degradation and remodeling of the cardiac collagenous ECM. Indeed, gene knockout studies in mice revealed a pivotal role of MMP-2 and MMP-9 in cardiac inflammation and dilatation after acute myocardial infarction (Heymans et al. 1999; Ducharme et al. 2000; Hayashidani et al. 2003; Schroen et al. 2004) and in inflammation, dilatation, and cardiac hypertrophy during pressure overload (Heymans et al. 2005). It remains, however, unknown whether these MMPs may also be implicated in inflammation and cardiac dilatation during viral myocarditis.

12.2.2 Regulation of MMP Expression and Activity

Because MMPs, once activated, are collectively capable of degrading the complete ECM, it is important that the activity of these enzymes is kept under tight control. The activity of MMPs is controlled at the following three levels: transcription, activation of latent proenzymes, and inhibition by their endogenous inhibitors, the TIMPs.

Regulation of MMP Expression by Inflammatory Cytokines at the Transcriptional Level

The expression of MMPs is low in normal adult tissue but is upregulated during certain physiological and pathological remodeling processes. Inflammatory cytokines, hormones, and growth factors are crucially involved in mediating MMP expression at the transcriptional level. Interleukin (IL)-1, IL-6, tumor necrosis factor-α (TNF-α) (Ito et al. 1990), transforming growth factor-β (TGF-β) (Fabunmi et al. 1996; Fabunmi et al. 1998) epidermal growth factor (EGF), platelet-derived growth factor (PDGF), basic fibroblast growth factor (bFGF), and CD40 (Malik et al. 1996) not only mediate MMP gene expression, but also regulate the transcriptional regulation of their inhibitors, TIMPs (Schonbeck et al. 1997; Creemers et al. 2001). Interferon (IFN)-γ may both increase or decrease MMP expression depending on cell-type (Mauviel 1993; Makela et al. 1998), whereas its effect on TIMP expression is unknown.

Both IL-1β (Siwik et al. 2000) and TNF-α (Brenner et al. 1989; Li et al. 1999) stimulate both collagenase and gelatinase activity by c-jun mediated activation of MMP-1, -2, and -9 activity. Recent studies in transgenic mice with targeted over-expression of TNF-α have shown that these mice develop progressive LV dilatation, associated with TNF-α-induced activation of MMPs and progressive degradation of fibrillar collagen (Li et al. 2000; Sivasubramanian et al. 2001). Whereas short-term over-expression of TNF-α resulted in dissolution of the fibrillar collagen surrounding the cardiomyocytes accompanied by increased MMP-expression, long-term stimulation with TNF-α resulted in an increase in fibrillar collagen that was accompanied by increased expression of TIMPs and the resulting decreased MMP-activity (Li et al. 2000). These observations suggest that sustained myocardial inflammation provokes time-dependent changes in the balance between MMP and TIMP activity. During the early acute inflammation, there is an increase in the ratio of MMP activity to TIMP levels that fosters LV dilatation. However, with chronic inflammatory signaling, there is a time-dependent increase in TIMP levels, with a resultant decrease in the ratio of MMP to TIMP activity and a subsequent increase in myocardial fibrillar collagen content. Such a time-dependent course may also explain the more pronounced cardiac dilatation and dysfunction in acute fulminant myocarditis as compared to the extensive fibrosis in chronic autoimmune myocarditis.

TGF-β is also an important regulator of MMP gene expression. It stimulates MMP-2 and MMP-9 but inhibits MMP-1 and MMP-3 synthesis (Siwik et al. 2000) and also increases TIMP expression (Wahl et al. 1993). Therefore, depending on cell type and space-specific expression, TGF-β may be either pro-fibrotic by reducing proteinase activity, or facilitate LV dilatation by increasing MMP-2 and -9 activity. Whereas TGF-β over-expression in mice results in increased cardiac fibrosis and dysfunction over the long-term by inhibition of MMP-1 activity, it protects against short-term ischemia-induced reperfusion injury, also by inhibition of MMP-1 activity, indicating a dual and time-dependent effect of TGFβ on cardiac remodeling. In the case of IFN-γ, the impact on MMP expression is cell-type specific. Whereas IFN-γ increases MMP-1 expression in keratinocytes, it decreases MMP-1 expression in macrophages and fibroblasts (Makela et al. 1998).

In conclusion, regulation of MMP activity by cytokines is diverse and acts on a time-, cell type-, and space-dependent way.

Activation of Latent Pro-enzymes

MMPs are synthesized and secreted as proenzymes (zymogens). After secretion, the proMMPs bind to various ECM components, which might serve as a means of extracellular storage for rapid activation and mobilization upon stimulation. Interestingly, MMPs can display selective affinity with discrete components of the ECM. MMP-2 appears to be associated with elastin-containing structures, MMP-3 with basement membrane and occasionally with collagen fibers, and MMP-13 with proteoglycans, collagen, and elastin (Lijnen et al. 1999). Importantly, proMMP-9 forms a high-affinity complex with α2 chains on the cell surface, promoting localized activity of MMP-9 at the cell surface/matrix (Olson et al. 1998), thereby facilitating invasion of cells into the degraded interstitial matrix. In addition, binding of MMP-9 to α-1 chains of type I collagen (Makela et al. 1998) results in a means to achieve collagen-targeted degradation.

Once pro-MMPs are secreted and stored in their inactive pro-form in the matrix, proteolytic degradation must activate them to their active form, which can occur by both plasmin-independent and plasmin-dependent pathways (Creemers et al. 2001; Lijnen 2001). Plasmin, one of the serine proteinases, is the active enzyme of the plasminogens system and degrades a variety of ECM components. However, most of the proteolytic features of plasmin are achieved by activating several MMPs, including proMMP-1, proMMP-3, proMMP-7, proMMP-9, proMMP-10, and proMMP-13 (Carmeliet et al. 1997; Lijnen 2001). The generation of plasmin is primarily controlled by the balance between the plasminogen activators [tissue-type plasminogen activator (tPA) and urokinase plasminogen activator (uPA)] and their physiological inhibitors, the plasminogen activators inhibitors (PAIs). uPA-mediated plasmin appears to play a pivotal role in MMP activation and inflammation in the heart. After acute myocardial infarction, total proteolytic activity of uPA increases, whereas tPA activity remains unchanged (Heymans et al. 1999). In concordance, influx of inflammatory cells and wound healing was virtually abolished in plasminogen-defideficient (Creemers et al. 2000) and uPA-deficient mice (Heymans et al. 1999), but was unaffected in tPA-deficient mice (Heymans et al. 1999). Together with increased uPA activity in the infarcted heart, both MMP-2 and MMP-9 activity increased, whereas uPA deficiency and plasminogen deficiency reduced the proteolytic activity of both MMP-2 and MMP-9. tPA deficiency did not alter MMP activity in the heart (Heymans et al. 1999; Creemers et al. 2000), indicating that uPA-mediated plasmin increases MMP activity. Two other observations also strongly suggested that plasminogen and uPA are crucial in activating MMPs. First, uPA was coexpressed with MMP-9 in infiltrating leukocytes. Second, MMP inhibition by TIMP-1 gene overexpression had comparable, although less-pronounced, effects on inflammation and infarct healing vs uPA and plasminogen deficiency.

Additionally, other serine proteases have been reported (i.e., serine elastase, trypsin, and cathepsin G) to be able to activate MMPs and destroy the inhibitory activity of TIMPs (Okada et al. 1988; Shamamian et al. 2000). The role of serine elastase has been investigated in relation to myocardial ischemia/reperfusion injury and viral myocarditis, and it was demonstrated that inhibitors of these enzymes are protective against inflammation-related myocardial injury (Murohara et al. 1995; Lee et al. 1998; Zaidi et al. 1999; Ohta et al. 2004)

Finally, activation of latent MMPs can be achieved through autocleavage by other MMPs (Murphy and Knauper 1997; Woessner 1999). Besides digesting ECM components, MMP-3 activates a number of proMMPs, and its action on a partially processed proMMP-1 is critical for the generation of fully active MMP-1 (Suzuki et al. 1990). Membrane-type MMPs (MT-MMPs), with the exception of MT4-MMP, are all capable of activating proMMP-2. Thus, rapid amplification of MMP activity can occur after an initial enzymatic step. Once activated, the MMP will degrade the matrix element to which it was bound. Indeed, the proMMPs bind to specific ECM components and, therefore, are in close juxtaposition to the future proteolytic substrate.

In conclusion, a pool of recruitable MMPs exists within the ECM and provides a means for the rapid induction of proteolytic activity. Furthermore, because these soluble MMPs bind to specific proteins and at specific protein sequences, the activation of MMPs can occur in specific patterns within the ECM.

Inhibition of MMPs

TIMPs are specific inhibitors that bind MMPs in a 1:1 stoichiometry (Visse and Nagase 2003). Four TIMPs (TIMP-1, TIMP-2, TIMP-3, and TIMP-4) have been identified in vertebrates, (Brew et al. 2000) and their expression is regulated during development and tissue remodeling. Under pathological conditions associated with unbalanced MMP activities, changes of TIMP levels are considered important because they directly affect the level of MMP activity. Inhibition is accomplished by their ability to interact with the zinc-binding site within the catalytic domain of active MMPs. There is a certain degree of specificity in the activity of different TIMPs toward distinct members of the MMP family. Whereas TIMP-1 potently inhibits the activity of most MMPs, with the exception of MMP-2 and MT1-MMP, TIMP-2 is a potent inhibitor of most MMPs, except MMP-9. This may be important for MMP-associated inflammation, since MMP-9, mainly produced by inflammatory cells, also stimulates further influx of new inflammatory cells in the injured heart (Heymans et al. 1999), resulting in cardiac dilatation and dysfunction (Ducharme et al. 2000). Indeed, TIMP-1 overexpression prevented inflammation and subsequent cardiac dilatation after myocardial infarction (Heymans et al. 1999) and cardiac overload (Heymans et al. 2005). Furthermore, TIMP-3, which is insoluble and binds to the ECM, has been shown to bind MMP-1, -2, -3, -9, and -13. TIMP-4 inhibits MMP-1, -3, -7, and -9 and shows a high level of expression in adult human cardiac tissue (Greene et al. 1996; Liu et al. 1997). Expression of TIMP, similar to MMPs, is also mediated by different cytokines, such as IL-1 and IL-6 (Mauviel 1993). Since these factors stimulate both TIMP and MMP expression, their net effect on total proteolytic activity might be minimal.

In addition to MMP-inhibiting activities, TIMPs have other biological functions. Both TIMP-1 and TIMP-2 may have growth factor activities (Murphy et al. 1993) and regulate tumor growth (Gomez et al. 1997), angiogenesis (Thorgeirsson et al. 1996; Gomez et al. 1997), and embryonic development (Barasch et al. 1999), independent of their MMP-inhibiting activity. Finally, TIMP-3 was shown to have proapoptotic activities, possibly through the stabilization of TNF-α cell receptor 1, Fas, and TNF-related apoptosis, inducing ligand receptor 1, as shown for tumor cells (Bond et al. 2002; Ahonen et al. 2003; Calabrese et al. 2004; Mohammed et al. 2004).

12.2.3 Importance of Proteinases in Cardiac Inflammation

Role of Plasminogen System and the Metalloproteinases in Cardiac Inflammation

Two recent studies clearly demonstrated that uPA-mediated plasmin is pivotal in infiltration of inflammatory cells in the injured heart, but also contributes to cardiac rupture after myocardial infarction (Heymans et al. 1999; Creemers et al. 2000). Absence of uPA or plasminogen prevented infiltration of inflammatory cells into the infarcted area, and completely aborted healing and scar formation of the infarct. By degrading matrix components, uPA-mediated proteolysis allows inflammatory cells to infiltrate the infarct and to disrupt the collagen network, a prerequisite for cardiac rupture. MMP-activity was reduced in both uPA and plasminogen-deficient mice, whereas MMP-9 gene inactivation and TIMP-1 gene overexpression, to a lesser extent than uPA inactivation, prevented cardiac rupture, but also impaired cardiac inflammation and wound healing. Furthermore, uPA was coexpressed with MMP-9 in infiltrating neutrophils. Together, these data indicate that uPA-mediated MMP-9 activation may play a central role in matrix degradation after cardiac injury, allowing further infiltration of inflammatory and other wound healing cells, but predisposing to cardiac rupture. uPA and MMPs may also promote inflammation through other mechanisms. First, PA/MMPs can produce matrix degradation fragments that affect growth and migration of inflammatory cells (MacKenna et al. 1994). In this perspective, shedding of glycoproteins or proteolytic activation of matricellular proteins, such as syndecans, tenascin-C, or osteonectin might strongly influence inflammation, angiogenesis, and matrix remodeling (Endo et al. 2003; Schellings et al. 2004; Schroen et al. 2004). uPA and MMPs may also activate and mobilize growth factors in the extracellular matrix (Kleiner and Stetler-Stevenson 1999; Rifkin et al. 1999). In particular, uPA is able to activate TGF-β1 (Plow et al. 1999); and decreased TGF-β activity in absence of uPA contributed to decreased inflammation and collagen deposition after myocardial infarction (Heymans et al. 1999).

It remains, however, unclear whether uPA or MMP may also affect inflammation during viral myocarditis. Acute viral myocarditis is characterized by viral infection of cardiomyocytes followed by a pronounced infiltration of inflammatory cells, including macrophages, natural killer cells, and T lymphocytes. Many pro-inflammatory cytokines are involved in inflammation during myocarditis, including TNF-α and IL-1β. TNF-α overexpressing (transgenic) mice develop increased inflammation, associated with severe LV dilatation and progressive cardiac failure (Kubota et al. 1997; Bryant et al. 1998; Sivasubramanian et al. 2001; Mann 2002). TNF-α plays an important role in the process of cardiac remodeling, including myocyte hypertrophy (Yokoyama et al. 1997), alterations in fetal gene expression (Kubota et al. 1997), and progressive myocyte loss through apoptosis (Krown et al. 1996). In addition, progressive LV dilatation in TNF-α transgenic mice was associated with increased activation of MMPs and progressive loss of fibrillar collagen in the cardiac matrix, predisposing to myocyte slippage and dilatation (Li et al. 2000; Sivasubramanian et al. 2001). However, long-term stimulation with TNF-α resulted in an increase in fibrillar collagen content that was accompanied by decreased MMP activity and increased expression of the tissue inhibitors of MMPs. Together, these observations suggest that sustained myocardial inflammation provokes time-dependent changes in the balance between MMP activity and TIMP activity. During the early stages of inflammation, there is an increase in the ratio of MMP activity to TIMP levels that fosters LV dilation. However, with chronic inflammatory signaling, there is a time-dependent increase in TIMP levels, with a resultant decrease in the ratio of MMP to TIMP activity and a subsequent increase in myocardial fibrillar content.

As for acute viral myocarditis, MMP upregulation was demonstrated in mice infected with a human coxsackievirus B3 on day 10 after infection, associated with alteration in cardiac collagen content and impaired LV function (Li et al. 2002). MMP upregulation was accompanied by T lymphocyte infiltration and increased expression of inflammatory cytokines such as TNF-α, IL-1β, IL-4, and TGF-β1. The myocardial inflammatory reaction resulted in a significant upregulation in MMP-3 and MMP-9, and downregulation in TIMP-1 and TIMP-4, both at messenger RNA and protein levels. This balance favored collagen degradation, as revealed by an elevated soluble fraction of myocardial collagen type I. These findings indicate that an imbalance in the myocardial ratio of MMs to TIMPs in acute viral myocarditis might lead to degradation of the collagen network in the heart, resulting in LV dilatation and dysfunction. Whether inactivation of MMPs may also prevent cardiac dilatation or dysfunction during viral myocarditis, as shown for myocardial infarction (Heymans et al. 1999; Creemers et al. 2000) or pressure overload (Heymans et al., in press), will be addressed in experiments using uPA or MMP gene-inactivated mice.

In conclusion, all these studies clearly point to a concerted interaction between inflammatory cells, cytokines, and proteinases in the process of myocardial inflammation, injury, and dysfunction during viral myocarditis. Inflammatory cytokines, i.e., IL1β, IL6, TNF-α, and TGF-β, may regulate myocardial collagen remodeling by mediating the expression and balance of the plasminogen and MMP/TIMP system (Creemers et al. 2000; Visse and Nagase 2003; Pauschinger et al. 2004b), whereas proteinases may regulate cytokine activity, as demonstrated by uPA-mediated increase in TGF-β activity (Plow et al. 1999). Together, cytokines, proteinases, and inflammatory cells during viral myocarditis result in a damaging response, resulting in cardiac dilatation and dysfunction.

Involvement of Other Proteinases in Cardiac Inflammation

Other proteinases that have been implicated in cardiac inflammation, remodeling, or wound healing are serine elastase (Lee et al. 1998; Zaidi et al. 1999; Tiede et al. 2003; Ohta et al. 2004), cathepsins (Sabri et al. 2003) and mast cell chymases and tryptases (Kitaura-Inenaga et al. 2003).

The importance of serine elastase activity in the pathophysiology of acute viral myocarditis and the therapeutic efficacy of an elastase inhibitor was demonstrated in DBA/2 mice inoculated with the encephalomyocarditis virus. Infection resulted in an increase exceeding 150% in myocardial serine elastase activity, which was suppressed efficiently by a selective serine elastase inhibitor, ZD0892, that is biologically effective after oral administration. Mice treated with this compound had little evidence of microvascular constriction and obstruction associated with myocarditis-induced ischemia reperfusion injury, much less inflammation and necrosis. Only mild fibrosis and myocardial collagen deposition, and normal ventricular function, compared with the infected non-treated group, was found (Lee et al. 1998). Concordantly, in transgenic mice over-expressing elafin, which is a serine elastase inhibitor, injection of a encephalomyocarditis virus resulted in reduced inflammation, improved cardiac function, and reduced mortality as compared to wild-type mice (Zaidi et al. 1999). Reduced inflammation by inhibition of serine elastase was also demonstrated in ischemia/reperfusion (Tiede et al. 2003) and after myocardial infarction (Ohta et al. 2004), and it protected against cardiac dilatation and dysfunction (Ohta et al. 2004).

Increased gene expression of mast cell chymase resulted in overexpression of mouse mast cell protease (mMCP)-4, -5, and -6, matrix metalloproteinase (MMP)-9, and type I procollagen 5–14 days after inoculation with a encephalomyocarditis virus, coinciding with a prominent inflammatory reaction and extensive myocardial necrosis and fibrosis (Kitaura-Inenaga et al. 2003). Increased expression of MMP-9 significantly correlated with upregulation of mast cell proteases. The gene expression of type I procollagen was increased at 5 days and continued to increase to day 14, suggesting that a fibrotic process had already begun during the acute stage of viral myocarditis (Kitaura-Inenaga et al. 2003). These findings suggest that mast cell chymase and tryptase may participate in the acute inflammation and remodeling process of viral myocarditis, but clear evidence that inhibition of these proteinases may reduce inflammation and injury during viral myocarditis is lacking.

12.3 Prospect of Therapy Based on Inflammation and Proteinases

Since myocardial inflammation appears to play a central role in the pathogenesis of viral myocarditis and dilated cardiomyopathy, and whereas direct infection of myocytes by the virus might be less important in myocardial damage, several therapeutic investigations have been conducted by modulating the immune response following viral infection. Each different presentation of viral myocarditis – including acute fulminant myocarditis with frequent complete recovery, and chronic autoimmune myocarditis mostly progressing to "idiopathic" dilated cardiomyopathy – represent pathogenetically distinct phases and therefore requires different diagnostic and therapeutic strategies (Frustaci et al. 2003). Several studies have been done in human patients using corticosteroids or cyclosporine to reduce cardiac inflammation during viral myocarditis. However, these studies did not distinguish acute fulminant or chronic autoimmune viral myocarditis, and these treatment strategies may even enhance the viral titer during the early phase of viral infection, and may result in increased mortality and cardiac insufficiency (Monrad et al. 1986; Tomioka et al. 1986; Mason et al. 1995).

Additionally, cytokine therapy may be either protective or deleterious, depending on the cellular and molecular states of the inflammatory phase. Administration of IFN-β was beneficial in patients with long-term LV dysfunction and myocardial viral persistence despite the failure of conventional therapy (Kuhl et al. 2003). IFN-β treatment resulted in the disappearance of viral genomes in the myocardium and improvement of LV function. These findings stress the importance of inflammation, induced by viral infected myocytes, in the development of matrix degradation and myocyte loss, resulting in cardiac dilatation, and in progressive systolic dysfunction.

These inflammatory cells are the major source of proteinases, including MMPs and uPA, and produce cytokines that mediate T lymphocytes but also modulate the expression and activity of MMPs. Increased activity of uPA (S. Heymans, unpublished data), MMP-2, -3, and -9 (Li et al. 2002) during experimental viral myocarditis in association with collagen degradation and LV dilatation suggest that increased proteolytic degradation of the interstitial matrix may be involved in cardiac dilatation and dysfunction during acute fulminant myocarditis, whereas viral infection and myocyte loss may be less important in cardiac dysfunction. This hypothesis also explains why patients with acute fulminant myocarditis may completely recover after clearance of the virus, since reversed remodeling of the matrix may result in normalization of the cardiac matrix, geometry, and function. Future direction in treating patients with acute myocarditis should therefore take into account the proteolytic degradation of the matrix mediated by inflammatory cells. But care should be taken to distinguish patients with acute fulminant myocarditis from patients with chronic autoimmune myocarditis resulting in so-called "idiopathic" dilated cardiomyopathy, since the latter is associated with a dynamic process of myocyte loss, increasing – reparative – interstitial fibrosis, and concurrent matrix remodeling resulting in LV dilatation. Therefore, the temporal and local expression of specific proteinases should be determined during different presentations of viral myocarditis in order to allow time-dependent inhibition of specific proteinases to prevent cardiac dilatation and dysfunction. Since inhibition of MMP-2/MMP-9 or uPA has proved to reduce cardiac inflammation and hinder deleterious matrix remodeling, thus preventing either cardiac rupture (Heymans et al. 1999; Hayashidani et al. 2003) or dilatation (Rohde et al. 1999; Ducharme et al. 2000), future experimental studies should focus on these candidates as novel therapeutic tools, together with cytokine therapy, to treat dilated cardiomyopathy during viral myocarditis.

Acknowledgements. The present work was supported by a Dr. Dekkers grant of the Netherlands Heart Foundation (NHS, 2003T036) to Dr. S. Heymans and a research grant of the Leuven University, Belgium (OT-0346) to Dr. S. Heymans and Prof. Dr. F. Van de Werf.

References

Ahonen M, Poukkula M, Baker AH, Kashiwagi M, Nagase H, Eriksson JE, Kahari VM (2003) Tissue inhibitor of metalloproteinases-3 induces apoptosis in melanoma cells by stabilization of death receptors. Oncogene 22:2121–2134

Aimes RT, Quigley JP (1995) Matrix metalloproteinase-2 is an interstitial collagenase. Inhibitor-free enzyme catalyzes the cleavage of collagen fibrils and soluble native type I collagen generating the specific 3/4- and 1/4-length fragments. J Biol Chem 270:5872–5876

Angelini A, Crosato M, Boffa GM, Calabrese F, Calzolari V, Chioin R, Daliento L, Thiene G (2002) Active versus borderline myocarditis: clinicopathological correlates and prognostic implications. Heart 87:210–215

Barasch J, Yang J, Qiao J, Tempst P, Erdjument-Bromage H, Leung W, Oliver JA (1999) Tissue inhibitor of metalloproteinase-2 stimulates mesenchymal growth and regulates epithelial branching during morphogenesis of the rat metanephros. J Clin Invest 103:1299–1307

Bond M, Murphy G, Bennett MR, Newby AC, Baker AH (2002) Tissue inhibitor of metalloproteinase-3 induces a Fas-associated death domain-dependent type II apoptotic pathway. J Biol Chem 277:13787–13795

Brenner DA, O'Hara M, Angel P, Chojkier M, Karin M (1989) Prolonged activation of jun and collagenase genes by tumour necrosis factor-alpha. Nature 337:661–663

Brew K, Dinakarpandian D, Nagase H (2000) Tissue inhibitors of metalloproteinases: evolution, structure and function. Biochim Biophys Acta 1477:267–283

Bryant D, Becker L, Richardson J, Shelton J, Franco F, Peshock R, Thompson M, Giroir B (1998) Cardiac failure in transgenic mice with myocardial expression of tumor necrosis factor-alpha. Circulation 97:1375–1381

Calabrese F, Carturan E, Chimenti C, Pieroni M, Agostini C, Angelini A, Crosato M, Valente M, Boffa GM, Frustaci A, Thiene G (2004) Overexpression of tumor necrosis factor (TNF)alpha and TNFalpha receptor I in human viral myocarditis: clinicopathologic correlations. Mod Pathol 17:1108–1118

Carmeliet P, Moons L, Lijnen R, Baes M, Lemaitre V, Tipping P, Drew A, Eeckhout Y, Shapiro S, Lupu F, Collen D (1997) Urokinase-generated plasmin activates matrix metalloproteinases during aneurysm formation. Nat Genet 17:439–444

Cleutjens JP, Creemers EE (2002) Integration of concepts: cardiac extracellular matrix remodeling after myocardial infarction. J Card Fail 8:S344–S348

Creemers E, Cleutjens J, Smits J, Heymans S, Moons L, Collen D, Daemen M, Carmeliet P (2000) Disruption of the plasminogen gene in mice abolishes wound healing after myocardial infarction. Am J Pathol 156:1865–1873

Creemers EE, Cleutjens JP, Smits JF, Daemen MJ (2001) Matrix metalloproteinase inhibition after myocardial infarction: a new approach to prevent heart failure? Circ Res 89:201–210

Ducharme A, Frantz S, Aikawa M, Rabkin E, Lindsey M, Rohde LE, Schoen FJ, Kelly RA, Werb Z, Libby P, Lee RT (2000) Targeted deletion of matrix metalloproteinase-9 attenuates left ventricular enlargement and collagen accumulation after experimental myocardial infarction. J Clin Invest 106: 55–62

Endo K, Takino T, Miyamori H, Kinsen H, Yoshizaki T, Furukawa M, Sato H (2003) Cleavage of syndecan-1 by membrane type matrix metalloproteinase-1 stimulates cell migration. J Biol Chem 278:40764–40770

Fabunmi RP, Baker AH, Murray EJ, Booth RF, Newby AC (1996) Divergent regulation by growth factors and cytokines of 95 kDa and 72 kDa gelatinases and tissue inhibitors or metalloproteinases-1, -2, and -3 in rabbit aortic smooth muscle cells. Biochem J 315:335–342

Fabunmi RP, Sukhova GK, Sugiyama S, Libby P (1998) Expression of tissue inhibitor of metalloproteinases-3 in human atheroma and regulation in lesion-associated cells: a potential protective mechanism in plaque stability. Circ Res 83:270–278

Fairweather D, Kaya Z, Shellam GR, Lawson CM, Rose NR (2001) From infection to autoimmunity. J Autoimmun 16:175–186

Fedak PW, Altamentova SM, Weisel RD, Nili N, Ohno N, Verma S, Lee TY, Kiani C, Mickle DA, Strauss BH, Li RK (2003) Matrix remodeling in experimental and human heart failure: a possible regulatory role for TIMP-3. Am J Physiol Heart Circ Physiol 284:H626–H634

Frustaci A, Chimenti C, Calabrese F, Pieroni M, Thiene G, Maseri A (2003) Immunosuppressive therapy for active lymphocytic myocarditis: virological and immunologic profile of responders versus nonresponders. Circulation 107:857–863

Gomez DE, Alonso DF, Yoshiji H, Thorgeirsson UP (1997) Tissue inhibitors of metalloproteinases: structure, regulation and biological functions. Eur J Cell Biol 74:111–122

Greene J, Wang M, Liu YE, Raymond LA, Rosen C, Shi YE (1996) Molecular cloning and characterization of human tissue inhibitor of metalloproteinase 4. J Biol Chem 271:30375–30380

Hayashidani S, Tsutsui H, Ikeuchi M, Shiomi T, Matsusaka H, Kubota T, Imanaka-Yoshida K, Itoh T, Takeshita A (2003) Targeted deletion of matrix metalloproteinase-2 attenuates early left ventricular rupture and late remodeling after experimental myocardial infarction. Am J Physiol Heart Circ Physiol 285:H1229–H1235

Heymans S, Luttun A, Nuyens D, Theilmeier G, Creemers E, Moons L, Dyspersin GD, Cleutjens JP, Shipley M, Angellilo A, Levi M, Nube O, Baker A, Keshet E, Lupu F, Herbert JM, Smits JF, Shapiro SD, Baes M, Borgers M, Collen D, Daemen MJ, Carmeliet P (1999) Inhibition of plasminogen activators or matrix metalloproteinases prevents cardiac rupture but impairs therapeutic angiogenesis and causes cardiac failure. Nat Med 5: 1135–1142

Heymans S, Lupu F, Terclavers S, Vanwetswinkel B, Herbert JM, Baker A, et al (2005) Loss or inhibition of uPA or MMP-9 attenuates LV remodeling and dysfunction after acute pressure overload in mice. Am J Pathol 166:15–25

Hill SL, Rose NR (2001) The transition from viral to autoimmune myocarditis. Autoimmunity 34:169–176

Ito A, Sato T, Iga T, Mori Y (1990) Tumor necrosis factor bifunctionally regulates matrix metalloproteinases and tissue inhibitor of metalloproteinases (TIMP) production by human fibroblasts. FEBS Lett 269:93–95

Kearney MT, Cotton JM, Richardson PJ, Shah AM (2001) Viral myocarditis and dilated cardiomyopathy: mechanisms, manifestations, and management. Postgrad Med J 77:4–10

Kitaura-Inenaga K, Hara M, Higuchi K, Yamamoto K, Yamaki A, Ono K, Nakano A, Kinoshita M, Sasayama S, Matsumori A (2003) Gene expression of cardiac mast cell chymase and tryptase in a murine model of heart failure caused by viral myocarditis. Circ J 67:881–884

Kleiner DE, Stetler-Stevenson WG (1999) Matrix metalloproteinases and metastasis. Cancer Chemother Pharmacol 43:S42–51

Knowlton KU (2000) The immune response following myocardial infarction: a role for T-cell-mediated myocyte damage. J Mol Cell Cardiol 32:2107–2110

Krown KA, Page MT, Nguyen C, Zechner D, Gutierrez V, Comstock KL, Glembotski CC, Quintana PJ, Sabbadini RA (1996) Tumor necrosis factor alpha-induced apoptosis in cardiac myocytes. Involvement of the sphingolipid signaling cascade in cardiac cell death. J Clin Invest 98:2854–2865

Kubota T, McTiernan CF, Frye CS, Slawson SE, Lemster BH, Koretsky AP, Demetris AJ, Feldman AM (1997) Dilated cardiomyopathy in transgenic mice with cardiac-specific overexpression of tumor necrosis factor-alpha. Circ Res 81:627–635

Kuhl U, Pauschinger M, Schwimmbeck PL, Seeberg B, Lober C, Noutsias M, Poller W, Schultheiss HP (2003) Interferon-beta treatment eliminates cardiotropic viruses and improves left ventricular function in patients with myocardial persistence of viral genomes and left ventricular dysfunction. Circulation 107:2793–2798

Lee JK, Zaidi SH, Liu P, Dawood F, Cheah AY, Wen WH, Saiki Y, Rabinovitch M (1998) A serine elastase inhibitor reduces inflammation and fibrosis and preserves cardiac function after experimentally-induced murine myocarditis. Nat Med 4:1383–1391

Li J, Schwimmbeck PL, Tschope C, Leschka S, Husmann L, Rutschow S, Reichenbach F, Noutsias M, Kobalz U, Poller W, Spillmann F, Zeichhardt H, Schultheiss HP, Pauschinger M (2002) Collagen degradation in a murine myocarditis model: relevance of matrix metalloproteinase in association with inflammatory induction. Cardiovasc Res 56:235–247

Li YY, McTiernan CF, Feldman AM (1999) Proinflammatory cytokines regulate tissue inhibitors of metalloproteinases and disintegrin metalloproteinase in cardiac cells. Cardiovasc Res 42:162–172

Li YY, Feng YQ, Kadokami T, McTiernan CF, Draviam R, Watkins SC, Feldman AM (2000) Myocardial extracellular matrix remodeling in transgenic mice overexpressing tumor necrosis factor alpha can be modulated by anti-tumor necrosis factor alpha therapy. Proc Natl Acad Sci U S A 97:12746–12751

Lijnen HR (2001) Plasmin and matrix metalloproteinases in vascular remodeling. Thromb Haemost 86:324–333

Lijnen HR, Lupu F, Moons L, Carmeliet P, Goulding D, Collen D (1999) Temporal and topographic matrix metalloproteinase expression after vascular injury in mice. Thromb Haemost 81:799–807

Lindsey ML (2004) MMP induction and inhibition in myocardial infarction. Heart Fail Rev 9:7–19

Liu PP, Mason JW (2001) Advances in the understanding of myocarditis. Circulation 104:1076–1082

Liu YE, Wang M, Greene J, Su J, Ullrich S, Li H, Sheng S, Alexander P, Sang QA, Shi YE (1997) Preparation and characterization of recombinant tissue inhibitor of metalloproteinase 4 (TIMP-4). J Biol Chem 272:20479–20483

MacKenna DA, Omens JH, McCulloch AD, Covell JW (1994) Contribution of collagen matrix to passive left ventricular mechanics in isolated rat hearts. Am J Physiol 266:H1007–H1018

Mahrholdt H, Goedecke C, Wagner A, Meinhardt G, Athanasiadis A, Vogelsberg H, Fritz P, Klingel K, Kandolf R, Sechtem U (2004) Multimedia article. Cardiovascular magnetic resonance assessment of human myocarditis: a comparison to histology and molecular pathology. Circulation 109:1250–1258

Makela M, Salo T, Larjava H (1998) MMP-9 from TNF alpha-stimulated keratinocytes binds to cell membranes and type I collagen: a cause for extended matrix degradation in inflammation? Biochem Biophys Res Commun 253:325–335

Malik N, Greenfield BW, Wahl AF, Kiener PA (1996) Activation of human monocytes through CD40 induces matrix metalloproteinases. J Immunol 156:3952–3960

Mann DL (2002) Inflammatory mediators and the failing heart: past, present, and the foreseeable future. Circ Res 91:988–998

Mason JW (2003) Myocarditis and dilated cardiomyopathy: an inflammatory link. Cardiovasc Res 60:5–10

Mason JW, O'Connell JB, Herskowitz A, Rose NR, McManus BM, Billingham ME, Moon TE (1995) A clinical trial of immunosuppressive therapy for myocarditis. The Myocarditis Treatment Trial Investigators. N Engl J Med 333:269–275

Mauviel A (1993) Cytokine regulation of metalloproteinase gene expression. J Cell Biochem 53:288–295

Mohammed FF, Smookler DS, Taylor SE, Fingleton B, Kassiri Z, Sanchez OH, English JL, Matrisian LM, Au B, Yeh WC, Khokha R (2004) Abnormal TNF activity in Timp3–/– mice leads to chronic hepatic inflammation and failure of liver regeneration. Nat Genet 36:969–977

Monrad ES, Matsumori A, Murphy JC, Fox JG, Crumpacker CS, Abelmann WH (1986) Therapy with cyclosporine in experimental murine myocarditis with encephalomyocarditis virus. Circulation 73:1058–1064

Murohara T, Guo JP, Lefer AM (1995) Cardioprotection by a novel recombinant serine protease inhibitor in myocardial ischemia and reperfusion injury. J Pharmacol Exp Ther 274:1246–1253

Murphy AN, Unsworth EJ, Stetler-Stevenson WG (1993) Tissue inhibitor of metalloproteinases-2 inhibits bFGF-induced human microvascular endothelial cell proliferation. J Cell Physiol 157:351–358

Murphy G, Knauper V (1997) Relating matrix metalloproteinase structure to function: why the "hemopexin" domain? Matrix Biol 15:511–518

Noutsias M, Pauschinger M, Poller WC, Schultheiss HP, Kuhl U (2003) Current insights into the pathogenesis, diagnosis and therapy of inflammatory cardiomyopathy. Heart Fail Monit 3:127–135

Ohta K, Nakajima T, Cheah AY, Zaidi SH, Kaviani N, Dawood F, You XM, Liu P, Husain M, Rabinovitch M (2004) Elafin-overexpressing mice have improved cardiac function after myocardial infarction. Am J Physiol Heart Circ Physiol 287:H286–292

Okada Y, Watanabe S, Nakanishi I, Kishi J, Hayakawa T, Watorek W, Travis J, Nagase H (1988) Inactivation of tissue inhibitor of metalloproteinases by neutrophil elastase and other serine proteinases. FEBS Lett 229:157–160

Okada Y, Naka K, Kawamura K, Matsumoto T, Nakanishi I, Fujimoto N, Sato H, Seiki M (1995) Localization of matrix metalloproteinase 9 (92-kilodalton gelatinase/type IV collagenase=gelatinase B) in osteoclasts: implications for bone resorption. Lab Invest 72:311–322

Olson MW, Toth M, Gervasi DC, Sado Y, Ninomiya Y, Fridman R (1998) High affinity binding of latent matrix metalloproteinase-9 to the alpha2 (IV) chain of collagen IV. J Biol Chem 273:10672–10681

Pauschinger M, Bowles NE, Fuentes-Garcia FJ, Pham V, Kuhl U, Schwimmbeck PL, Schultheiss HP, Towbin JA (1999) Detection of adenoviral genome in the myocardium of adult patients with idiopathic left ventricular dysfunction. Circulation 99:1348–1354

Pauschinger M, Chandrasekharan K, Noutsias M, Kuhl U, Schwimmbeck LP, Schultheiss HP (2004a) Viral heart disease: molecular diagnosis, clinical prognosis, and treatment strategies. Med Microbiol Immunol (Berl) 193: 65–69

Pauschinger M, Chandrasekharan K, Schultheiss HP (2004b) Myocardial remodeling in viral heart disease: possible interactions between inflammatory mediators and MMP-TIMP system. Heart Fail Rev 9:21–31

Plow EF, Ploplis VA, Busuttil S, Carmeliet P, Collen D (1999) A role of plasminogen in atherosclerosis and restenosis models in mice. Thromb Haemost 82 Suppl 1:4–7

Rifkin DB, Mazzieri R, Munger JS, Noguera I, Sung J (1999) Proteolytic control of growth factor availability. Apmis 107:80–85

Rohde LE, Ducharme A, Arroyo LH, Aikawa M, Sukhova GH, Lopez-Anaya A, McClure KF, Mitchell PG, Libby P, Lee RT (1999) Matrix metalloproteinase inhibition attenuates early left ventricular enlargement after experimental myocardial infarction in mice. Circulation 99:3063–3070

Sabri A, Alcott SG, Elouardighi H, Pak E, Derian C, Andrade-Gordon P, Kinnally K, Steinberg SF (2003) Neutrophil cathepsin G promotes detachment-induced cardiomyocyte apoptosis via a protease-activated receptor-independent mechanism. J Biol Chem 278:23944–23954

Schellings MW, Pinto YM, Heymans S (2004) Matricellular proteins in the heart: possible role during stress and remodeling. Cardiovasc Res 64:24–31

Schonbeck U, Mach F, Sukhova GK, Murphy C, Bonnefoy JY, Fabunmi RP, Libby P (1997) Regulation of matrix metalloproteinase expression in human vascular smooth muscle cells by T lymphocytes: a role for CD40 signaling in plaque rupture? Circ Res 81:448–454

Schroen B, Heymans S, Sharma U, Blankesteijn WM, Pokharel S, Cleutjens JP, Porter JG, Evelo CT, Duisters R, van Leeuwen RE, Janssen BJ, Debets JJ, Smits JF, Daemen MJ, Crijns HJ, Bornstein P, Pinto YM (2004) Thrombospondin-2 is essential for myocardial matrix integrity: increased expression identifies failure-prone cardiac hypertrophy. Circ Res 95:515–522

Shamamian P, Pocock BJ, Schwartz JD, Monea S, Chuang N, Whiting D, Marcus SG, Galloway AC, Mignatti P (2000) Neutrophil-derived serine proteinases enhance membrane type-1 matrix metalloproteinase-dependent tumor cell invasion. Surgery 127:142–147

Sivasubramanian N, Coker ML, Kurrelmeyer KM, MacLellan WR, DeMayo FJ, Spinale FG, Mann DL (2001) Left ventricular remodeling in transgenic mice with cardiac restricted overexpression of tumor necrosis factor. Circulation 104:826–831

Siwik DA, Chang DL, Colucci WS (2000) Interleukin-1beta and tumor necrosis factor-alpha decrease collagen synthesis and increase matrix metalloproteinase activity in cardiac fibroblasts in vitro. Circ Res 86:1259–1265

Spinale FG (2002) Matrix metalloproteinases: regulation and dysregulation in the failing heart. Circ Res 90:520–530

Spinale FG, Gunasinghe H, Sprunger PD, Baskin JM, Bradham WC (2002) Extracellular degradative pathways in myocardial remodeling and progression to heart failure. J Card Fail 8:S332–S338

Suzuki K, Enghild JJ, Morodomi T, Salvesen G, Nagase H (1990) Mechanisms of activation of tissue procollagenase by matrix metalloproteinase 3 (stromelysin). Biochemistry 29:10261–10270

Tao ZY, Cavasin MA, Yang F, Liu YH, Yang XP (2004) Temporal changes in matrix metalloproteinase expression and inflammatory response associated with cardiac rupture after myocardial infarction in mice. Life Sci 74:1561–1572

Thorgeirsson UP, Yoshiji H, Sinha CC, Gomez DE (1996) Breast cancer; tumor neovasculature and the effect of tissue inhibitor of metalloproteinases-1 (TIMP-1) on angiogenesis. In Vivo 10:137–144

Tiede K, Stoter K, Petrik C, Chen WB, Ungefroren H, Kruse ML, Stoll M, Unger T, Fischer JW (2003) Angiotensin II AT (1)-receptor induces biglycan in neonatal cardiac fibroblasts via autocrine release of TGFbeta in vitro. Cardiovasc Res 60:538–546

Tomioka N, Kishimoto C, Matsumori A, Kawai C (1986) Effects of prednisolone on acute viral myocarditis in mice. J Am Coll Cardiol 7:868–872

Towbin JA (2001) Myocarditis and pericarditis in adolescents. Adolesc Med 12:47–67

Tyagi SC (1998) Extracellular matrix dynamics in heart failure: a prospect for gene therapy. J Cell Biochem 68:403–410

Vallejo J, Mann DL (2003) Antiinflammatory therapy in myocarditis. Curr Opin Cardiol 18:189–193

Visse R, Nagase H (2003) Matrix metalloproteinases and tissue inhibitors of metalloproteinases: structure, function, and biochemistry. Circ Res 92:827–839

Wahl SM, Allen JB, Weeks BS, Wong HL, Klotman PE (1993) Transforming growth factor beta enhances integrin expression and type IV collagenase secretion in human monocytes. Proc Natl Acad Sci U S A 90:4577–4581

Wilson EM, Moainie SL, Baskin JM, Lowry AS, Deschamps AM, Mukherjee R, Guy TS, St John-Sutton MG, Gorman JH 3rd, Edmunds LH Jr, Gorman RC, Spinale FG (2003) Region- and type-specific induction of matrix metalloproteinases in post-myocardial infarction remodeling. Circulation 107:2857–2863

Woessner JF Jr (1999) Matrix metalloproteinase inhibition. From the Jurassic to the third millennium. Ann N Y Acad Sci 878:388–403

Yokoyama T, Nakano M, Bednarczyk JL, McIntyre BW, Entman M, Mann DL (1997) Tumor necrosis factor-alpha provokes a hypertrophic growth response in adult cardiac myocytes. Circulation 95:1247–1252

Zaidi SH, Hui CC, Cheah AY, You XM, Husain M, Rabinovitch M (1999) Targeted overexpression of elafin protects mice against cardiac dysfunction and mortality following viral myocarditis. J Clin Invest 103:1211–1219

13 Inflammatory Cardiomyopathy: There Is a Specific Matrix Destruction in the Course of the Disease

J.A. Towbin

Abstract. Cardiomyopathies are responsible for a high proportion of cases of congestive heart failure and sudden death, as well as for the need for transplantation. Understanding of the causes of these disorders has been sought in earnest over the past decade. We hypothesized that DCM is a disease of the cytoskeleton/sarcolemma, which affects the sarcomere. Evaluation of the sarcolemma in DCM and other forms of systolic heart failure demonstrates membrane disruption; and, secondarily, the extracellular matrix architecture is also affected. Disruption of the links from the sarcolemma to ECM at the dystrophin C-terminus and those to the sarcomere and nucleus via N-terminal dystrophin interactions could lead to a "domino effect" disruption of systolic function and development of arrhythmias. We also have suggested that dystrophin mutations play a role in idiopathic DCM in males. The T-cap/MLP/α-actinin/titin complex appears to stabilize Z-disc function via mechanical stretch sensing. Loss of elasticity results in the primary defect in the endogenous cardiac muscle stretch sensor machinery. The over-stretching of individual myocytes leads to activation of cell death pathways, at a time when stretch-regulated survival cues are diminished due to defective stretch sensing, leading to progression of heart failure. Genetic DCM and the acquired disorder viral myocarditis have the same clinical features including heart failure, arrhythmias, and conduction block, and also similar mechanisms of disease based on the proteins targeted. In dilated cardiomyopathy, the process of progressive ventricular dilation and changes of the shape of the ventricle to a more spherical shape, associated with changes in ventricular function and/or hypertrophy, occurs without known initiating disturbance. In those cases in which resolution of cardiac dysfunction does not occur, chronic DCM results. It has been unclear what the underlying etiology of this long-term sequela could be, but viral persistence and autoimmunity have been widely speculated.

13.1 Introduction

Cardiomyopathies are heart muscle disorders associated with significant morbidity and mortality. These disorders are classified by the World Health Organization (WHO) into four forms (WHO 1996): (1) dilated cardiomyopathy (DCM), (2) hypertrophic cardiomyopathy (HCM), (3) restrictive cardiomyopathy (RCM), and (4) arrhythmogenic right ventricular cardiomyopathy (ARVC). Recently, another cardiomyopathy, left ventricular noncompaction (LVNC), has gained attention, although it does not meet criteria for a separate classification currently (Pignatelli et al. 2003).

The most common cardiomyopathy is DCM, accounting for approximately 55% of cases (WHO 1996). HCM is the second most common, approximately 35%, with the remaining forms accounting for approximately 5% or less in each case (WHO 1996). The importance of these disorders lies in the fact that they are responsible for a high proportion of cases of congestive heart failure and sudden death, as well as the need for transplantation. The mortality rate in the United States due to cardiomyopathy is greater than 10,000 deaths per annum, with DCM being the major contributor to this death rate (Abelman and Lorrell 1989). The total cost of health care in the United States focused on cardiomyopathies is in the billions of dollars and only limited success has been achieved (O'Connell and Bristow 1994). In order to improve care and outcomes in children and adults, understanding of the causes of these disorders has been sought in earnest over the past decade.

13.2 Basic Research

13.2.1 Normal Cardiac Structure

In order to understand the mechanisms responsible for the development of the clinical disease, an understanding of normal cardiac structure is necessary.

Cardiac muscle fibers comprise separate cellular units (myocytes) connected in series (Schwartz et al. 2001). In contrast to skeletal muscle fibers, cardiac fibers do not assemble in parallel arrays but bifurcate and recombine to form a complex three-dimensional network. Cardiac myocytes are joined at each end to adjacent myocytes at the intercalated disc, the specialized area of interdigitating cell membrane (Fig. 1). The intercalated disc contains gap junctions (containing connexins), and mechanical junctions, comprising adherens junctions (con-

▶

Fig. 1. Cardiac myocyte cytoarchitecture. Schematic of the interactions between dystrophin and the dystrophin-associated proteins in the sarcolemma and intracellular cytoplasm (dystroglycans, sarcoglycans, syntrophins, dystrobrevin, sarcospan) at the C-terminal end of the dystrophin. The integral membrane proteins interact with the extracellular matrix via α-dystroglycan-laminin α2 connections. The N-terminus of dystrophin binds actin and connects dystrophin with the sarcomere intracellularly, the sarcolemma and extracellular matrix. *N*, amino terminus; *C*, carboxy terminus; *MLP*, muscle *LIM* protein

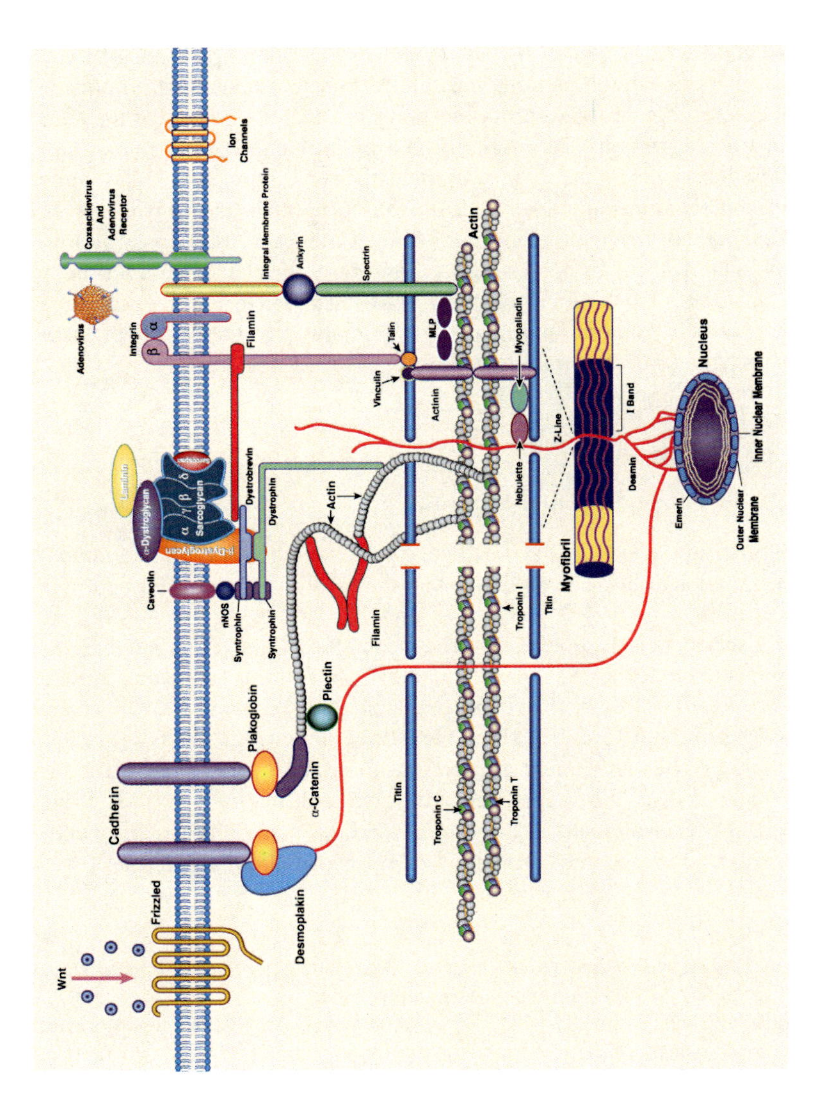
Wnt
Frizzled
Cadherin
Desmoplakin
Plakoglobin
α-Catenin
Plectin
Caveolin
nNOS
Syntrophin
Syntrophin
α-Dystroglycan
β-Dystroglycan
α γ β δ
Sarcoglycan
Sarcospan
Laminin
Dystrobrevin
Dystrophin
Actin
Filamin
Adenovirus
Integrin
β
α
Filamin
Coxsackievirus
And
Adenovirus
Receptor
Integral Membrane Protein
Ankyrin
Spectrin
Ion
Channels
Talin
Vinculin
Actinin
MLP
Actin
Myopalladin
Nebulette
Titin
Titin
Titin
Troponin C
Troponin T
Troponin I
Z-Line
I Band
Myofibril
Desmin
Emerin
Nucleus
Outer Nuclear
Membrane
Inner Nuclear Membrane

taining N-cadherin, catenins, and vinculin) and desmosomes (containing desmin, desmoplakin, desmocollin, and desmoglein). Cardiac myocytes are surrounded by a thin membrane (sarcolemma) and the interior of each myocyte contains bundles of longitudinally arranged myofibrils. The myofibrils are formed by repeating sarcomeres, the basic contractile units of cardiac muscle comprising interdigitating thin (actin) and thick (myosin) filaments (Fig. 1), that give the muscle its characteristic striated appearance (Gregorio and Antin 2000; Squire 1997). The thick filaments are composed primarily of myosin but additionally contain myosin-binding proteins C, H, and X. The thin filaments are composed of cardiac actin, α-tropomyosin (α-TM), and troponins T, I, and C (cTnT, cTnI, cTnC). In addition, myofibrils contain a third filament formed by the giant filamentous protein, titin, which extends from the Z-disc to the M-line and acts as a molecular template for the layout of the sarcomere. The Z-disc at the borders of the sarcomere is formed by a lattice of interdigitating proteins that maintain myofilament organization by cross-linking antiparallel titin and thin filaments from adjacent sarcomeres (Fig. 2). Other proteins in the Z-disc include α-actinin, nebulette, telethonin/T-cap, capZ, MLP, myopalladin, myotilin, cypher/ZASP, filamin, and FATZ (Clark et al. 2002; Gregorio and Antin 2000; Pyle and Solaro 2004; Squire 1997; Vigoreaux 1994).

Finally, the extrasarcomeric cytoskeleton, a complex network of proteins linking the sarcomere with the sarcolemma and the extracellular matrix (ECM), provides structural support for subcellular structures and transmits mechanical and chemical signals within and between cells. The extrasarcomeric cytoskeleton has intermyofibrillar and subsarcolemmal components, with the intermyofibrillar cytoskeleton composed of intermediate filaments (IFs), microfilaments, and microtubules (Barth et al. 1997; Burridge and Chrzanowska-Wodnicka 1996; Capetanaki 2002; Clark et al. 2002; Pyle and Solaro 2004; Stewart 1993; Vigoreaux 1994). Desmin IFs form a three-dimensional scaffold throughout the extra-sarcomeric cytoskeleton with desmin filaments surrounding the Z-disc, allowing for longitudinal connections to adjacent Z-discs and lateral connections to subsarcolemmal costameres (Capetanaki 2002; Stewart 1993). Microfilaments composed of non-sarcomeric actin (mainly γ-actin) also form complex networks linking the sarcomere (via α-actinin) to various components of the costameres.

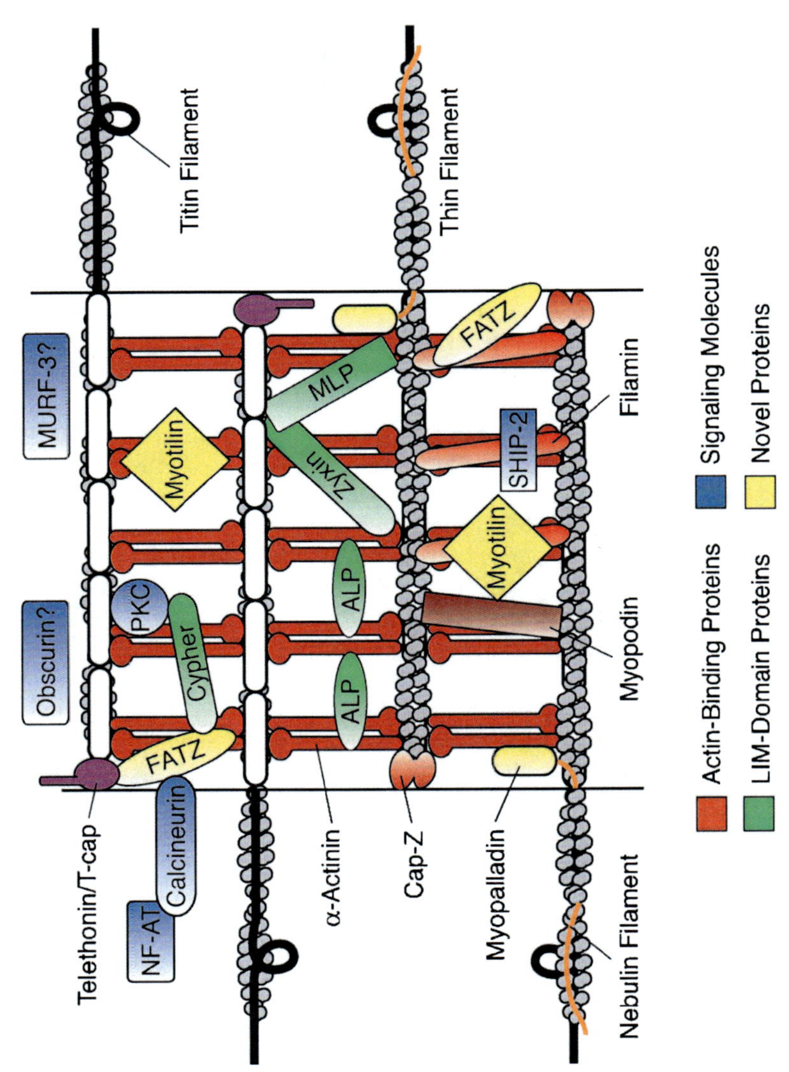

Fig. 2. Z-disc architecture. The Z-disc of the sarcomere comprises multiple interacting proteins that anchor the sarcomere. Reported with permission from Clark et al. (2002)

Costameres are subsarcolemmal domains located in a periodic, grid-like pattern, flanking the Z-discs and overlying the I bands, along the cytoplasmic side of the sarcolemma. These costameres are sites of interconnection between various cytoskeletal networks linking sarcomere and sarcolemma and are thought to function as anchor sites for stabilization of the sarcolemma and for integration of pathways involved in mechanical force transduction. Costameres contain three principal components: the focal adhesion-type complex, the spectrin-based complex, and the dystrophin/dystrophin-associated protein complex (DAPC) (Sharp et al. 1997; Straub and Campbell 1997). The focal adhesion-type complex, comprising cytoplasmic proteins (i.e., vinculin, talin, tensin, paxillin, zyxin), connects with cytoskeletal actin filaments and with the transmembrane proteins α-, β-dystroglycan, α-, β-, γ-, δ-sarcoglycans, dystrobrevin, and syntrophin (Barth et al. 1997; Burridge and Chrzanowska-Wodnicka 1996). Several actin-associated proteins are located at sites of attachment of cytoskeletal actin filaments with costameric complexes, including α-actinin and the muscle LIM protein, MLP. The C-terminus of dystrophin binds β-dystroglycan (Fig. 1), which in turn interacts with α-dystroglycan to link to the ECM (via α-2-laminin). The N-terminus of dystrophin interacts with actin. Also notable, voltage-gated sodium channels co-localize with dystrophin, β-spectrin, ankyrin, and syntrophins while potassium channels interact with the sarcomeric Z-disc and intercalated discs (Furukawa et al. 2001; Kucera et al. 2002; Ribaux et al. 2001). Since arrhythmias and conduction system diseases are common in children and adults with DCM, this could play an important role. Hence, disruption of the links from the sarcolemma to ECM at the dystrophin C-terminus and those to the sarcomere and nucleus via N-terminal dystrophin interactions could lead to a "domino effect" disruption of systolic function and development of arrhythmias.

13.2.2 Z-Disc Organization

The precise organization of Z-discs, the borders of individual sarcomeres in vertebrate striated muscle, defines a supramolecular assembly of eukaryotic cell structure (Fig. 2). Z-discs contain the barbed ends of actin thin filaments, the N-terminal ends of titin filaments, and the C-terminal ends of nebulette/nebulin in cardiac and skeletal muscle, respectively, as well as a variety of other regulatory and structural proteins. In addition to being a boundary between successive sarcomeres, Z-discs are responsible for transmitting tension generated by individual sarcomeres along the length of the myofibril, allowing for efficient contractile activity (i.e., the primary conduits of the force generated by contraction) (Clark et al. 2002; Gregorio and Antin 2000; Squire 1997; Vigoreaux 1994). Z-disc associated proteins also appear to be crucial for early stages of myofibril assembly since I-Z-I structures (i.e., Z-disc precursors) form the earliest identifiable protein assemblies observed during muscle differentiation (Ehler et al. 1999; Epstein and Fischman 1991; Gregorio and Antin 2000; Holtzer et al. 1997). Detailed ultrastructural and biochemical investigations of the Z-disc and its various components have yielded valuable information concerning its structural architecture. The width of the Z-disc can vary from roughly 30 nm in fish skeletal muscle to greater than 1 μm in patients with Nemaline myopathy (Franzini-Armstrong 1973; Rowe 1973; Vigoreaux 1994). The thin filaments from adjacent sarcomeres fully overlap within the Z-disc and are cross-linked by the direct interaction of actin filaments and the actin filament barbed end-capping protein, CapZ, with α-actinin (Goll et al. 1991; Papa et al. 1999). Titin filaments from adjacent sarcomeres also fully overlap in the Z-lines and are cross-linked by α-actinin (Clark et al. 2002). In vitro studies have identified two distinct α-actinin-binding sites within titin's N-terminal, 80-kDa, Z-disc integral segment. These binding sites may link together titin and α-actinin filaments both inside the Z-disc and around its periphery (Granzier and Labeit 2004; Gregorio et al. 1998; Ohtsuka et al. 1997; Pyle and Solaro 2004; Sorimachi et al. 1997; Young et al. 1998). A direct interaction between the Z-disc peripheral region of nebulin/nebulette and the intermediate filament protein desmin has been identified, and it is believed that this link contributes to the lateral connections between adjacent Z-discs (Bang et al. 2001). Another protein, myopalladin, interacts with nebulette and α-actinin, as well as with tinin and the thin filaments, forming intra-Z-disc meshworks. Myopalladin complex and α-actinin appear to form a linking system – tethering the barbed ends of actin–titin filaments, the N-terminal ends of titin filaments, and the C-terminal ends of nebulin/nebulette within the Z-disc – and provide the anchor to stabilize these structures against stress and other forces.

Another important interaction within the Z-disc occurs between titin and telethonin (T-cap), an interaction critical for sarcomeric function (Gregorio et al. 1998). Telethonin has recently been shown to interact with α-actinin as well as MLP, another Z-disc protein (Knoll et al. 2002). This complex (T-cap/MLP/α-actinin/titin) appears to stabilize Z-disc function via mechanical stretch sensing. Loss of the MLP/T-cap/titin-dependent mechanical stress sensor pathway (due to MLP mutations) is thought to result in destabilization of the anchoring of the Z-disc to the proximal end of the T-cap/titin complex, which leads to conformational alteration of the intrinsic titin molecular spring elements (Towbin 2002). In turn, this loss of elasticity results in the primary defect in the endogenous cardiac muscle stretch sensor machinery. The over-stretching of individual myocytes leads to activation of cell death pathways, at a time when stretch-regulated survival cues are diminished due to defective stretch sensing, leading to myocyte cell death and progression of heart failure (Chien 1999).

13.2.3 Dystrophin

Dystrophin is a cytoskeletal protein which provides structural support to the myocyte by creating a lattice-like network to the sarcolemma (Hoffman et al. 1987). In addition, dystrophin plays a major role in linking the sarcomeric contractile apparatus to the sarcolemma and extracellular matrix (Campbell 1995; Cox and Kunkel 1997; Kaprielian et al. 2000; Meng et al. 1996). Furthermore, dystrophin is involved in cell signaling, particularly through its interactions with nitric oxide synthase (Chang et al. 1996). The dystrophin gene is responsible for Duchenne and Becker muscular dystrophy (DMD/BMD) when mutated as well (Koenig et al. 1987). These skeletal myopathies present early in life (DMD is diagnosed before age 12 years while BMD is seen in teenage males older than 16 years of age), and the vast majority of patients develop DCM before the 25th birthday. In most patients, creatine kinase muscle type (CK-MM) is elevated, the same is seen in X-linked dilated cardiomyopathy (XLCM) (Berko and Swift 1987; Feng et al. 2002a; Ferlini et al. 1998; Franz et al. 2000; Milasin et al. 1996; Muntoni et al. 1993; Ortiz-Lopez et al. 1997; Towbin et al. 1993b; Yoshida et al. 1998). In addition, manifesting female carriers develop disease late in life, similar to XLCM. Furthermore, immunohistochemical analysis demonstrates reduced levels (or absence) of dystrophin, similar to that seen in the hearts of patients with XLCM.

Murine models of dystrophin deficiency demonstrate abnormalities of muscle physiology based on membrane structural support abnormalities. In addition to the dysfunction of dystrophin, mutations in dystrophin secondarily affect proteins which interact with dystrophin. At the amino-terminus (N-terminus), dystrophin binds to the sarcomeric protein actin, a member of the thin filament of the contractile apparatus. At the carboxy-terminus (C-terminus), dystrophin interacts with α-dystroglycan, a dystrophin-associated membrane-bound protein which is involved in the function of DAPC, which includes β-dystroglycan, the sarcoglycan subcomplex (α, β, γ, δ, and ϵ sarcoglycan), syntrophins, and dystrobrevins (Fig. 1). In turn, this complex interacts with α2-laminin and the ECM (Emery 2002). Like dystrophin, mutations in these genes lead to muscular dystrophies with or without cardiomyopathy, supporting the contention that this group of proteins are important to the normal function of the myocytes of the heart and skeletal muscles (Campbell 1995; Emery 2002; Klietsch et al. 1993; Ozawa et al. 1995). In both cases, mechanical stress (Petrof et al. 1993) appears to play a significant role in the age-onset dependent dysfunction of these muscles. The information gained from the studies on XLCM, DMD, and BMD led us to hypothesize that DCM is a disease of the cytoskeleton/sarcolemma which affects the sarcomere (Danialou et al. 2001; Kyoi et al. 2003; Towbin 1998; Wilding et al. 2005). We also have suggested that dystrophin mutations play a role in idiopathic DCM in males. Recently we showed that 3/22 boys with DCM studied for dystrophin mutations using a rapid DNA mutation screening method had mutations and all had elevated CK-MM as well (Danialou et al. 2001; Kyoi et al. 2003; Towbin 1998; Wilding et al. 2005). In addition, 8 families with DCM and possible X-linked inheritance were also screened and in 3/8 families, dystrophin mutations were noted. Again, CK-MM was elevated in all subjects carrying mutations (Feng et al. 2002a).

Disruption of cardiac cytoarchitecture by genetic mutations or acquired abnormalities leads to clinical cardiomyopathic disorders. These disorders will be discussed later on.

13.2.4 Cytoskeletal Disruption in Inflammatory Cardiomyopathy

Vatta and colleagues demonstrated that N-terminal dystrophin is disrupted and lost in myocarditis and other forms of DCM (Vatta et al. 2002, 2004). Using immunohistochemistry and Western blotting, myocardial specimens from patients were evaluated using antibodies against the N-terminus, rod domain, and C-terminus of dystrophin. In nearly all cases, the C-terminus and rod domain staining was normal while N-terminal dystrophin staining was reduced or absent. This abnormality was shown to be reversible using mechanical unloading therapy (ventricular assist device, VAD) after 3 months of therapy (Fig. 3). The authors suggest that proteases, caspases, and apoptotic mechanisms are at play.

Badorff, Knowlton, and colleagues have shown that coxsackievirus (CV)B_3 myocarditis results in DCM and heart failure due to disruption of dystrophin (Badorff et al. 1999, 2000; Badorff and Knowlton 2004; Lee et al. 2000; Xiong et al. 2002). This viral genome encodes for a variety of structural proteins and proteases. One such protease, enteroviral protease 2A cleaves dystrophin in the third huge region and causes loss of function

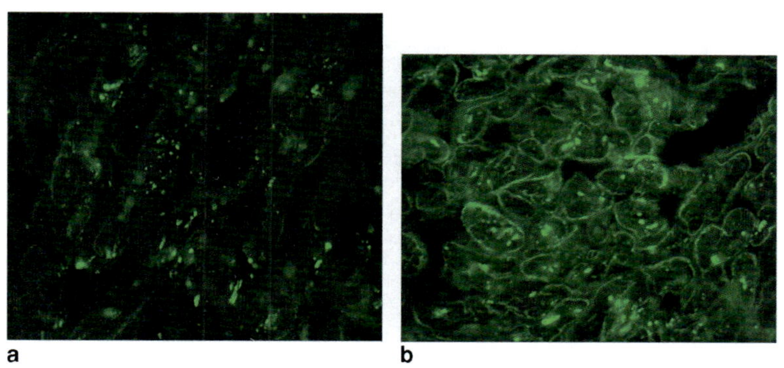

Fig. 3a,b. Dystrophin loss in dilated cardiomyopathy/myocarditis and normalization after LVAD therapy. **a** Amino-terminus: (N-terminus) dystrophin staining demonstrates loss of dystrophin in the sarcolemma of cardiomyocytes in the heart of a patient with myocarditis-induced DCM. **b** Amino-terminus: (N-terminus) dystrophin staining demonstrates normalization of dystrophin in the sarcolemma of cardiomyocytes in the heart of a patient with myocarditis-induced DCM treated with LVAD

(Badorff et al. 2000; Badorff and Knowlton 2004; Badorff et al. 1999; Lee et al. 2000; Xiong et al. 2002).

Evaluation of the sarcolemma in DCM and other forms of systolic heart failure demonstrates membrane disruption. Evans blue dye in mouse models has been shown to extravasate in these cases, consistent with membrane disruption; and, secondarily, the extracellular matrix architecture is also affected. These findings have been confirmed by others (Heydemann et al. 2001; Lapidos et al. 2004; McNally et al. 2003).

13.2.5 Cytokines

Over the last few years, there has been considerable interest in the role of cytokines in the pathogenesis of myocarditis and DCM. Animal studies have suggested that a relationship may exist between subclinical viral infection and later development of DCM. This process is presumed to occur by an autoimmune-like mechanism triggered by the initial viral insult (Kubota et al. 1997; Lane et al. 1993; Neumann et al. 1993). Several murine models have been studied which suggest that cytokine-mediated modulation of the immune response to viral infection may lead to induction of chronic autoimmune myocarditis (Huber 1997; Kroemer and Martinez 1991; Lane et al. 1992; Smith and Allen 1992). Among their many immunomodulatory activities, cytokines contribute to regulation of antibody production and maintenance of "self-tolerance." Certain susceptible murine strains, when infected with CVB3, are known to develop myocyte necrosis and an acute inflammatory response consisting mainly of neutrophils and macrophages within the heart. After the initial viral infection, resolution of inflammation eventually occurs. In other strains, however, a second autoimmune phase of myocarditis appears later with findings of diffuse mononuclear cell infiltrates within the heart. These mononuclear cells are a significant source of the cytokines interleukin-1 (IL-1) and tumor necrosis factor-α (TNF-α), and work by Henke et al. (1991) demonstrated that release of large amounts of TNF-α and IL-1β by human monocytes exposed to CVB3 occurs. Both of these cytokines are known to participate in leukocyte activation that may be beneficial in promoting a specific lymphocyte response to viral infection. However, these cytokines may also promote cardiac fibroblast activity, and therefore it has been speculated that local secretion of cytokines in the myocardium perpetuates the inflammatory process, which secondarily leads to the fibrosis associated with DCM and resultant deterioration of cardiac function. Studies by Gulick et al.(1989) initially implicated IL-1 and TNF-α as potential inhibitors of cardiac myocyte β-adrenergic responsiveness, and further studies have shown IL-1 and TNF-α to be the macrophage factors mediating this effect. In particular, TNF-α has been studied in some detail, resulting in reports of elevated TNF-α levels in the serum of patients with chronic heart disease (Torre-Amione et al. 1996), including a subset of patients with myocarditis or DCM (Matsumori et al. 1994). TNF-α is able to potentiate the immune response and induce apoptosis in cells, both of which appear to hold special importance in the pathogenesis of myocarditis. Other inflammatory mediators, including granulocyte colony-stimulating factor (GCSF), are also elevated in myocarditis patients (Matsumori et al. 1994) and have received attention as well. Other studies have suggested that inflammatory cytokines may actually cause a direct negative inotropic response (Bowles and Vallejo 2003).

13.2.6 Cell Adhesion Molecules

Now known to play a major role in many processes of inflammation, the distinct classes of cell adhesion molecules (CAMs) may also play a role in the pathogenesis of myocarditis. One molecule which is well known to play a major role in cell–cell adhesion, particularly leukocyte adherence and transendothelial migration, is intercellular adhesion molecule-1 (ICAM-1). ICAM-1 is a member of the immunoglobulin supergene family of CAMs and is a single chain glycoprotein of 80–115 kDa with an extracellular domain made up of five immunoglobulin-like repeats (Seko et al. 1993; Springer 1990). ICAM-1 is predominantly expressed on endothelial cells, but also on fibroblasts, epithelial cells, mucosal cells, lymphocytes, monocytes, and cardiac myocytes after inflammatory injury. Expression of ICAM-1 on endothelial cells is known to be upregulated by cytokines such as IL-1 and TNF-α. A well-established binding ligand of ICAM-1 is lymphocyte function-associated antigen-1 (LFA-1), a molecule which is part of the β-2 integrin family and consists of a 180-kDa α-subunit (CD11a) and a 95-kDa β-subunit (CD18). LFA-1 is expressed on virtually all leukocytes, including monocytes. The adhesive interaction between LFA-1 and ICAM-1 is known to mediate adhesion-dependent helper T cell, cytotoxic T cell, and natural killer (NK) cell functions. Antibody to LFA-1 has been used for therapy in animal models of myocarditis with resultant blockade of the inflammatory response.

13.2.7 Apoptosis

Apoptosis, or programmed cell death, has an important role in embryogenesis, tissue homeostasis, and regulation of immunologic responses, among normal physiologic processes, and is associated with the growth and regression of tumors (Cohen 1993). Cells undergoing apoptosis exhibit characteristic morphologic and biochemical features, including chromatin aggregation, nuclear and cytoplasmic aggregation, and formation of apoptotic bodies resulting from the partition of the cytoplasm and nucleus into membrane bound-vesicles. These apoptotic bodies are rapidly phagocytosed by adjacent macrophages or epithelial cells, without resulting in an inflammatory response. Apoptotic cells are detectable by terminal transferase labeling [terminal deoxynucleotide transferase-mediated biotin-deoxyuridine triphosphate nick end labeling (TUNEL)] in myocardial tissue samples from patients with DCM. It has been shown that up to 0.1% of cells stain positive by this technique. A number of viruses have been implicated in the induction of apoptosis, including human immunodeficiency virus (HIV), Epstein-Barr virus (EBV), and adenovirus. Apoptotic cells have been detected in myocardial sections from patients with adenovirus-associated myocarditis and DCM, the areas of staining are usually focal, and a number of positively staining areas may be detected within each section. Within such areas, up to 1% of cells may stain positive, including myocytes, infiltrating inflammatory cells, and endothelial cells. In the tissue sections from control patients, either unstained or sporadic (one or two per section) stained cells may be detected. These data suggest a relationship between infection of the myocardium by adenovirus and the onset of apoptosis, which could result in pathologic processes associated with myocarditis and DCM (White 1993). Further, a number of inflammatory cells may be seen to be undergoing apoptosis. Although this could reflect the natural defense mechanism of the host against the virus, it also raises the possibility of virus-induced apoptosis as a mechanism of immune system avoidance. Strand et al. (1996) reported that in tumors, infiltrating immune cells are destroyed by the induction of apoptosis through the expression of Fas ligand on the tumor cell that binds Fas on the lymphocyte.

13.3 Clinical Aspects

13.3.1 Dilated Cardiomyopathy

DCM has become a popular target of research over the past 7–8 years, with multiple genes identified during that time period (Seidman and Seidman 2001; Towbin and Bowles 2002). These genes appear to encode two major subgroups of proteins, cytoskeletal and sarcomeric proteins (Seidman and Seidman 2001; Towbin and Bowles 2002). The cytoskeletal proteins identified to date include dystrophin, desmin, lamin A/C, δ-arcoglycan, metavinculin, muscle LIM protein, and α-actinin. In the case of sarcomere-encoding genes, the same genes identified for HCM appear to be culprits in select patients. A new gene, cypher/ZASP, a Z-line protein, has been identified recently as well (Vatta et al. 2003).

13.3.2 Viral Myocarditis

Another form besides genetic DCM, the acquired disorder viral myocarditis, has the same clinical features as DCM including heart failure, arrhythmias, and conduction block (Towbin 2002). Evidence currently exists that suggests that viral myocarditis and genetic DCM have similar mechanisms of disease based on the proteins targeted (Bowles and Towbin 2000; Bowles and Vallejo 2003).

Most cases of myocarditis in the United States and western Europe result from viral infections (Berkovich et al. 1968; Bowles et al. 2003; Dec et al. 1985), including infections with the enteroviruses (CVA and CVB, echovirus, poliovirus), particularly CVB (Bowles and Towbin 1998; Hirschman and Hammer 1974; Rosenberg and McNamara 1964; Woodruff 1980), adenoviruses [particularly serotype 2 and 5, (Bowles et al. 2003; Bowles and Towbin 1998; Dec et al. 1985)], and parvovirus B19 (Schowengerdt et al. 1997; Tschope et al. 2005), among other cardiotropic viruses (Bowles et al. 2003; Tschope et al. 2005) (see Table 1).

Table 1. Viral PCR analysis of myocarditis and dilated cardiomyopathy (DCM). The detection of viruses by PCR in myocardial samples

Diagnosis	No. of samples	No. of PCR+ samples	PCR amplimer (#)
Myocarditis	624	239 (38%)	Adenovirus 142 (23%) Enterovirus 85 (14%) CMV 18 (3%) Parvovirus 6 (< 1%) Influenza A 5 (< 1%) HSV 5 (< 1%) EBV 3 (< 1%) RSV 1 (< 1%)
DCM	149	30 (20%)	Adenovirus 18 (12%) Enterovirus 12 (8%)
Controls	215	3 (1.4%)	Enterovirus 1 (< 1%) CMV 2 (< 1%)

However, a wide variety of other viral causes of myocarditis (Bowles et al. 2003; Bowles and Towbin 1998; Perez-Pulido 1984) in children and adults have been described, including influenza (Karjalainen et al. 1980; Proby et al. 1986), cytomegalovirus (CMV) (Schonian et al. 1995; Tiula and Leinikki 1972), herpes simplex virus (HSV) (Lowry et al. 1983), hepatitis C (HCV) (Okabe et al. 1997), rubella (Ainger et al. 1966), varicella (Lorber et al. 1988; Osama et al. 1979), mumps (Chaudary and Jaski 1989; Noren et al. 1963), EBV (Frishman et al. 1977), HIV (Barbaro et al. 1998; Herskowitz et al. 1994; Lipshultz et al. 1998), and respiratory syncytial virus (RSV) (Puchkov and Minkovich 1972), among others.

Usually sporadic, viral myocarditis can also occur as an epidemic. Epidemics usually are seen in newborns, most commonly in association with CVB. Intrauterine myocarditis has also been seen during epidemics as well as sporadically (Towbin et al. 1993a; Van den Veyver et al. 1998). Postnatal spread of coxsackievirus is via the fecal/oral or airborne route (Friedman et al. 1998; Wynn and Braunwald 1997). The World Health Organization (WHO) reports that this ubiquitous family of viruses results in cardiovascular sequelae in less than 1% of infections, although this increases to 4% when CVB is considered (Grist and Reid 1962). Other important viral causes, such as adenovirus and influenza A, are transmitted through the air (Friedman et al. 1998; Shimizu et al. 1995; Sterner 1962). Although the disease can occur equally throughout the year, the exact etiology is probably season-dependent [in other words, certain viral causes are seasonal (i.e., coxsackievirus) while others are year-round (i.e., adenovirus)].

Viral Diagnostics

The diagnostic gold standard of the viral etiology is positive viral culture from the myocardium; this is rare, however. Viral cultures of peripheral specimens, such as blood, stool, or urine are commonly performed but are unreliable at identifying the causative infection. Other studies used to delineate the viral etiology include serologic studies in which a fourfold rise in antibody titer is required (Friedman et al. 1998). Antibody studies commonly performed include type-specific neutralizing, hemagglutination inhibiting, or complement-fixing antibody studies. However, these studies are nonspecific, since prior infection with the causative virus is commonplace.

Molecular analysis using in situ hybridization has been used to identify CVB sequences in myocardial samples (Archard et al. 1987; Bowles et al. 1986; Bowles et al. 1989), but this method never gained popularity. Currently, polymerase chain reaction (PCR) (Bowles and Towbin 1998; Martin et al. 1994; Woodruff 1980) has been used to rapidly and specifically amplify viral sequences from cardiac tissue samples. In 1986, the first molecular diagnostic approach was reported by Bowles et al. (1986) in which in situ hybridization was performed on myocardial tissue using probes for coxsackievirus. Subsequently, other reports noted the utility of this method in identifying coxsackievirus RNA within cardiac tissue specimens (Archard et al. 1987; Bowles et al. 1989). However, the difficulty of using this technique in the hospital-based setting reduced the interest in pursuing this technique. In 1990, Jin and colleagues (1990) first reported the use of PCR in identification of viral genome within the myocardium. This amplification process allows for specific portions of the viral genome of interest to be identified on an agarose gel after electrophoresis and is quite sensitive and specific (Fig. 4). During the

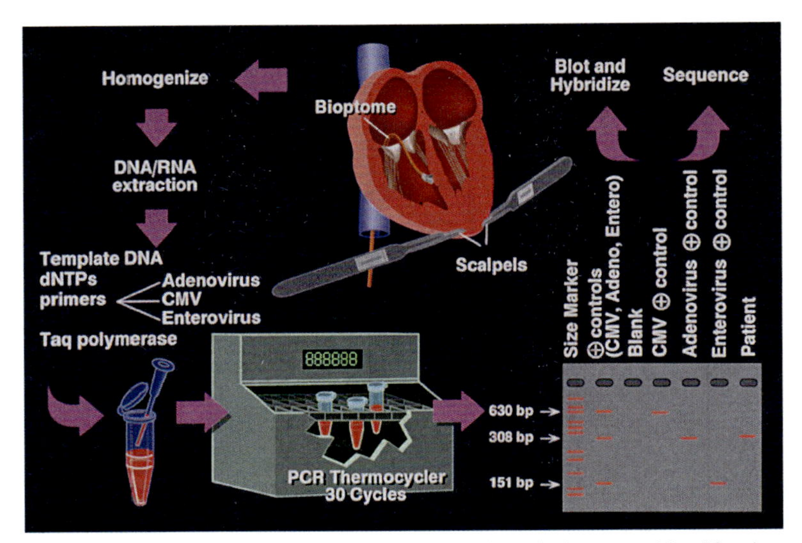

Fig. 4. Polymerase chain reaction (PCR) approach to viral genome identification in myocarditis

past decade, a large number of investigators have demonstrated the ability to identify enteroviral genome by PCR (Archard et al. 1998; Griffin et al. 1995; Hilton et al. 1993; Kyu et al. 1992) and, in fact, 20%–50% of cases are reportedly identified as enterovirus PCR-positive in these studies. PCR has also been used to screen for other viral genomes within cardiac tissue specimens and, using this method, Towbin and colleagues have showed adenovirus to be at least as common as enterovirus in heart tissue specimens of patients with myocarditis or DCM (Griffin et al. 1995; Martin et al. 1994; Woodruff 1980). This was confirmed by other laboratories as well (Calabrese et al. 2002; Lozinski et al. 1994; Pauschinger et al. 1999). In addition, other viral genomes have been identified using PCR, including cytomegalovirus, parvovirus, RSV, EBV, HSV, and influenza A virus (Ni et al. 1997; Schowengerdt et al. 1997; Woodruff 1980). Further, this method has been used to identify mumps virus as the responsible agent in endocardial fibroelastosis (EFE) (Ni et al. 1997), a previously important cause of heart failure in children which has disappeared over the past 20 years (Chaudary and Jaski 1989; Hutchins and Vie 1972; Noren et al. 1963).

PCR analysis usually does not identify viral genome in peripheral blood of patients with myocarditis; however, Akhtar et al. (1996) demonstrated the ability of this method to identify viral genome in tracheal aspirates of intubated children with myocarditis, potentially reducing the need for EMB.

13.3.3 Non-viral Myocarditis

Non-viral etiologies include infectious agents such as rickettsiae, bacteria, protozoa and other parasites, fungi, and yeast (Table 2; Seko et al. 1993); various drugs including antimicrobial medications (Wynn and Braunwald 1997); hypersensitivity, autoimmune or collagen-vascular diseases, such as systemic lupus erythematosus, mixed connective tissue disease, rheumatic fever, rheumatoid arthritis, and scleroderma; toxic reactions to infectious agents (e.g., diphtheria); or other disorders such as Kawasaki disease and sarcoidosis (Table 2). In most cases, however, "idiopathic" myocarditis is encountered (Wynn and Braunwald 1997).

13.3.4 Syndrome of Heart Failure

Traditionally, heart failure has been viewed as a constellation of clinical findings resulting from inadequate systolic function ("pump function"). However, over the past decade this view has been altered by a variety of clinical and basic data including the continued poor outcomes of patients despite therapies designed to improve systolic function (Hunt et al. 2001). In addition, information regarding inflammatory mediators, apoptosis, structure–function studies, and genetics has resulted in a concept that the syndrome of heart failure occurs due to a complex interaction of structural, functional, and biologic disturbances of the heart. The current concepts regarding the syndrome of heart failure integrate a series of models of heart failure, including the hemodynamic model, neurohormonal model, structural model, and autocrine-paracrine model of heart failure (Francis and Wilson-Tang 2003; Jessup and Brozena 2003). The structural model focuses on the necessity of proper interactions between the sarcomere and sarcolemma via cytoskeletal and other

Table 2. Causes of myocarditis

Viral
- Coxsackievirus A
- Coxsackievirus B
- Echoviruses
- Rubella virus
- Measles virus
- Parvovirus
- Adenovirus
- Polio viruses
- Vaccinia virus
- Mumps Virus
- Herpes simplex virus
- Epstein-Barr virus
- Cytomegalovirus
- Rhinoviruses
- Hepatitis viruses
- Arboviruses
- Influenza viruses
- Varicella virus

Rickettsial
- *Rickettsia rickettsii*
- *Rickettsia tsutsugamushi*

Bacterial
- Meningococcus
- Klebsiella
- Leptospira
- Diphtheria
- Salmonella
- Clostridia
- Tuberculosis
- Brucella
- *Legionella pneumophila*
- Streptococcus

Protozoal
- *Trypanosoma cruzi*
- Toxoplasmosis
- Amebiasis

Other Parasites
- *Toxocara canis*
- Schistosomiasis
- Heterophyiasis
- Cysticercosis
- Echinococcus
- Visceral larva migrans

Fungi and yeasts
- Actinomycosis
- Coccidioidomycosis
- Echinococcus
- Histoplasmosis
- Candida

Toxic
- Scorpion
- Diphtheria

Drugs
- Sulfonamides
- Phenylbutazone
- Cyclophosphamide
- Neomercazole
- Acetazolamide
- Amphotericin B
- Indomethacin
- Tetracycline
- Isoniazid
- Methyldopa
- Phenytoin
- Penicillin

Hypersensitivity/autoimmune
- Rheumatoid arthritis
- Rheumatic fever
- Ulcerative colitis
- Systemic lupus erythematosus

Other
- Sarcoidosis
- Scleroderma
- Idiopathic
- Cornstarch

proteins. These, as well as models relying on inflammatory mediators and apoptosis, have all been used to explain the features of heart failure; it is likely that all play a role in the clinical disorder. Despite the probable complexity of interactions resulting in heart failure, a major abnormality at the center of the disorder and its reversal is the process of remodeling. Improved outcomes appear to be linked to the ability to reverse this process ("reverse-remodeling"). Remodeling of the left ventricle is a process in which ventricular size, shape, and function is altered due to mechanical, genetic, and neurohormonal factors that lead to hypertrophy, myocyte loss, and interstitial fibrosis (Sutton and Sharpe 2000). In dilated cardiomyopathy, the process of progressive ventricular dilation and changes of the shape of the ventricle to a more spherical shape, associated with changes in ventricular function and/or hypertrophy, occurs without known initiating disturbance except in patients with myocardial infarction. Due to this remodeling event, mitral regurgitation may develop as the geometric relation between the mitral valve apparatus, mitral ring, and papillary muscles are altered. The presence of mitral regurgitation results in increasing volume overload on an already compromised ventricle, further contributing to remodeling, heart-failure symptoms, disease progression, and left atrial dilation.

13.3.5 Chronic DCM

In those cases in which resolution of cardiac dysfunction does not occur, chronic DCM results (Martin et al. 1994; Woodruff 1980). It has been unclear what the underlying etiology of this long-term sequela could be, but viral persistence and autoimmunity have been widely speculated. Recently, Badorff et al. (Badorff et al. 1999, 2000; Badorff and Knowlton 2004) demonstrated that enteroviral protease 2A directly cleaves the cytoskeletal protein dystrophin, resulting in dysfunction of this protein. Since mutations in dystrophin are known to cause an inherited form of DCM (Danialou et al. 2001; Feng et al. 2002b; Kyoi et al. 2003; Rimessi et al. 2005; Towbin et al. 1993b; Towbin 1998; Wilding et al. 2005), as well as the DCM associated with the neuromuscular diseases Duchenne and Becker muscular dystrophy, it is likely that these are to a large extent responsible for the chronic DCM seen in enteroviral myocarditis. Other viruses, such as adenoviruses, also have enzymes which cleave membrane structural proteins or result in activation or inactivation of transcription factors, cytokines, or adhesion molecules to cause chronic DCM. Therefore, it appears as if a complex interaction between the viral genome and the heart occurs and results in the long-term outcome of affected patients.

There are some reports which suggest myocarditis could be inherited (O'Connell et al. 1984; Shapiro et al. 1983). Support for this includes the frequent finding of myocardial lymphocytic infiltrate in patients with familial and sporadic DCM, as well as the few reports of families in which two or more individuals have been diagnosed with myocarditis on EMB. The recent finding of a common receptor for the four most common viral causes of myocarditis (CVB_3 and B_4 and adenovirus 2 and 5), namely the human coxsackie and adenovirus receptor (hCAR) (Fig. 1) (Bergelson et al. 1997; Tomko et al. 1997) – which if mutated could result in responsible host differences leading to myocarditis – is intriguing and requires study.

References

Abelman WH, Lorrell BH (1989) The challenge of cardiomyopathy. J Am Coll Cardiol 13:1219

Ainger LE, Lawyer NG, Fitch CW (1966) Neonatal rubella myocarditis. Br Heart J 28:691–697

Akhtar N, Ni J, Langston C, et al.(1996) PCR diagnosis of viral pneumonitis from fixed-lung tissue in children. Biochem Mol Med 58:66–76

Archard LC, Bowles NE, Olsen EGJ, Ricardson PJ (1987) Detection of persistent coxsackie B virus in dilated cardiomyopathy and myocarditis. Eur Heart J 8:437–440

Archard LC, Khan MA, Soteriou BA, et al. (1998) Characterization of coxsackie B virus RNA in myocardium from patients with dilated cardiomyopathy by nucleotide sequencing of reverse transcription-nested polymerase chain reaction products. Hum Pathol 29:578–584

Badorff C, Knowlton KU (2004) Dystrophin disruption in enterovirus-induced myocarditis and dilated cardiomyopathy: from bench to bedside. Med Microbiol Immunol (Berl) 193:121–126

Badorff C, Lee GH, Lamphear BJ, Martone ME, Campbell KP, Rhoads RE, Knowlton KU (1999) Enteroviral protease 2A cleaves dystrophin: evidence of cytoskeletal disruption in an acuiqred cardiomyopathy. Nat Med 5:266–267

Badorff C, Berkely N, Mehrotra S, Talhouk JW, Rhoads RE, Knowlton KU (2000) Enteroviral protease 2A directly cleaves dystrophin and is inhibited by a dystrophin-based substrate analogue. J Biol Chem 275:11191–11197

Bang ML, Mudry RE, McElhinny AS (2001) Myopalladin, a novel 145-kilo dalton sarcomeric protein with multiple roles in Z-disc and I-band protein assemblies. J Cell Biol 153:413–427

Barbaro G, Di Lorenzo G, Grisorio B, et al. (1998) Incidence of dilated cardiomyopathy and detection of HIV in myocardial cells of HIV-positive patients. N Engl J Med 339:1093–1099

Barth AL, Nathke IS, Nelson WJ (1997) Cadherins, catenins and APC protein; interplay between cytoskeletal complexes and signaling pathways. Curr Opin Cell Biol 9:683–690

Bergelson JM, Cunningham JA, Drouguett G, et al.(1997) Isolation of a common receptor for coxsackie B viruses and adenoviruses 2 and 5. Science 275:1320–1323

Berko BA, Swift M (1987) X-linked dilated cardiomyopathy. N Engl J Med 316:1186–1191

Berkovich S, Rodriguez-Torres R, Lin JS (1968) Virologic studies in children with acute myocarditis. Am J Dis Child 115:207–221

Bowles NE, Towbin JA (1998) Molecular aspects of myocarditis. Curr Opin Cardiol 13:179–184

Bowles NE, Towbin JA (2000) Molecular aspects of myocarditis. Curr Infect Dis Rep 2:308–314

Bowles NE, Vallejo J (2003) Viral causes of cardiac inflammation. Curr Opin Cardiol 18:182–188

Bowles NE, Richardson PJ, Olsen EGJ, Archard LC (1986) Detection of coxsackie-B virus specific RNA sequences in myocardial biopsy samples from patients with myocarditis and dilated cardiomyopathy. Lancet 1:1120–1123

Bowles NE, Rose ML, Taylor P (1989) End-stage dilated cardiomyopathy: persistence of enterovirus RNA in myocardium at cardiac transplantation and lack of immune response. Circulation 80:1128–1136

Bowles NE, Bowles NE, Ni J, et al.(2003) Detection of viruses in myocardial tissues by polymerase chain reaction: evidence of adenovirus as a common cause of myocarditis in children and adults. J Am Coll Cardiol 42:466–472

Burridge K, Chrzanowska-Wodnicka M (1996) Focal adhesions, contractility, and signaling. Annu Rev Cell Dev Biol 12:463–518

Calabrese F, Rigo E, Milanesi O, et al. (2002) Molecular diagnosis of myocarditis and dilated cardiomyopathy in children: clinicopathologic features and prognostic implications. Diagn Mol Pathol 11:212–221

Campbell KP (1995) Three muscular dystrophies: Loss of cytoskeleton-extracellular matrix linkage. Cell 80:675–679
Capetanaki Y (2002) Desmin cytoskeleton: a potential regulator of muscle mitochondrial behaviour and function. Trends Cardiovasc Med 12:339–348
Chang WJ, Iannaccone ST, Lau KS, et al. (1996) Neuronal nitric oxide synthase and dystrophin-deficient muscular dystrophy. Proc Natl Acad Sci USA 93:9142–9147
Chaudary S, Jaski BE (1989) Fulminant mumps myocarditis. Ann Intern Med 110:569–570
Chien KR (1999) Stress pathways and heart failure. Cell 98:555–558
Clark KA, McElhinny AS, Beckerle MC, Gregorio CC (2002) Striated muscle cytoarchitecture: an intricate web of form and function. Annu Rev Cell Dev Biol 18:637–706
Cohen JJ (1993) Apoptosis. Immunol Today 14:126–130
Cox GF, Kunkel LM (1997) Dystrophies and heart disease. Curr Opin Cardiol 12:329–343
Danialou G, Comtois AS, Dudley R, Karpatt G, Vincent G, Des Rosiers C, Petrof BJ (2001) Dystrophin-deficient cardiomyocytes are abnormally vulnerable to mechanical stress-induced contractile failure and injury. FASEB J 15:1655–12657
Dec GW, Palacios IF, Fallon JT, et al.(1985) Active myocarditis in the spectrum of acute dilated cardiomyopathies: clinical features, histologic correlates, and clinical outcome. N Engl J Med 312:885–890
Ehler E, Rothen BM, Hammerle SP, Komiyama M, Perriard JC (1999) Myofibrillogenesis in the developing chicken heart: assembly of Z-disk, M-line and the thick filaments. J Cell Sci 112:1529–1539
Emery AE (2002) The muscular dystrophies. Lancet 359:687–695
Epstein HF, Fischman DA (1991) Molecular analysis of protein assembly in muscle development. Science 251:1039–1044
Feng J, Yan J, Buzin CH, Sommer SS, Towbin JA (2002a) Comprehensive mutation scanning of the dystrophin gene in patients with nonsyndromic X-linked dilated cardiomyopathy. J Am Coll Cardiol 40:1120–1124
Feng J, Yan J, Buzin CH, Towbin JA, Sommer SS (2002b) Mutations in the dystrophin gene are associated with sporadic dilated cardiomyopathy. Mol Gen Metab 77:119–126
Ferlini A, Galie N, Merlini L, et al.(1998) A novel Alu-like element rearranged in the dystrophin gene causes a splicing mutation in a family with X-linked dilated cardiomyopathy. Am J Hum Genet 63:436–460
Francis GS, Wilson-Tang WH (2003) Pathophysiology of congestive heart failure. Rev Cardiovasc Med 4(Supp 2):S14–S20

Franz W-M, Muller M, Muller AJ, et al.(2000) Association of nonsense mutation of dystrophin gene with disruption of sarcoglycan complex in X-linked dilated cardiomyopathy. Lancet 355:1781–1785

Franzini-Armstrong C (1973) The structure of a simple Z line. J Cell Biol 58:630–642

Friedman RA, Schowengerdt KO, Towbin JA (1998) Myocarditis. In: Bricker JT, Garson A Jr, Fisher DJ, Neish SR (eds) The science and practice of pediatric cardiology, 2nd edn. Williams and Wilkins Publishers, Baltimore, pp 1777–1794

Frishman W, Kraus ME, Zabkar J, et al.(1977) Infectious mononucleosis and fatal myocarditis. Chest 72:535–538

Furukawa T, Ono Y, Tsuchiya H, et al.(2001) Specific interaction of the potassium channel beta-subunit iph with the sarcomeric protein T-cap suggests a T-tubule-myofibril linking system. J Mol Biol 313:775–784

Goll DE, Dayton WR, Singh I, Robson RM (1991) Studies of the alpha-actinin/actin interaction in the Z-disk by using calpain. J Biol Chem 266:8501–8510

Granzier H, Labeit S (2004) The grant protein titin: a major player in myocardial mechanics, signaling, and disease. Circ Res 94:284–295

Gregorio CC, Antin PB (2000) To the heart of myofibril assembly. Trends Cell Biol 10:355–362

Gregorio CC, Trombitas K, Centner T, et al. (1998) The NH2 terminus of titin spans the Z-disc: its interaction with a novel 19-kD ligand (T-cap) is required for sarcomeric integrity. J Cell Biol 143:1013–1027

Griffin LD, Kearney D, Ni J, et al.(1995) Analysis of formalin-fixed and frozen myocardial autopsy samples for viral genome in childhood myocarditis and dilated cardiomyopathy with endocardial fibroelastosis using polymerase chain reaction (PCR). Cardiovasc Pathol 4:3–11

Grist NR, Reid D (1962) General pathogenicity and epidemiology. In: Bendinelli M, Friedman H (eds) Coxsackieviruses: a general update. Plenum Press, New York, pp 241–252

Gulick T, Chung MK, Pieper SJ, et al.(1989) Interleukin-1 and tumor necrosis factor inhibit cardiac myocyte β-adrenergic responsiveness. Proc Natl Acad Sci USA 86:6753–6757

Henke A, Nain M, Stelzner A, et al.(1991) Induction of cytokine release from human monocytes by coxsackievirus infection. Eur Heart J 12(Suppl D):134–136

Herskowitz A, Wu T-C, Willoughby SB, et al. (1994) Myocarditis and cardiotropic viral infection associated with severe left ventricular dysfunction in late-stage infection with human immunodeficiency virus. J Am Coll Cardiol 24:1025–1032

Heydemann A, Wheeler MT, McNally EM (2001) Cardiomyopathy in animal models of muscular dystrophy. Curr Opin Cardiol 16:211–217
Hilton DA, Variend S, Pringle JH (1993) Demonstration of coxsackie virus RNA in formalin-fixed tissue sections from childhoold myocarditis by in situ hybridization and the polymerase chain reaction. J Pathol 170:45–51
Hirschman ZS, Hammer SG (1974) Coxsackie virus myopericarditis: a microbiological and clinical review. Am J Cardiol 34:224–232
Hoffman EP, Brown RH, Kunkel LM (1987) Dystrophin: the protein product of the Duchenne muscular dystrophy locus. Cell 51:919–928
Holtzer H, Hijikata T, Lin ZX, et al.(1997) Independent assembly of 1.6 microns long bipolar MHC filaments and I-Z-I bodies. Cell Struct Funct 22:83–93
Huber SA (1997) Autoimmunity in myocarditis: relevance of animal models. Clin Immunol Immunopathol 83:93–102
Hunt SA, Baker DW, Chin MH, et al.(2001) ACC/AHA guidelines for the evaluation and management of chronic heart failure in the adult: executive summary: a report of the American College of Cardiology/American Heart Association Task Force on Practice Guidelines. J Am Coll Cardiol 38:2101–2213
Hutchins GM, Vie SA (1972) The progression of interstitial myocarditis to idiopathic endocardial fibroelastosis. Am J Pathol 66:483–496
Jessup M, Brozena S (2003) Heart failure. N Engl J Med 348:2007–2018
Jin O, Sole M, Butany J (1990) Detection of enterovirus RNA in myocardial biopsies from patients with myocarditis and cardiomyopathy using gene amplification by polymerase chain reaction. Circulation 82:8–16
Kaprielian RR, Stevenson S, Rothery SM, Cullen MJ, Severs NJ (2000) Distinct patterns of dystrophin organization in myocyte sarcolemma and transverse tubules of normal and diseased human myocardium. Circulation 101:2586–2594
Karjalainen J, Nieminen MS, Heikkila J (1980) Influenza A1 myocarditis in conscripts. Acta Med Scand 20:27–30
Klietsch R, Ervasti JM, Arnold W, Campbell KP, Jorgensen AO (1993) Dystrophin-glycoprotein complex and laminin colocalize to the sarcolemma and transverse tubules of cardiac muscle. Circ Res 72:349–360
Knoll R, Hoshijima M, Hoffman HM, et al. (2002) The cardiac mechanical stretch sensor machinery involves a Z disc complex that is defective in a subset of human dilated cardiomyopathy. Cell 11:943–955
Koenig M, Hoffman EP, Bertelson CJ, et al. (1987) Complete cloning of the Duchenne muscular dystrophy (DMD) cDNA and preliminary genomic organization of the DMD gene in normal and affected individuals. Cell 50:509–517

Kroemer G, Martinez AC (1991) Cytokines and autoimmune disease. Clin Immunol Immunopathol 61:275–295
Kubota T, McTiernan CF, Frye CS, et al. (1997) Cardiospecific overexpression of tumor necrosis factor-alpha causes lethal myocarditis. J Card Fail 3:117–124
Kucera JP, Rohr S, Rudy Y (2002) Localization of sodium channels in intercalated disks modulates cardiac conduction. Circ Res 91:1176–1182
Kyoi S, Otani H, Sumida T, Okada T, Osako M, Imamura H, Kamihata H, Matsubara H, Iwasaka T (2003) Loss of intracellular dystrophin-A potential mechanism for myocardial reperfusion injury. Circ J 67:725–727
Kyu BS, Matsumori A, Sato Y (1992) Cardiac persistence of enteroviral RNA detected by polymerase chain reaction in a murine model of dilated cardiomyopathy. Circulation 86:522–530
Lane JR, Neumann DA, Lafond-Walker A, et al. (1992) Interleukin 1 or tumor necrosis factor can promote coxsackie B3-induced myocarditis in resistant B10.A mice. J Exp Med 175:1123–1129
Lane JR, Neumann DA, Lafond-Walker A, et al. (1993) Role of IL-I and tumor necrosis factor in coxsackie virus-induced autoimmune myocarditis. J Immunol 151:1682–1690
Lapidos KA, Kakkar R, McNally EM (2004) The dystrophin glycoprotein complex: signaling strength and integrity for the sarcolemma. Circ Res 94:1023–1031
Lee GH, Badorff C, Knowlton KU (2000) Dissociation of sarcoglycans and the dystrophin carboxyl terminus from the sarcolemma in enteroviral cardiomyopathy. Circ Res 87:489–495
Lipshultz SE, Easley KA, Orav EJ (1998) Left ventricular structure and function in children infected with human immunodeficiency virus. Circulation 97:1246–1256
Lorber A, Zonis A, Maisuls E, et al.(1988) The scale of myocardial involvement in varicella myocarditis. Int J Cardiol 20:257–262
Lowry PJ, Thompson RA, Littler WA (1983) Humoral immunity in cardiomyopathy. Br Heart J 50:390–394
Lozinski GM, Davis GG, Krous HF, Billman GF, Shimizu H, Burns JC (1994) Adenovirus myocarditis: retrospective diagnosis by gene amplification from formalin-fixed, paraffin-embedded tissues. Hum Pathol 25:831–834
Martin AB, Webber S, Fricker FJ, et al. (1994) Acute myocarditis: rapid diagnosis by PCR in children. Circulation 90:330–333
Matsumori A, Yamada T, Suzuki H, et al. (1994) Increased circulating cytokines in patients with myocarditis and cardiomyopathy. Br Heart J 72:561–566
McNally E, Allikian M, Wheeler MT, Mislow JM, Heydemann A (2003) Cytoskeletal defects in cardiomyopathy. J Mol Cell Cardiol 35:231–241

Meng H, Leddy JJ, Frank J, Holland P, Tuana BS (1996) The association of cardiac dystrophin with myofibrils/z-discs regions in cardiac muscle suggests a novel role in the contractile apparatus. J Biol Chem 271:12364–12371

Milasin J, Muntoni F, Severini CM, et al. (1996) A point mutation in the 5′ splice site of the dystrophin gene first intron responsible for X-linked dilated cardiomyopathy. Hum Mol Genet 5:73–79

Muntoni F, Cau M, Ganau A, et al.(1993) Brief report: deletion of the dystrophin muscle-specific promoter region associated with X-linked dilated cardiomyopathy. N Engl J Med 329:921–925

Neumann DA, Lane JR, Allen GS, et al.(1993) Viral myocarditis leading to cardiomyopathy: do cytokines contribute to pathogenesis? Clin Immunol Immunopathol 68:181–191

Ni J, Bowles NE, Kim Y-H, et al.(1997) Viral infection of the myocardium in endocardial fibroelastosis: molecular evidence for the role of mumps virus as an etiological agent. Circulation 95:133–139

Noren GR, Adams P Jr, Anderson RC (1963) Positive skin reactivity to mumps virus antigen in endocardial fibroelastosis. J Pediatr 62:604–606

O'Connell JB, Bristow MR (1994) Economic impact of heart failure in the United States: time for a different approach. J Heart Lung Transplant 13:S107-S112

O'Connell JB, Fowles RE, Robinson JA, et al. (1984) Clinical and pathologic findings of myocarditis in two families with dilated cardiomyopathy. Am Heart J 167:127–135

Ohtsuka H, Yajima H, Maruyama K, Kimura S (1997) Binding of the N-terminal 63 kDa portion of connectin/titin to alpha-actinin as revealed by the yeast two-hybrid system. FEBS Lett 401:65–67

Okabe M, Fukuda K, Arakawa K, Kikuchi M (1997) Chronic variant myocarditis associated with hepatitis C virus infection. Circulation 96:22–24

Ortiz-Lopez R, Li H, Su J, Goytia V, Towbin JA (1997) Evidence for a dystrophin missense mutation as a cause of X-linked dilated cardiomyopathy. Circulation 95:2434–2440

Osama SM, Krishnamurti S, Gupta DN (1979) Incidence of myocarditis in varicella. Indian Heart J 31:315–320

Ozawa E, Yoshida M, Suzuki A, Mizuno Y, Hagiwara Y, Noguchi S (1995) Dystrophin-associated proteins in muscular dystrophy. Hum Mol Genet 4:1711–1716

Papa I, Astier C, Kwiatek O, et al.(1999) Alpha-actinin-CapZ, an anchoring complex for thin filaments in Z-line. J Muscle Res Cell Motil 20:187–197

Pauschinger M, Bowles NE, Fuentes-Garcia FJ, et al.(1999) Detection of adenoviral genome in the myocardium of adult patients with idiopathic left ventricular dysfunction. Circulation 99:1348–1354

Perez-Pulido S (1984) Acute and subacute myocarditis. Cardiovasc Rev Rep 5:912–926
Petrof BJ, Shrager JB, Stedman HH, Kelly AM, Sweeny HL (1993) Dystrophin protects the sarcolemma from stresses developed during muscle contraction. Proc Natl Acad Sci USA 90:3710–3714
Pignatelli RH, McMahon CJ, Dreyer WJ, et al. (2003) Clinical characterization of left ventricular noncompaction in children. A relatively common form of cardiomyopathy. Circulation 108:2672–2678
Proby CM, Hackett S, Gupta S, Cox TM (1986) Acute myopericarditis in influenza A infection. Q J Med 60:887–892
Puchkov GF, Minkovich BM (1972) A case of respiratory syncytial infection in a child by interstitial myocarditis with lethal outcome. Arkh Patol 34:70–73
Pyle WG, Solaro RJ (2004) At the crossroads of myocardial signaling: the role of Z-discs in intracellular signaling and cardiac function. Circ Res 94:296–305
Ribaux P, Bleicher F, Couble ML, et al. (2001) Voltage-gated sodium channel (SkM1) content in dystrophin-deficient muscle. Pflugers Arch 441:746–755
Rimessi P, Gualandi F, Duprez L, Spitali P, Neri M, Merlini L, Calzolari E, Muntoni F, Ferlini A (2005) Genomic and transcription studies as diagnostic tools for a prenatal detection of X-linked dilated cardiomyopathy due to a dystrophin gene mutation. Am J Med Genet 132:391–394
Rosenberg HS, McNamara DG (1964) Acute myocarditis in infancy and childhood. Prog Cardiovasc Dis 7:179–197
Rowe RW (1973) The ultrastructure of Z disks from white, intermediate, and red fibers of mammalian striated muscles. J Cell Biol 57:261–277
Schonian U, Crombach M, Maser S, Maisch B (1995) Cytomegalovirus associated heart muscle disease. Eur Heart J 16(Suppl 0):46–49
Schowengerdt KO, Ni J, Denfield SW, et al. (1997) Parvovirus B19 infection as a cause of myocarditis and cardiac allograft rejection: diagnosis using the polymerase chain reaction (PCR). Circulation 96:3549–3554
Schwartz SM, Duffy JY, Pearl JM, Nelson DP (2001) Cellular and molecular aspects of myocardial dysfunction. Crit Care Med 29:S214–S219
Seidman JG, Seidman C (2001) The genetic basis for cardiomyopathy from mutation identification to mechanistic paradigms. Cell 108:557–567
Seko Y, Matsuda H, Kato K, et al.(1993) Expression of intercellular adhesion molecule-1 in murine hearts with acute myocarditis caused by coxsackievirus B3. J Clin Invest 91:1327–1336
Shapiro LM, Rozkovec A, Cambridge G, et al. (1983) Myocarditis in siblings leading to chronic heart failure. Eur Heart J 4:742–746
Sharp WW, Simpson DG, Borg TK, Samarel AM, Terracio L (1997) Mechanical forces regulate focal adhesion and costamere assembly in cardiac myocytes. Am J Physiol 273:H546–H556

Shimizu C, Rambaud C, Cheron G, et al.(1995) Molecular identification of viruses in sudden infant death associated with myocarditis and pericarditis. Pediatr Infect Dis J 14:584–588

Smith SC, Allen PM (1992) Neutralization of endogenous tumor necrosis factor ameliorates the severity of myosin-induced myocarditis. Circ Res 70:856–863

Sorimachi H, Freiburg A, Kolmerer B, et al. (1997) Tissue-specific expression and alpha-actinin binding properties of the Z-disc titin: implications for the nature of vertebrate Z-discs. J Mol Biol 270:688–695

Springer TA (1990) Adhesion receptors of the immune system. Nature 346:425–434

Squire JM (1997) Architecture and function in the muscle sarcomere. Curr Opin Struct Biol 7:247–257

Sterner G (1962) Adenovirus infections in childhood: an epidemiological and clinical survey among Swedish children. Acta Paediatr 142:1–30

Stewart M (1993) Intermediate filament structure and assembly. Curr Opin Cell Biol 5:3–11

Strand S, Hofmann WJ, Hug H, et al.(1996) Lymphocyte apoptosis induced by CD95 (APO-1/Fas) ligand-expressing tumor cells – a mechanism of immune evasion? Nat Med 2:1361–1366

Straub V, Campbell KP (1997) Muscular dystrophies and the dystrophin-glycoprotein complex. Curr Opin Neurol 10:168–175

Sutton MG, Sharpe N (2000) Left ventricular remodeling after myocardial infarction: pathophysiology and therapy. Circulation 101:2981–2988

Tiula E, Leinikki P (1972) Fatal cytomegalovirus infection in a previously healthy boy with myocarditis and consumption coagulopathy as a presenting sign. Scand J Infect Dis 4:57–60

Tomko RP, Xu R, Philipson L (1997) HCAR and MCAR: the human and mouse cellular receptors for subgroup C adenoviruses and group B coxsackieviruses. Proc Natl Acad Sci USA 94:3352–3356

Torre-Amione G, Kapadia S, Lee J, et al. (1996) Tumor necrosis factor-alpha and tumor necrosis factor receptors in the failing human heart. Circulation 93:704–711

Towbin JA (1998) The role of cytoskeletal proteins in cardiomyopathies. Curr Opin Cell Biol 10:131–139

Towbin JA (2002) Myocarditis. In: Finberg L, Kleinman R (eds) Saunders manual of pediatric practice, 2nd edn. WB Saunders, Philadelphia, pp 660–663

Towbin JA, Bowles NE (2002) The failing heart. Nature 415:227–233

Towbin JA, Griffin LD, Martin AB, et al. (1993a) Intrauterine adenoviral myocarditis presenting as non-immune hydrops fetalis: diagnosis by polymerase chain reaction. Pediatr Infect Dis J 13:144–150

Towbin JA, Hejtmancik JF, Brink P, et al. (1993b) X-linked dilated cardiomyopathy (XLCM): molecular genetic evidence of linkage to the Duchenne muscular dystrophy gene at the Xp21 locus. Circulation 87:1854–1865

Tschope C, Bock CT, Kasner M, Noutsias M, Westermann D, Schwimmbeck PL, Paulschinger M, Poller WC, Kuhl U, Kandolf R, Schultheiss HP (2005) High prevalence of cardiac parvovirus B19 infection in patients with isolated left ventricular diastolic dysfunction. Circulation 111:879–886

Van den Veyver IB, Ni J, Bowles N, et al. (1998) Detection of intrauterine viral infection using the polymerase chain reaction (PCR). Mol Gen Metab 63:85–95

Vatta M, Stetson SJ, Perez-Verdra A, et al. (2000) Molecular remodeling of dystrophin in patients with end-stage cardiomyopathies and reversal for patients on assist device therapy. Lancet 359:936–941

Vatta M, Mohapatra B, Jimenez S, et al. (2003) Mutations in cypher/ZASP in patients with dilated cardiomyopathy and left ventricular non-compaction. J Am Coll Cardiol 42:2014–2027

Vatta M, Stetson SJ, Jimenez S, et al.(2004) Molecular normalization of dystrophin in the failing left and right ventricle of patients treated with either pulsatile or continuous flow-type ventricular assist devices. J Am Coll Cardiol 43:811–817

Vigoreaux JO (1994) The muscle Z band: lessons in stress management. J Muscle Res Cell Motil 15:237–255

White E (1993) Regulation of apoptosis by the transforming genes of the DNA tumor virus adenovirus. Proc Soc Exp Biol Med 204:30–39

WHO (1996) Report of the 1995 World Health Organization/International Society and Federation of Cardiology Task Force on the Definition and Classification of Cardiomyopathies. Circulation 93:841–842

Wilding JR, Schneider JE, Sang AE, Davies KE, Neubauer S, Clarke K (2005) Dystrophin- and MLP-deficient mouse hearts: marked differences in morphology and function, but similar accumulation of cytoskeletal proteins. FASEB J 19:79–81

Woodruff JF (1980) Viral myocarditis: a review. Am J Pathol 101:427–484

Wynn J, Braunwald E (1997) The cardiomyopathies and myocarditides. In: Braunwald E (ed) Heart disease: a textbook of cardiovascular medicine. WB Saunders, Philadelphia, pp 1404–1463 Xiong D, Lee GH, Badorff C, Dorner A, Lee S, Wolf P, Knowlton KU (2002) Dystrophin deficiency markedly increases enterovirus-induced cardiomyopathy: a genetic predisposition to viral heart disease. Nat Med 8:872–877

Yoshida K, Nakamura A, Yazak M, et al.(1998) Insertional mutation by transposable element, L1, in the DMD gene results in X-linked dilated cardiomyopathy. Hum Mol Genet 7:1129–1132

Young P, Ferguson C, Banuelos S, Gautel M (1998) Molecular structure of the sarcomeric Z-disk: two types of titin interactions lead to an asymmetrical sorting of alpha-actinin. EMBO J 17:1614–1624

14 *Invited for Debate: Is There a Virus-Specific Matrix Destruction in the Course of Disease in Dilated Cardiomyopathy?*

F. Waagstein

In the normal myocardium there is a equilibrium between degrading and rebuilding of the matrix (Fig. 1) which during the remodeling process in chronic heart failure is displaced by an increase in degrading of matrix in the early phase leading to slipping of myocytes and dilatation of ventricles (Fig. 2) later followed by increase in fibrosis and myocardial stiffness (Tsuruda et al. 2004). In inflammatory viral cardiomyopathy, the question is whether this process is mediated by a specific virus-induced process due to loss of myocytes and depression of contractility or mediated by an increase in inflammatory cytokines, which indirectly leads to a cascade of events resulting in upregulation of metalloproteinases.

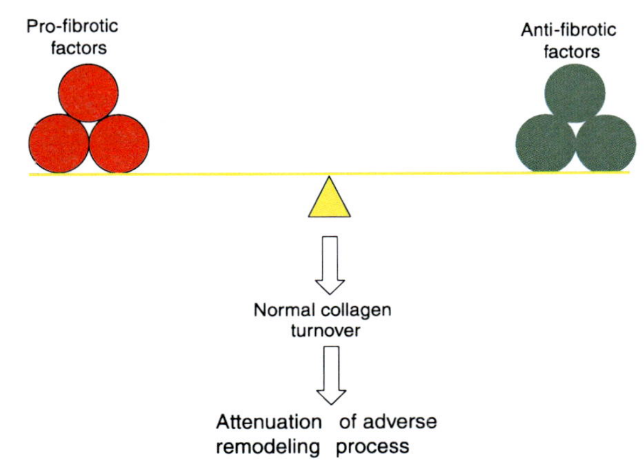

Fig. 1. Adequate extracellular matrix metabolism

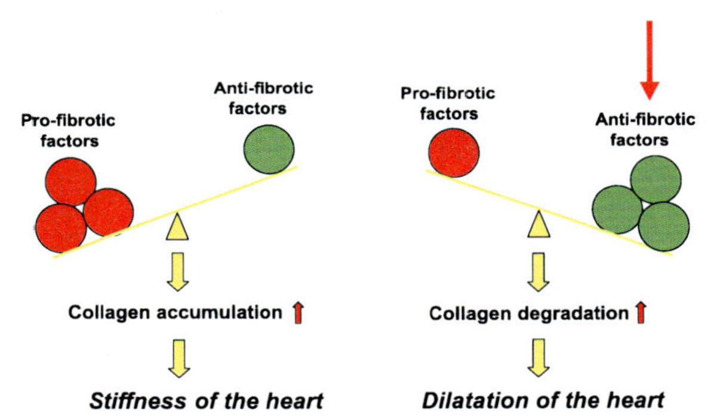

Fig. 2. Inadequate extracellular matrix metabolism

Figure 3 illustrates a hypothetical series of events leading to matrix destruction during chronic virus infection.

When the time course of matrix destruction is studied in an animal model it is obvious that a bimodal pattern is present. Early during stretch – after a few hours – there is an activation of angiotensin II that slows down after 24 h (Wang et al. 2004; Fig. 4).

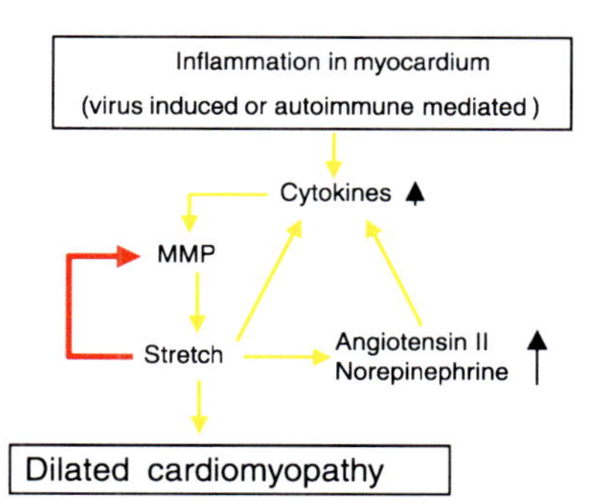

Fig. 3. Possible mechanism for developing heart dilatation in inflammatory myocardial disease

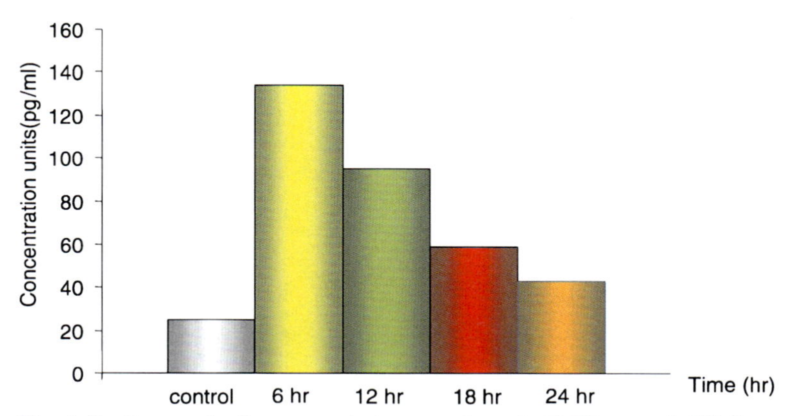

Fig. 4. Cyclic stretch of myocytes induces angiotensin II (Wang et al. 2004)

In a volume overload model of heart failure, there is an early activation of metalloproteinases followed by downregulation, and then upregulation again, during the phase of decompensation (Fig. 5). In the early phase, downregulation of collagen volume fraction is seen, followed by upregulation in the late phase of decompensation (Janicki et al. 2004; Fig. 6).

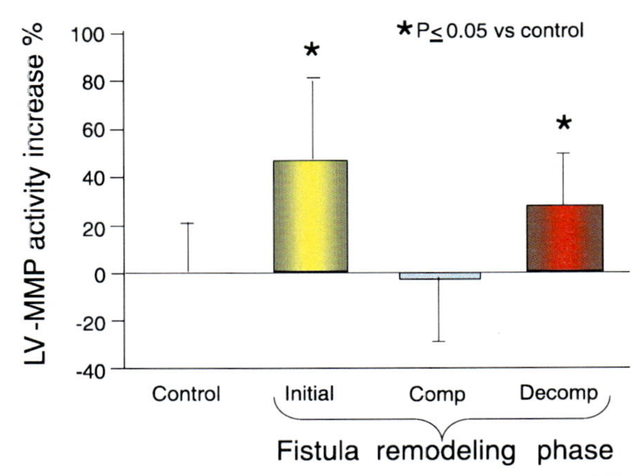

Fig. 5. Activation of matrix metalloproteinase (MMP) in a heart failure volume overload model (Janicki et al. 2004)

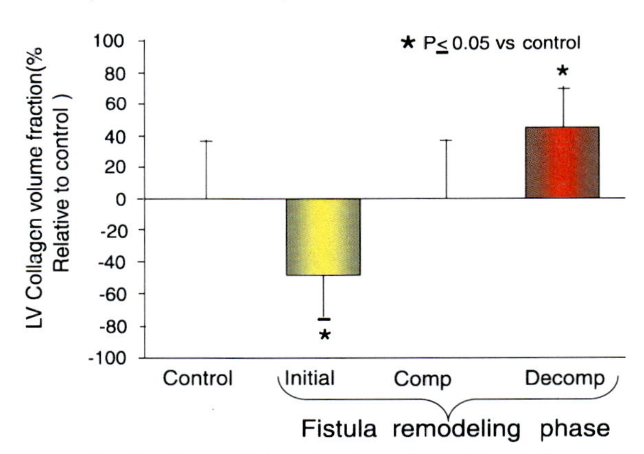

Fig. 6. Effects of activation and deactivation of MMP on left ventricular (LV) collagen in a volume overload model of heart failure (Janicki et al. 2004)

This process could be reversed by unloading the failing heart by left ventricular assist, which downregulates metalloproteinases (Fig. 7) and upregulates tissue inhibitors of matrix metalloproteinase (TIMP) (Li et al. 2001; Fig. 8).

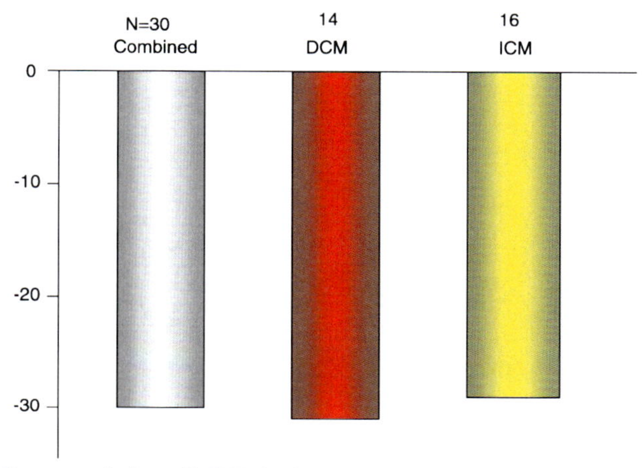

Fig. 7. Downregulation of MMP-9 after LV assist device (LVAD) in severe heart failure (Li et al. 2001)

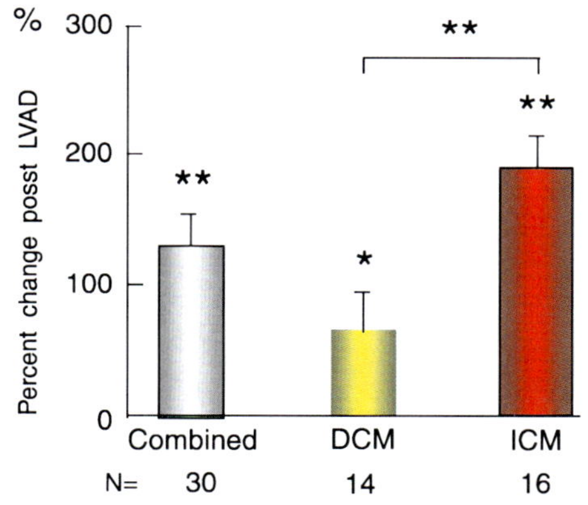

Fig. 8. Effect of LVAD on TIMP-3 expression in dilated cardiomyopathy (DCM) and ischemic cardiomyopathy (ICM) (Lin et al. 2001)

Additionally, pharmacotherapy, such as beta-blockade, could reverse the inflammatory response (Fig. 9) and downregulate metalloproteinases (Ohtsuka et al. 2003; Fig. 10).

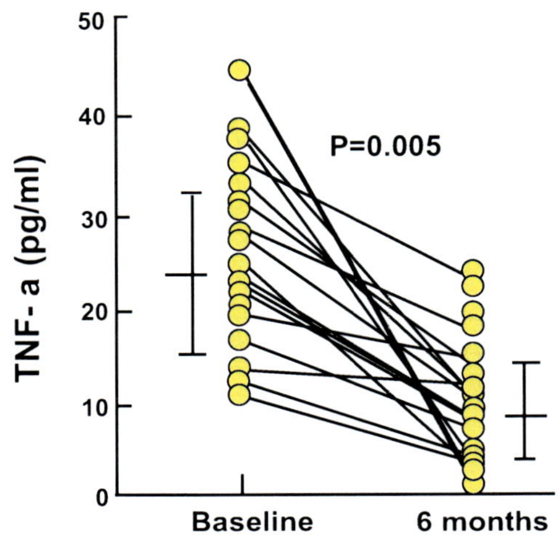

Fig. 9. Serum tumor necrosis factor (TNF)-α in DCM treated with carvedilol (Ohtsuka et al. 2003)

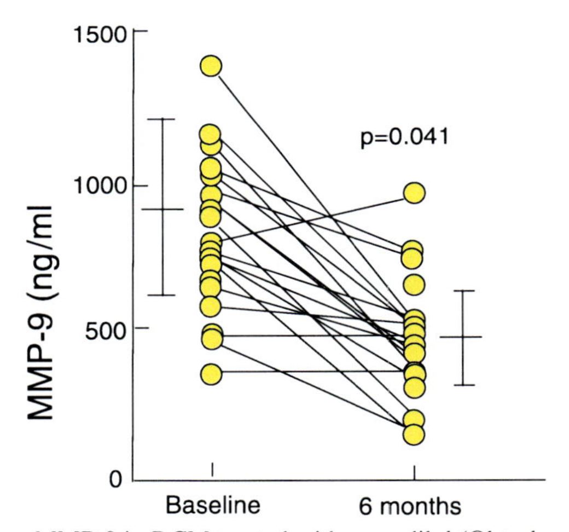

Fig. 10. Serum MMP-9 in DCM treated with carvedilol (Ohtsuka et al. 2003)

14.1 Conclusion

There is more evidence that matrix destruction in inflammatory cardiomyopathy is a secondary event to virus infection caused by a cascade of processes induced by the inflammatory response to virus infection and loss of myocytes than matrix destruction is a specific process caused by virus proteases.

It is, however, possible that both processes, a direct virus effect and an indirect effect from virus infection, act in cooperation. Future treatment with specific antiviral therapy may show which mechanism is the dominating factor in inflammatory cardiomyopathy causing matrix destruction.

References

Janicki JS, Brower GL, Gardner JD, Chancey AL, Stewart JA Jr (2004) The dynamic interaction between matrix metalloproteinase activity and adverse myocardial remodeling. Heart Fail Rev 9:33–42

Li YY, Feng Y, McTiernan CF, Pei W, Moravec CS, Wang P, Rosenblum W, Kormos RL, Feldman AM (2001) Down regulation of matrix metalloproteinases and reduction in collagen damage in the failing human heart after support with left ventricular assist devices. Circulation 104:1147–1152

Ohtsuka T, Hamada M, Saeki H, Ogimoto A, Hara Y, Shigematsu Y, Higaki J (2003) Serum levels of serum metalloproteinases and tumour necrosis factor-alpha in patients with idiopathic dilated cardiomyopathy and effect of carvedilol on these levels. J Am Coll Cardiol 91:1024–1027

Tsuruda T, Costello-Boerrigter LC, Burnett JC Jr (2004) Matrix metalloproteinases: pathways of induction by bioactive molecules. Heart Fail Rev 9:53–61

Wang TL, Yang YH, Chang H, Hung CR (2004) Angiotensin II signals mechanical stretch-induced cardiac matrix metalloproteinase expression via JAK-STAT pathway. J Mol Cell Cardiol 37:785–794

V Diagnosis and Treatment

15 New Non-invasive Approaches for the Diagnosis of Cardiomyopathy: Magnetic Resonance Imaging

U. Sechtem, H. Mahrholdt, S. Hager, H. Vogelsberg

Abstract. Cardiac magnetic resonance imaging (CMR) permits a detailed look at the myocardium in patients with recent onset heart failure. Late-enhancement CMR provides information that is similar to that obtained by the naked eye of a pathologist. Myocardial scarring is endocardial in myocardial infarction, but it is epicardial in myocarditis and intramyocardial in hypertrophic cardiomyopathy. Thus, the distinction between these entities is possible by depicting scar via late-enhancement CMR and observing myocardial function by cine magnetic resonance imaging. Moreover, non-invasive follow-up – and hence observation of the healing or remodelling process – can be achieved using CMR. New CMR pulse sequences also permit depiction of myocardial oedema, which may occur early in patients with myocarditis and may be the only sign of the disease in the absence of necrosis. It is anticipated that cardiac MRI will become a standard diagnostic technique in patients with new onset of heart failure, left-ventricular hypertrophy or clinical symptoms suggestive of myocarditis.

As clinicians we are frequently confronted with patients who present with symptoms and signs of heart failure of recent onset. Echocardiography reveals an enlarged, poorly contractile left ventricle (LV) and sometimes also signs of right ventricular (RV) dysfunction. As the patients often also report chest pain, the differential diagnosis of the aetiology of heart failure is broad. In order to distinguish between ischaemia, infection, inflammation or 'idiopathic' disease we need to look at the myocardium. Echocardiography is not well suited to provide detailed information about myocardial tissue. First, a substantial portion of patients has poor acoustic windows, which prevent full visualisation of the ventricles, especially the RV. Second, myocardial tissue differentiation by echocardiography is still in its infancy. As it may be difficult to rule out ischaemia as the underlying cause of ventricular dysfunction clinically or by echocardiography, coronary angiography is still routinely performed in these patients. Thus, although the final diagnosis will still mainly rely on the results of endomyocardial biopsy, other non-invasive means of distinguishing between the above-mentioned causes of the patient's problem would be desirable. Cardiovascular magnetic resonance (CMR) is becoming established in the initial functional assessment of patients with heart failure when echocardiography does not provide all the information needed. CMR is also assuming an important role for determining secondary causes of heart failure. The introduction of the so-called "late enhancement" technique by Kim and Judd and co-workers (Kim et al. 1999)provided a very accurate new CMR tool to identify areas of myocardial infarction (Wu et al. 2001; Ricciardi et al. 2001), and other forms of myocardial fibrosis (Varghese and Pennell 2004; Choudhury et al. 2002). CMR also permits imaging of tissue oedema using appropriate pulse sequences (Abdel-Aty et al. 2004a), which might be helpful in patients with suspected infectious of inflammatory myocardial disease (Hiramitsu et al. 2001). The purpose of this chapter is to review the usefulness and the indications for CMR in patients with systolic heart failure.

15.1 Differentiation of Ischaemic and Non-ischaemic Heart Failure

The most common cause of chronic heart failure is no longer hypertension or valvular heart disease but rather coronary artery disease (CAD) (Bourassa et al. 1993). In multi-centre heart failure trials, CAD is the underlying aetiology of heart failure in about 70% of the patients (Gheorghiade and Bonow 1998). Yet, this high number may still be an underestimation of the true prevalence of CAD, because not all patients with heart failure underwent coronary angiography in these trials.

The distinction between the causes of systolic LV dysfunction is clinically important given that the prognosis of patients with heart failure and CAD is considerably worse than that of patients without CAD (Bart et al. 1997). Moreover, patients with chronic ischaemic LV dysfunction benefit from revascularisation by improving their ejection fraction and exercise capacity (Di Carli et al. 1994). Non-invasive exclusion of coronary artery disease has been described using myocardial perfusion imaging (Iskandrian et al. 1986) including positron emission tomography, echocardiography (Duncan et al. 2003) and computed tomography (Le et al. 2000). However, perfusion defects may also occur in patients with dilated cardiomyopathy (DCM), and segmental wall motion abnormalities – the mainstay of the echocardiographic differentiation – may not always distinguish between the two entities. The presence of coronary calcification by multislice computed tomography suggests the presence of coronary artery disease but is not proving that the cause of heart failure is indeed ischaemic. Thus, most patients with dilated poorly contracting ventricles still undergo coronary angiography during the workup of their disease in order to exclude significant coronary artery stenoses. On the other hand, the diagnosis of DCM requires exclusion of secondary causes such as alcohol ingestion or iron overload (Anderson et al. 2002).

How reliable is CMR in distinguishing between patients with ischaemic and non-ischaemic heart failure? Using new inversion-recovery prepared T_1-weighted gradient-echo sequences after the administration of a gadolinium chelate with image acquisition between 10 and 20 min after contrast administration (Kim et al. 2003), delineation and quantification of ischaemic scar is possible with high accuracy (Kim et al. 1999) and reproducibility (Mahrholdt et al. 2002a) (Fig. 1). The high spatial resolution of CMR makes it possible to distinguish between subendocardial and transmural ischaemic necrosis (Wu et al. 2001). As some degree of ischaemic necrosis is virtually always present in patients with severely reduced LV ejection fraction caused by coronary artery disease and CMR is very sensitive in detecting even small amounts of subendocardial infarction (Wagner et al. 2003), one would expect a very high sensitivity of CMR for detecting ischaemic injury. On the other hand, subendocardial enhancement should be absent in patients with normal coronary arteries and LV dysfunction.

Indeed, initial experience suggested that late gadolinium enhancement which is found in almost all patients with a previously identified infarct-related artery does not occur in patients with nonischaemic car-

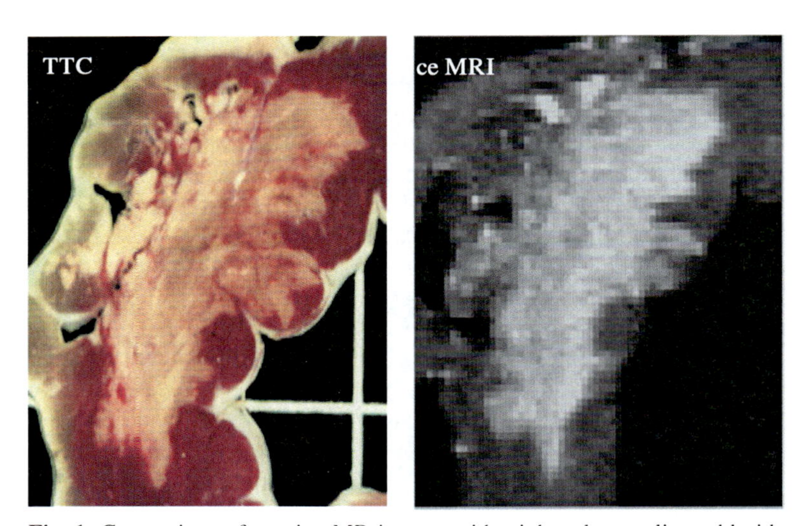

Fig. 1. Comparison of ex vivo MR images with triphenyltetrazolium chloride (TTC)-stained slices in a dog at 3 days after myocardial infarction. *Left*: TTC-stained slice shows pale necrotic myocardium and red viable myocardium. *Right*: the high signal intensity area represents enhancement following gadolinium-diethylenetriaminepentaacetic acid (DTPA). There is a very close spatial relationship between the MR image and infarct anatomy. Reproduced with permission from Kim et al. (1999)

diomyopathy or healthy volunteers (Wu et al. 2001). This finding was confirmed and extended in another study of 90 patients with heart failure and LV systolic dysfunction (McCrohon et al. 2003). All of the 27 patients with LV dysfunction and CAD had subendocardial or transmural enhancement by CMR. In contrast, patients without coronary artery disease (hence by definition DCM) showed three different findings: no enhancement in 59% of the 63 subjects with DCM, myocardial enhancement indistinguishable from the patients with coronary artery disease in 13% and patchy or longitudinal striae of midwall enhancement clearly different from the distribution in patients with coronary artery disease in 28% (Fig. 2). These findings show that using the coronary angiogram alone for identifying ischaemic LV dysfunction might lead to an incorrect assignment of the diagnosis of DCM in up to 13% of patients. In these patients, ischaemic damage was probably caused by transient occlusion of the coronary artery or by emboli originating in ruptured plaques (Kotani et al. 2002).

A new finding is the non-endocardial myocardial enhancement which is not associated with the perfusion bed of one coronary artery in patients with DCM. Myocardial fibrosis is usually seen at post-mortem exami-

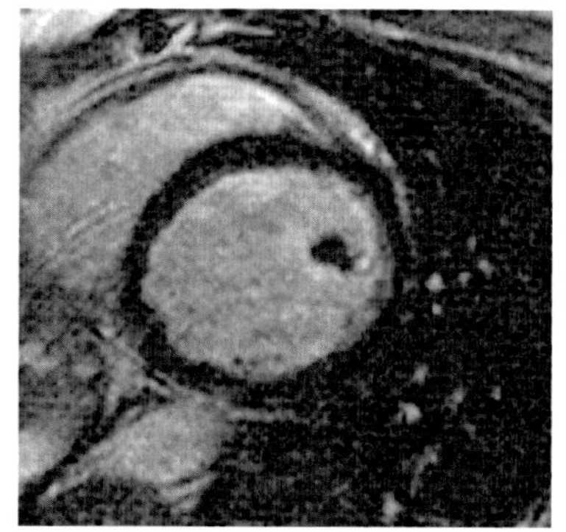

Fig. 2. Late enhancement images in two patients presenting with a history of several months of shortness of breath. Both had normal coronary arteries by coronary angiography and severely reduced LV function leading to the diagnosis of dilated cardiomyopathy. *Left*: midwall striae of enhancement in the interventricular septum. *Right*: no late enhancement of LV

nations in patients with DCM. Such fibrotic changes may be diffuse or patchy. The late gadolinium enhancement technique is unlikely to detect diffuse fibrotic changes as it is designed to optimise contrast between normal and necrotic myocardium in patients with ischaemic necrosis (Simonetti et al. 2001). Hence, no enhancement will be seen in patients with diffuse fibrosis. In contrast, CMR does have sufficient resolution to image small foci of myocardial fibrosis in vivo. It is currently unknown whether the finding of focal fibrosis is indicative of a specific aetiology or has prognostic value. As myocarditis is known to sometimes preferentially affect the interventricular septum (Hauck et al. 1989) and those enhancing streaks are sometimes seen just in this area (Fig. 2) one might speculate about a linkage between these two observations.

15.2 Hypertrophic Cardiomyopathy

CMR is a very useful technique in the management of patients with hypertrophic cardiomyopathy (HCM). Differentiating between HCM and other forms of hypertrophy has remained a substantial challenge for imaging techniques. Although most patients with HCM have a typical asymmetric pattern of hypertrophy affecting the interventricular septum more than the posterolateral wall, concentric, apical and other atypical distributions of hypertrophy do exist (Fig. 3). Especially these atypical forms of hypertrophy may be difficult to recognise by echocardiography. CMR may be helpful in these patients because image quality is often better than that of echocardiography, and unusual distributions of hypertrophy may be identified beyond that shade of doubt often left by suboptimal echocardiographic images.

CMR can also be used to assess myocardial tissue characteristics in vivo by applying the late enhancement technique to patients with HCM. Myocardial scarring is a common finding in patients with HCM and may have important prognostic impact (Thaman et al. 2004; Varnava et al. 2000). Scarring occurs mostly in hypertrophied regions and is usually patchy with multiple foci predominantly involving the midventricular wall (Choudhury et al. 2002) (Fig. 4). However, the location of scarring does not correspond to the perfusion territories of the epicardial coronary arteries. It is currently not clear whether increased

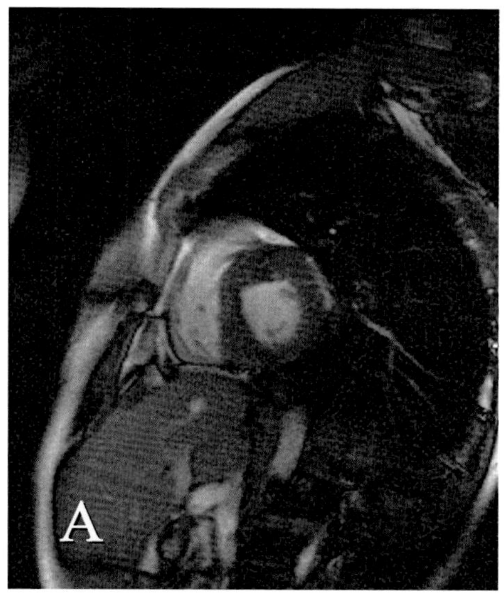

Fig. 3a. A 36-year-old lady with palpitations and angina pectoris. Short runs of ventricular tachycardia by Holter ECG. Normal echocardiogram. Anterior ischaemia by perfusion scintigraphy. Coronary angiography: normal coronary arteries. Short axis diastolic CMR image clearly demonstrates thickening of the inferior portion of the interventricular septum (**a**)

numbers of structurally abnormal intramural coronary arteries within areas of scarred myocardium have a causal role in producing myocardial ischaemia leading to scarring. In patients with small areas of scarring, the junctions of the interventricular septum and the RV walls are commonly involved (Fig. 5). The extent of scarring measured by CMR correlates with conventional risk markers (McKenna and Behr 2002; Mahrholdt et al. 2002b). It is currently not clear whether the extent of scarring in the CMR will be more predictive of serious arrhythmic events than the current risk stratification based on risk factors (Maron et al. 2003).

Some patients with HCM develop progressive LV impairment (Cannan et al. 1995). These patients have progressive interstitial fibrosis, myocardial disarray, small vessel disease and microscopic scarring leading to wall thinning (Thaman et al. 2004; Varnava et al. 2000). CMR

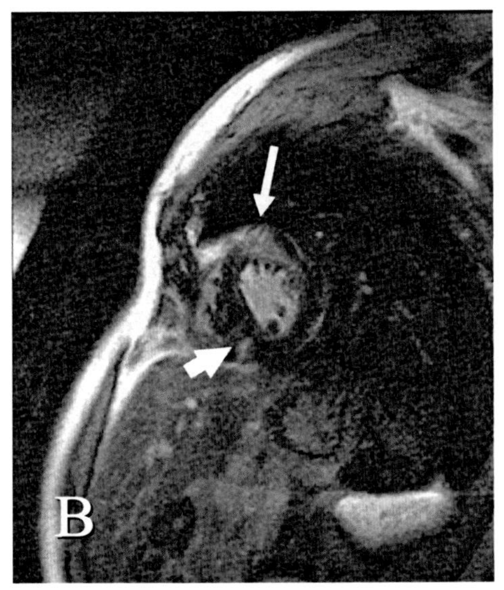

Fig. 3b. The inversion-recovery image (**b**, late enhancement) obtained 10 min after application of gadolinium-DTPA shows a small area of hyperenhancement in the hypertrophied portion of the interventricular septum (*short arrow*) and more extensive hyperenhancement in the anterior wall (*long arrow*)

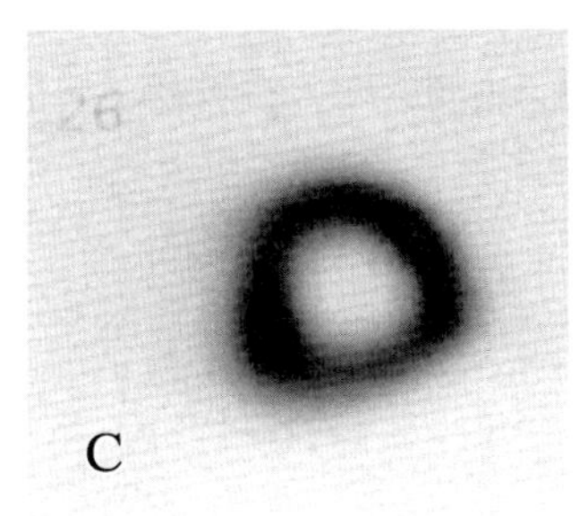

Fig. 3c. Retrospectively seen, the perfusion scintigram already demonstrates the localised hypertrophy of the inferior interventricular septum (**c**)

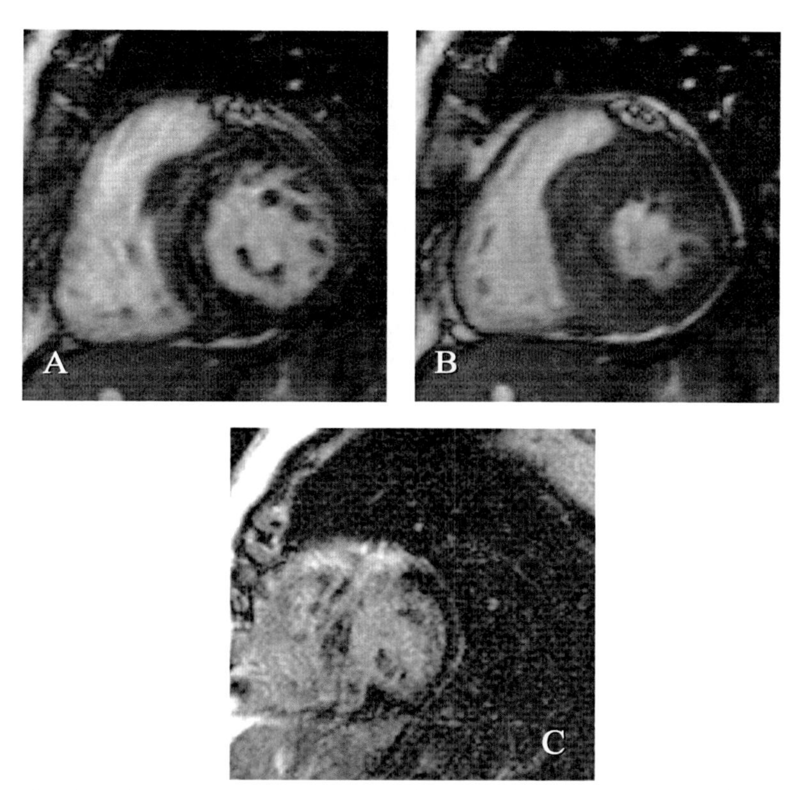

Fig. 4a–c. Asymptomatic 46-year-old patient with HCM. Family history of sudden cardiac death (father died at age 60). The short axis end diastolic image demonstrates the asymmetric hypertrophy of the interventricular septum (**a**). Septal thickness is 23 mm. The systolic image shows appropriate thickening of the entire circumference including the interventricular septum (**b**). The late enhancement image of the same slice (**c**) shows extensive scar of the interventricular septum and the anterior wall. Implantation of an ICD was discussed but the patient refused this form of prophylactic therapy

indeed demonstrates a greater extent of hyper-enhancement in patients with clinically progressive disease, and this is associated with thinning of previously grossly hypertrophied myocardium (Moon et al. 2003).

▶

Fig. 5a,b. *Left*: post-mortem specimen in a patient with HCM. There are two areas of scarring: one in the anterior wall near the junction of the right and the LV and one in the inferior wall also near the junction of the left and the right ventricles (*arrows*). *Right*: late enhancement MR image after application of gadolinium-DTPA at the corresponding sites indicating the presence of myocardial scarring in this patient. Images reproduced from Roberts and Roberts (1989) (*left*) and Choudhury et al. (2002) (*right*) with permission

Similar to colour Doppler echocardiography, CMR can depict the turbulent jet in the LV outflow tract (LVOT) in patients with the obstructive form of HCM. However, quantification of the subaortic gradient is easier performed by echocardiography than by velocity-encoded CMR techniques. However, CMR provides a unique opportunity for monitoring and quantifying the changes induced by septal ablation of the obstructive lesion (Fig. 6; van Dockum et al. 2004). The remodelling of the LVOT can be serially monitored during the healing process (Schulz-Menger et al. 2000).

Thus, CMR provides comprehensive information on anatomy, function and tissue composition in patients with HCM. Although the value of CMR as compared to echocardiography is not clearly defined yet (Kim and Judd 2003), CMR will be useful in patients in whom complete information cannot be obtained by echocardiography.

15.3 Myocarditis

Acute viral myocarditis may be difficult to recognise clinically. Even when the diagnosis is considered, there are currently no non-invasive tools to verify the diagnosis and assess the extent of myocardial involvement. The ultimate proof that the patient has myocarditis is provided by endomyocardial biopsy, which may demonstrate the typical inflammatory infiltrate in the myocardium. Using molecular hybridisation or polymerase chain reaction (PCR) methods (Pauschinger et al. 2004), it is possible to identify the causative viral agent. However, myocarditis is a patchy disease, which may explain the sampling error limiting the diagnostic value of endomyocardial biopsy (Hauck et al. 1989).

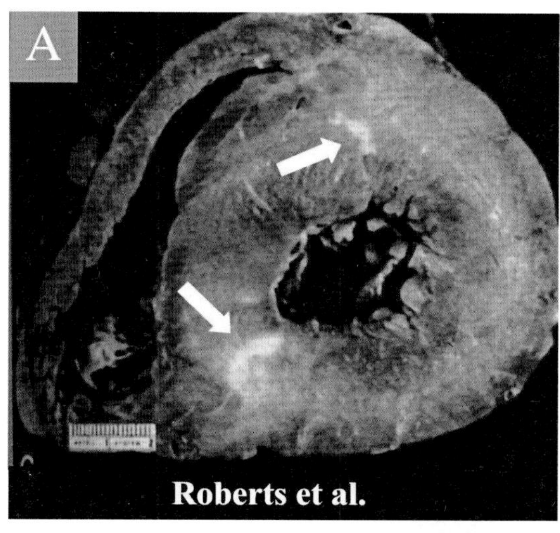
A
Roberts et al.

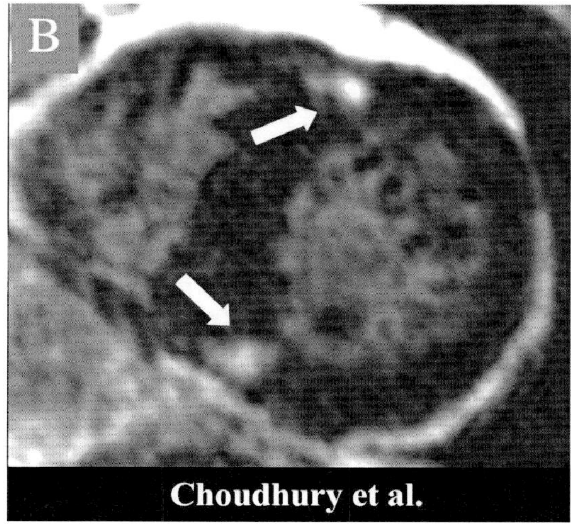
B
Choudhury et al.

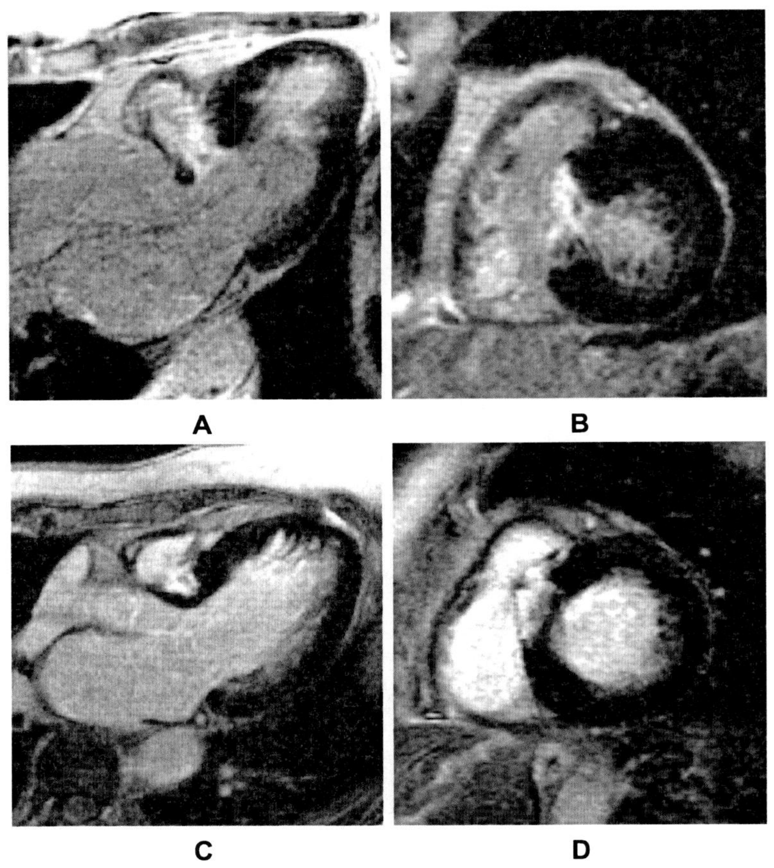

Fig. 6a–d. Contrast-enhanced images 20 min after administration of gadolinium-DTPA in two patients with hypertrophic obstructive cardiomyopathy 1 month after percutaneous transluminal septal myocardial ablation (**a**, **b**). Three-chamber view and short-axis view in a patient with transmural septal infarction (**c**, **d**). Comparable views in a patient with myocardial infarction located exclusively on the right ventricular side of the interventricular septum. Reproduced from van Dockum et al. (2004) with permission

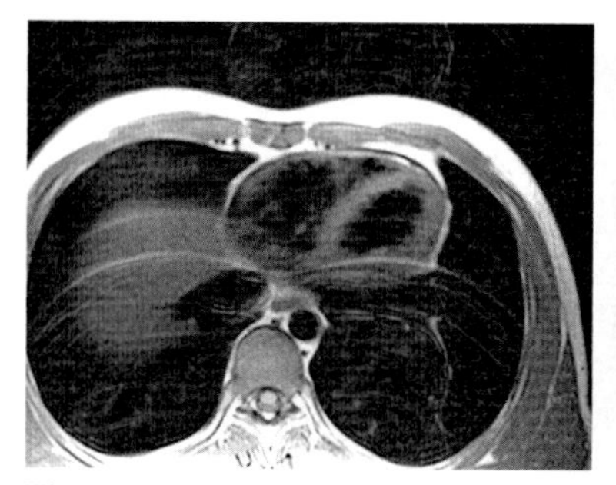
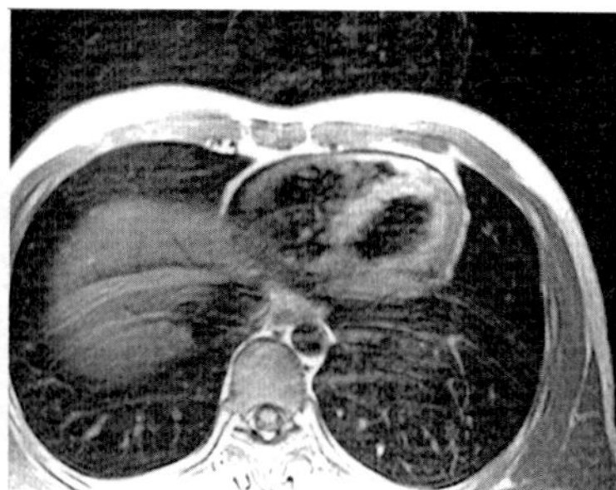

Fig. 7. Patient with biopsy-proven myocarditis. T_1-weighted spin echo imaging before contrast (*left*) shows a transverse section through the inferior portion of the left and right ventricles. *Right*: after the application of contrast material there is enhancement of predominantly the interventricular septum relative to the skeletal muscle. This enhancement is likely related to the presence of oedema

Friedrich et al. (1998) were the first to propose CMR for non-invasive diagnosis of myocarditis. They observed that the myocardium in patients with the clinical manifestations of myocarditis showed hyper-enhancement relative to skeletal muscle on T_1-weighted images (Fig. 7). However, the imaging protocol used in that study yielded a low contrast between inflamed and normal myocardium and suffered from image artefacts.

New contrast-enhanced CMR techniques such as the one employed for infarct imaging (Simonetti et al. 2001) provide an improvement in contrast between diseased and normal myocardium of up to 500% when compared with the protocol used by Friedrich et al. (1998). When these new inversion recovery gradient echo-techniques are used in patients with clinically suspected myocarditis, contrast enhancement is found in 90% of the patients (Mahrholdt et al. 2004). The regions of contrast enhancement have a patchy distribution throughout the LV. They are most frequently located in the lateral free wall (Fig. 8) and originate from the epicardial quartile of that wall. In the interventricular septum, patterns may occur which resemble those found in some patients with DCM. In contrast to ischaemic injury, areas of contrast enhancement do not originate from the subendocardial portion of the wall.

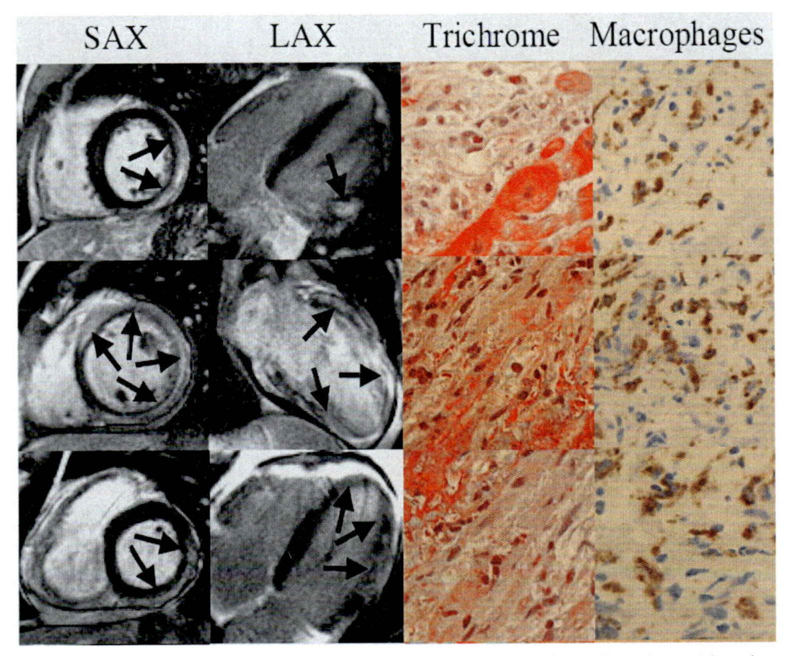

Fig. 8. CMR images and histopathology of typical patients in whom biopsies were obtained from area of contrast enhancement. Patients had active myocarditis with contrast enhancement extending from the epicardium towards the endocardium associated with myocyte damage and infiltration of macrophages. *SAX*, short axis; *LAX*, long axis. Reproduced with permission from Mahrholdt et al. (2004)

When it is possible to obtain biopsies from the area of contrast enhancement, which was the case in 21 of the 28 patients in the study of Mahrholdt et al., histopathologic analysis shows active acute or chronic myocarditis in a larger portion of patients than would be expected on the basis of historic controls (90% vs 63%) (Hauck et al. 1989). In our experience, parvovirus B19 (PVB19) is the most commonly encountered causative agent followed by human herpes virus type 6 (HHV6). In those patients in whom the biopsy cannot be obtained from the region of contrast enhancement, active myocarditis is seen less consistently (found in only one of 7 patients in the study by Mahrholdt et al.). Thus, although the number of patients was small, this study (Mahrholdt et al. 2004) suggests that CMR-guided biopsy may result in a higher yield of positive findings than routine RV biopsy.

Why should endomyocardial biopsy yield more frequent positive findings in patients who show hyper-enhancement on CMR images? The biopsies are taken from the endocardium, whereas the areas of hyper-enhancement are clearly located towards the epicardium. There are two possible explanations for this paradox. First, in many patients, epicardially located areas of hyper-enhancement extend almost transmurally towards the endocardium (see patients 7 and 14 in Fig. 9). Thus,

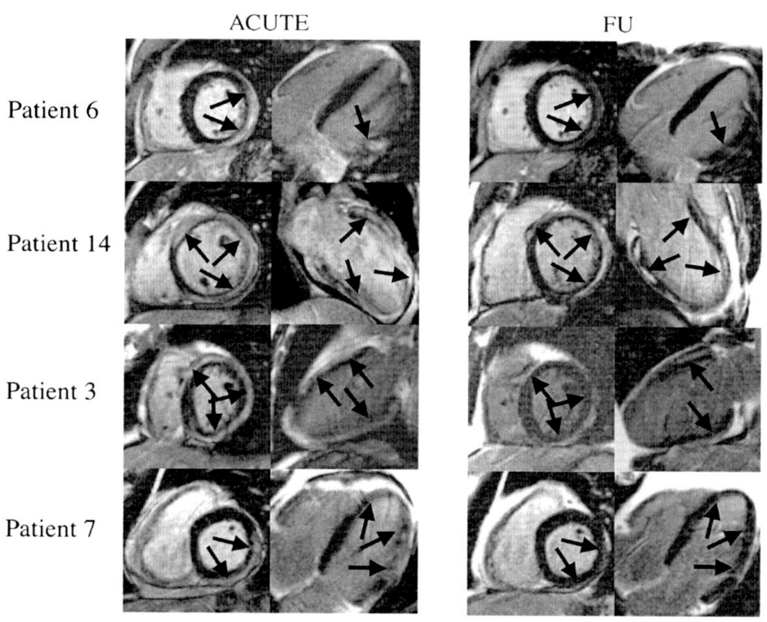

Fig. 9. CMR images in the acute (*left*) and sub-acute (*FU*; *right*) phase. Areas of contrast enhancement are smaller at follow-up (*right*) as mean ejection fraction and mean end diastolic volumes returned towards normal. Reproduced with permission from Mahrholdt et al. (2004)

the bioptome might have reached areas of ongoing inflammation in these patients. Second, it can be assumed that damage to the myocardium was most pronounced in the areas of contrast enhancement. Consequently, the nearby areas located endocardially may have a higher inflammatory burden than areas remote from the centres of inflammation. Therefore, a RV biopsy in patient 7 (see Fig. 9) would be less likely to yield positive results than an LV biopsy in the same patient.

Thus, it is now possible to depict the location and the extent of active myocarditis in vivo by using CMR. In addition, CMR gives insights regarding the clinical course in those patients, depending on the extent and location of areas of hyper-enhancement. In our initial series, 20 patients were available for a follow-up (Fig. 9; Mahrholdt et al. 2004). Interestingly, the average area of enhancing tissue decreased and enhancement disappeared completely in 3 patients. There was a moderate positive correlation between the decrease of enhancement and the improvement in ejection fraction that almost achieved statistical significance (Fig. 10).

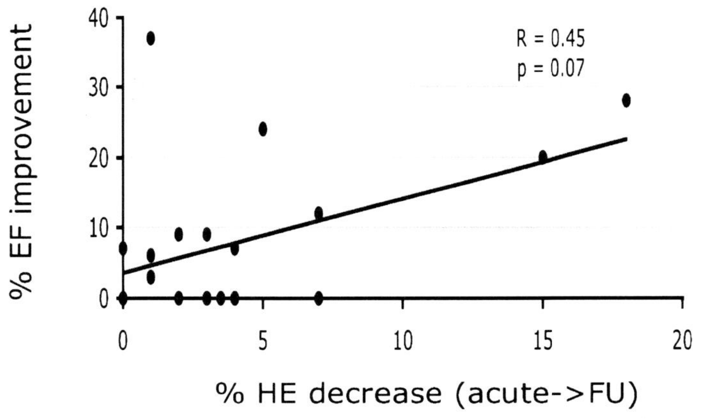

Fig. 10. Correlation between the decrease of contrast enhancement and improvement in ejection fraction at follow-up. Number of data points visible in graph is influenced by projection of data points on top of other data points ($n = 20$). *HE*, hyper-enhancement; *EF*, ejection fraction. Reproduced with permission from Mahrholdt et al. (2004)

A predominant involvement of the free lateral wall of the LV in the inflammatory process has not been previously described in vivo. However, this finding correlates nicely to previous reports from post-mortem examinations (Shirani et al. 1993). The peculiar pattern of a sandwiched stripe of late enhancement within the interventricular septum may correspond to previous post-mortem observations showing that active myocarditis was in fact occurring in isolation in the centre of the interventricular septum (Hauck et al. 1989).

The mechanism for contrast enhancement on inversion recovery gradient echo images is likely to follow the same principles as in myocardial infarction (Mahrholdt et al. 2002c) and HCM (Moon et al. 2004). The commonly accepted hypothesis is that the damage to the cellular membranes following myocyte death allows the contrast agent to diffuse into the cells, thus increasing the normally very small extracellular space in the myocardium. Normal myocardium is characterised by very densely packed myocytes which occupy about 90% of the available space. Scattered small foci of necrosis would thus lead to an increase in the proportion of a voxel occupied by extracellular space, which in turn would cause an increase in the concentration of extracellular contrast agents such as the gadolinium-based compounds. Thus, the signal enhancement seen in patients with acute myocarditis on late gadolinium enhancement images could either be the result of areas of scattered myocyte necrosis interspersed between viable myocardial cells or of larger confluent areas in which most of the myocytes were killed by the viral infection. The latter scenario would mimic the effects of myocardial infarction. Unfortunately, it is not possible to distinguish between these two possible scenarios based on the information contained in the CMR late enhancement images, as there is not a linear relationship between the proportion of extracellular space within a voxel and voxel signal. However, the brightness of myocarditis lesions is usually less intense than that of myocardial infarcts, suggesting a less-complete cell destruction.

How can one explain the shrinkage of lesions commonly observed with follow-up of myocarditis patients (Mahrholdt et al. 2004)? Shrinkage of myocardial infarcts is common and may lead to a reduction of the initial infarct mass to less than 1/3 of its initial value (Fieno et al. 2004; Reimer and Jennings 1979). Oedema, haemorrhage and inflammation within infarcts cause infarct sizes to increase by 20% to 30% during the first 4 days, a process which has been termed "early infarct expansion" (Richard et al. 1995). This magnitude of early increase in infarct size is also seen by CMR as an increase in MRI hyper-enhanced area in dogs between 2 h and 48 h following coronary occlusion (Rochitte et al. 1998). Infarct size decreases progressively following the phase of early infarct expansion, as necrotic tissue is replaced by collagenous scar (Richard et al. 1995). At 6 weeks, mean infarct size is only 35% to 40% of values at 4 days. Interestingly, smaller infarcts show a larger percentage of shrinkage at 6 weeks than larger infarcts (Richard et al. 1995).

If one translates these findings into the situation of cellular damage caused by myocarditis, shrinkage of the initial area of hyper-enhancement fits nicely with the observations in the healing process of myocardial infarcts. Contrast enhancement at 3 months was $3 \pm 4\%$ as compared to $9 \pm 11\%$ of the LV in the study of Mahrholdt et al. (2004) even though some patients presented weeks after onset of flu-like and cardiac symptoms. Thus, the full extent of shrinking may not have been observed in all patients.

As in myocardial infarction, oedema may be prominent in the initial stages of tissue injury caused by viral myocarditis (Hiramitsu et al. 2001). In the acute phase of myocarditis, ventricular wall thickening is sometimes observed and is believed to be caused by interstitial oedema (Hauser et al. 1983). Oedema may lead to a reversible 50% increase in LV wall thickness (Fig. 11). Such a prominent oedema would undoubtedly result in a massive expansion of the extracellular space without necessarily resulting in swelling of myocytes (Hiramitsu et al. 2001).

The late enhancement technique is optimised to detect differences between normal and necrotic myocardium. Therefore, even substantial diffuse increases in the interstitial space, as caused by the oedema accompanying acute myocarditis, would go unnoticed on late enhancement CMR images due to the diffuse nature of the process. More localised oedema of a lesser extent, in contrast, without accompanying myocyte death might not result in enough increase in extracellular space to cause late gadolinium enhancement. Indeed, the spectrum of myocardial injury caused by the disease ranges from mild inflammation with hyperaemia or oedema only to frank necrosis (Lieberman et al. 1991). The frequent occurrence of milder forms of myocarditis with minor cellular injury

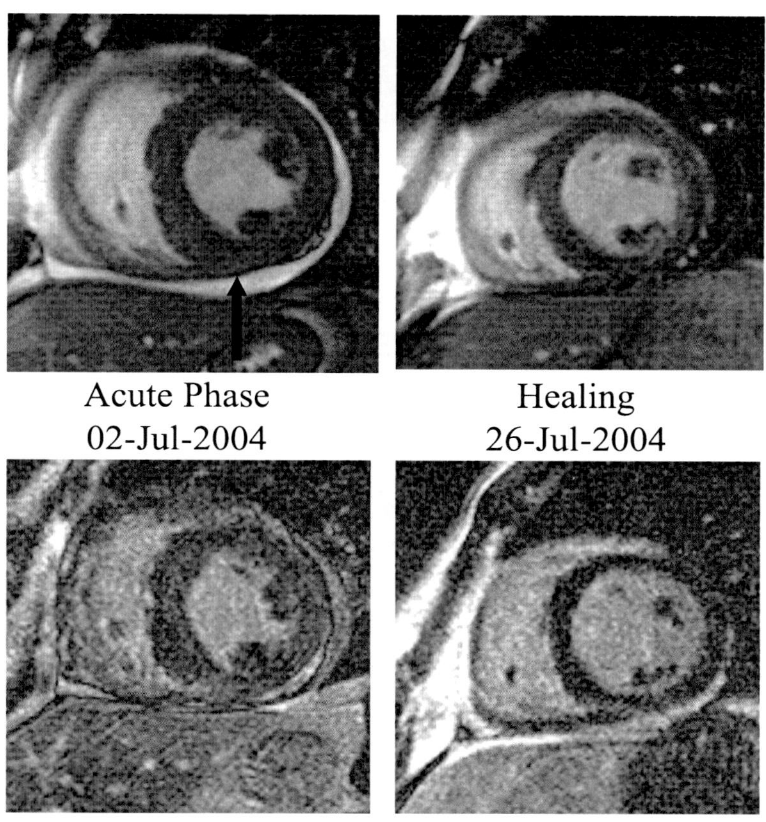

Fig. 11. Patient with severe upper respiratory tract infection and increasing shortness of breath. *Upper left panel*: end diastolic image during acute phase shows pericardial effusion (*arrow*) and diffusely hypertrophied LV. *Lower left panel*: late enhancement image shows diffuse increase in myocardial signal intensity. *Right upper panel*: end diastolic image 4 weeks later shows disappearance of the pericardial effusion and shrinking of the right and left ventricles. The hypertrophy is no longer present. This is indicative of fulminant myocarditis with reversible oedema. *Lower right panel*: late enhancement image demonstrates low signal intensity in the LV myocardium. The field of view (FOV) was identical for all images

but prominent oedema may explain the recent observation by Abdel-Aty et al. (2005) that CMR imaging of oedema may be more sensitive than late enhancement imaging in detecting acute myocarditis. They investigated 25 acute myocarditis patients a few days after the onset of clinical symptoms by CMR. CMR studies included: (1) T_2-weighted triple inversion recovery, (2) T_1-weighted spin echo before and shortly after contrast injection (as described by Friedrich et al. (1998) and (3) inversion recovery gradient echo 10 min after contrast injection (late gadolinium enhancement). Global myocardial T_2 signal intensity was significantly higher in patients than in volunteers (2.3 ± 0.4 vs 1.7 ± 0.4; $p < 0.0001$) although there was overlap. A cut-off value of 1.9 had a sensitivity of 84% and a specificity of 74% to identify the disease. Global myocardial relative enhancement measured from T_1-weighted spin-echo images was also significantly higher in patients compared to volunteers (6.8 ± 4.0 vs 3.7 ± 2.3; $p < 0.001$). A cut-off value of 4.0 had a sensitivity of 80% and a specificity of 73% to identify myocarditis. The sensitivity of late gadolinium enhancement was lower, with only 44%, but the specificity was high (100%). Combining the high sensitivity of global T_2 increases with the excellent specificity of focal late gadolinium enhancement may result in an even better diagnostic performance of CMR to identify acute myocarditis.

In patients with myocarditis (often PVB19) who present with a clinical picture mimicking that of acute myocardial infarction (Kühl et al. 2003) immediate multi-sequence CMR might result in the correct diagnosis without cardiac catheterisation. However, as missing acute coronary occlusion has such severe consequences for the patient, immediate catheterisation to exclude coronary disease with the highest possible specificity may still remain the preferred management strategy in these patients. CMR would be employed as the second step followed by endomyocardial biopsy to identify the causative agent.

15.4 Conclusions

The ability of CMR to precisely depict areas of myocardial necrosis expands the armamentarium provided by established imaging techniques to diagnose and characterise cardiomyopathy and myocarditis. As subendocardial necrosis is almost always due to ischaemia and the patterns of necrosis are different in dilated and hypertrophic cardiomyopathies as well as in myocarditis, late enhancement CMR provides a powerful tool to differentiate these disease entities. If the initial impressions as described in this chapter will be confirmed in more and larger studies, CMR may become mandatory in patient groups such as those with new onset heart failure, those with LV hypertrophy or those with clinical symptoms suggesting acute or chronic myocarditis.

References

Abdel-Aty H, Zagrosek A, Schulz-Menger J, Taylor AJ, Messroghli D, Kumar A, Gross M, Dietz R, Friedrich MG (2004a) Delayed enhancement and T2-weighted cardiovascular magnetic resonance imaging differentiate acute from chronic myocardial infarction. Circulation 109:2411–2416

Abdel-Aty H, Boyé P, Zagrosek A, Wassmuth R, Kumar A, Messroghli D, Bock P, Dietz R, Fiedrich MG, Schulz-Menger J (2005) The sensitivity and specificity of contrast-enhanced and T2-weighted cardiovascular magnetic resonance to detect acute myocarditis. J Am Coll Cardiol (in press)

Anderson LJ, Wonke B, Prescott E, Holden S, Walker JM, Pennell DJ (2002) Comparison of effects of oral deferiprone and subcutaneous desferrioxamine on myocardial iron concentrations and ventricular function in beta-thalassaemia. Lancet 360:516–520

Bart BA, Shaw LK, McCants CB Jr, Fortin DF, Lee KL, Califf RM, O'Connor CM (1997) Clinical determinants of mortality in patients with angiographically diagnosed ischaemic or nonischaemic cardiomyopathy. J Am Coll Cardiol 30:1002–1008

Bourassa MG, Gurne O, Bangdiwala SI, Ghali JK, Young JB, Rousseau M, Johnstone DE, Yusuf S (1993) Natural history and patterns of current practice in heart failure. The Studies of Left Ventricular Dysfunction (SOLVD) Investigators. J Am Coll Cardiol 22(Suppl A):14A–19A

Cannan CR, Reeder GS, Bailey KR, Melton LJ 3rd, Gersh BJ (1995) Natural history of hypertrophic cardiomyopathy. A population-based study, 1976 through 1990. Circulation 92:2488–2495

Choudhury L, Mahrholdt H, Wagner A, Choi KM, Elliott MD, Klocke FJ, Bonow RO, Judd RM, Kim RJ (2002) Myocardial scarring in asymptomatic or mildly symptomatic patients with hypertrophic cardiomyopathy. J Am Coll Cardiol 40:2156–2164

Di Carli MF, Davidson M, Little R, Khanna S, Mody FV, Brunken RC, Czernin J, Rokhsar S, Stevenson LW, Laks H (1994) Value of metabolic imaging with positron emission tomography for evaluating prognosis in patients with coronary artery disease and left ventricular dysfunction. Am J Cardiol 73:527–533

Duncan AM, Francis DP, Gibson DG, Henein MY (2003) Differentiation of ischaemic from nonischaemic cardiomyopathy during dobutamine stress by left ventricular long-axis function: additional effect of left bundle-branch block. Circulation 108:1214–1220

Fieno DS, Hillenbrand HB, Rehwald WG, Harris KR, Decker RS, Parker MA, Klocke FJ, Kim RJ, Judd RM (2004) Infarct resorption, compensatory hypertrophy, and differing patterns of ventricular remodeling following myocardial infarctions of varying size. J Am Coll Cardiol 43:2124–2131

Friedrich MG, Strohm O, Schulz-Menger J, Marciniak H, Luft FC, Dietz R (1998) Contrast media-enhanced magnetic resonance imaging visualizes myocardial changes in the course of viral myocarditis. Circulation 97:1802–1809

Gheorghiade M, Bonow RO (1998) Chronic heart failure in the United States: a manifestation of coronary artery disease. Circulation 97:282–289

Hauck AJ, Kearney DL, Edwards WD (1989) Evaluation of postmortem endomyocardial biopsy specimens from 38 patients with lymphocytic myocarditis: implications for role of sampling error. Mayo Clin Proc 64:1235–1245

Hauser AM, Gordon S, Cieszkowski J, Timmis GC (1983) Severe transient left ventricular 'hypertrophy' occurring during acute myocarditis. Chest 83:275–277

Hiramitsu S, Morimoto S, Kato S, Uemura A, Kubo N, Kimura K, Sugiura A, Itoh T, Hishida H (2001) Transient ventricular wall thickening in acute myocarditis: a serial echocardiographic and histopathologic study. Jpn Circ J 65:863–866

Iskandrian AS, Hakki AH, Kane S (1986) Resting thallium-201 myocardial perfusion patterns in patients with severe left ventricular dysfunction: differences between patients with primary cardiomyopathy, chronic coronary artery disease, or acute myocardial infarction. Am Heart J 111:760–767

Kim RJ, Judd RM (2003) Gadolinium-enhanced magnetic resonance imaging in hypertrophic cardiomyopathy: in vivo imaging of the pathologic substrate for premature cardiac death? J Am Coll Cardiol 41:1568–1572

Kim RJ, Fieno DS, Parrish TB, Harris K, Chen EL, Simonetti O, Bundy J, Finn JP, Klocke FJ, Judd RM (1999) Relationship of MRI delayed contrast enhancement to irreversible injury, infarct age, and contractile function. Circulation 100:1992–2002

Kim RJ, Shah DJ, Judd RM (2003) How we perform delayed enhancement imaging. J Cardiovasc Magn Reson 5:505–514

Kotani J, Nanto S, Mintz GS, Kitakaze M, Ohara T, Morozumi T, Nagata S, Hori M (2002) Plaque gruel of atheromatous coronary lesion may contribute to the no-reflow phenomenon in patients with acute coronary syndrome. Circulation 106:1672–1677

Kühl U, Pauschinger M, Bock T, Klingel K, Schwimmbeck CPL, Seeberg B, Krautwurm L, Poller W, Schultheiss H-P, Kandolf R (2003) Parvovirus B19 infection mimicking acute myocardial infarction. Circulation 108:945–950

Le T, Ko JY, Kim HT, Akinwale P, Budoff MJ (2000) Comparison of echocardiography and electron beam tomography in differentiating the etiology of heart failure. Clin Cardiol 23:417–420

Lieberman EB, Hutchins GM, Herskowitz A, Rose NR, Baughman KL (1991) Clinicopathologic description of myocarditis. J Am Coll Cardiol 18:1617–1626

Mahrholdt H, Wagner A, Holly TA, Elliott MD, Bonow RO, Kim RJ, Judd RM (2002a) Reproducibility of chronic infarct size measurement by contrast-enhanced magnetic resonance imaging. Circulation 106:2322–2327

Mahrholdt H, Choudhury L, Wagner A, Honold M, Bonow RO, Sechtem U, Judd RM, Kim RJ (2002b) Relation of myocardial scarring to clinical risk factors for sudden cardiac death in hypertrophic cardiomyopathy. Circulation 106(Suppl III):III-652 (abstr)

Mahrholdt H, Wagner A, Judd RM, Sechtem U (2002c) Assessment of myocardial viability by cardiovascular magnetic resonance imaging. Eur Heart J 23:602–619

Mahrholdt H, Goedecke C, Wagner A, Meinhardt G, Athanasiadis A, Vogelsberg H, Fritz P, Klingel K, Kandolf R, Sechtem U (2004) Cardiovascular magnetic resonance assessment of human myocarditis: a comparison to histology and molecular pathology. Circulation 109:1250–1258

Maron BJ, Estes NA 3rd, Maron MS, Almquist AK, Link MS, Udelson JE (2003) Primary prevention of sudden death as a novel treatment strategy in hypertrophic cardiomyopathy. Circulation 107:2872–2875

McCrohon JA, Moon JC, Prasad SK, McKenna WJ, Lorenz CH, Coats AJ, Pennell DJ (2003) Differentiation of heart failure related to dilated cardiomyopathy and coronary artery disease using gadolinium-enhanced cardiovascular magnetic resonance. Circulation 108:54–59

McKenna WJ, Behr ER (2002) Hypertrophic cardiomyopathy: management, risk stratification, and prevention of sudden death. Heart 87:169–176

Moon JC, McKenna WJ, McCrohon JA, Elliott PM, Smith GC, Pennell DJ (2003) Toward clinical risk assessment in hypertrophic cardiomyopathy with gadolinium cardiovascular magnetic resonance. J Am Coll Cardiol 41:1561–1567

Moon JC, Reed E, Sheppard MN, Elkington AG, Ho SY, Burke M, Petrou M, Pennell DJ (2004) The histologic basis of late gadolinium enhancement cardiovascular magnetic resonance in hypertrophic cardiomyopathy. J Am Coll Cardiol 43:2260–2264

Pauschinger M, Chandrasekharan K, Noutsias M, Kuhl U, Schwimmbeck LP, Schultheiss HP (2004) Viral heart disease: molecular diagnosis, clinical prognosis, and treatment strategies. Med Microbiol Immunol (Berl) 193: 65–69

Reimer KA, Jennings RB (1979) The changing anatomic reference base of evolving myocardial infarction. Underestimation of myocardial collateral blood flow and overestimation of experimental anatomic infarct size due to tissue edema, hemorrhage and acute inflammation. Circulation 60: 866–876

Ricciardi MJ, Wu E, Davidson CJ, Choi KM, Klocke FJ, Bonow RO, Judd RM, Kim RJ (2001) Visualization of discrete microinfarction after percutaneous coronary intervention associated with mild creatine kinase-MB elevation. Circulation 103:2780–2783

Richard V, Murry CE, Reimer KA (1995) Healing of myocardial infarcts in dogs. Effects of late reperfusion. Circulation 92:1891–1901

Roberts CS, Roberts WC (1989) Morphologic features. In: Zipes DP, Rowlands DJ (eds) Progress in cardiology, vol. 2/2. Lea and Febiger, Philadelphia, pp 3-32

Rochitte CE, Lima JA, Bluemke DA, Reeder SB, McVeigh ER, Furuta T, Becker LC, Melin JA (1998) Magnitude and time course of microvascular obstruction and tissue injury after acute myocardial infarction. Circulation 98:1006–1014

Schulz-Menger J, Strohm O, Waigand J, Uhlich F, Dietz R, Friedrich MG (2000) The value of magnetic resonance imaging of the left ventricular outflow tract in patients with hypertrophic obstructive cardiomyopathy after septal artery embolization. Circulation 101:1764–1766

Shirani J, Freant LJ, Roberts WC (1993) Gross and semiquantitative histologic findings in mononuclear cell myocarditis causing sudden death, and implications for endomyocardial biopsy. Am J Cardiol 72:952–957

Simonetti OP, Kim RJ, Fieno DS, Hillenbrand HB, Wu E, Bundy JM, Finn JP, Judd RM (2001) An improved MR imaging technique for the visualization of myocardial infarction. Radiology 218:215–223

Thaman R, Gimeno JR, Reith S, Esteban MT, Limongelli G, Murphy RT, Mist B, McKenna WJ, Elliott PM (2004) Progressive left ventricular remodeling in patients with hypertrophic cardiomyopathy and severe left ventricular hypertrophy. J Am Coll Cardiol 44:398–405

van Dockum WG, ten Cate FJ, ten Berg JM, Beek AM, Twisk JW, Vos J, Hofman MB, Visser CA, van Rossum AC (2004) Myocardial infarction after percutaneous transluminal septal myocardial ablation in hypertrophic obstructive cardiomyopathy: evaluation by contrast-enhanced magnetic resonance imaging. J Am Coll Cardiol 43:27–34

Varghese A, Pennell DJ (2004) Late gadolinium enhanced cardiovascular magnetic resonance in Becker muscular dystrophy. Heart 90:e59

Varnava AM, Elliott PM, Sharma S, McKenna WJ, Davies MJ (2000) Hypertrophic cardiomyopathy: the interrelation of disarray, fibrosis, and small vessel disease. Heart 84:476–482

Wagner A, Mahrholdt H, Holly TA, Elliott MD, Regenfus M, Parker M, Klocke FJ, Bonow RO, Kim RJ, Judd RM (2003) Contrast-enhanced MRI and routine single photon emission computed tomography (SPECT) perfusion imaging for detection of subendocardial myocardial infarcts: an imaging study. Lancet 361:374–379

Wu E, Judd RM, Vargas JD, Klocke FJ, Bonow RO, Kim RJ (2001) Visualisation of presence, location, and transmural extent of healed Q-wave and non-Q-wave myocardial infarction. Lancet 357:21–28

16 New Therapeutics Targets in Chronic Viral Cardiomyopathy

W. Poller, H. Fechner, U. Kühl, M. Pauschinger, H.-P. Schultheiss

Abstract. Dilated cardiomyopathy (DCM) is a prevalent heart muscle disease characterized by impaired contractility and dilation of the ventricles. Recent clinical research suggests that cardiotropic viruses are important environmental pathogenic factors in human DCM, which may therefore be considered as a chronic viral cardiomyopathy. All virus-positive DCM patients thus come into the focus of virological research and should be considered for antiviral strategies. Interferon-β therapy has been shown to mediate virus elimination in patients with adenovirus or coxsackievirus persistence. We discuss here several possible new molecular targets for patients infected with cardiotropic viruses in (1) the cellular virus uptake system, (2) virus-induced cellular signaling pathways, and (3) interactions between virus-encoded proteins with important cellular target proteins. The potential of these approaches in the setting of a chronic viral infection is significantly different from that in an acute viral infection. Specific problems encountered in a chronic situation and possible solutions are discussed.

16.1 Introduction

Dilated cardiomyopathy (DCM) is a prevalent heart muscle disease characterized by impaired contractility and dilation of the ventricles (Poller et al. 2005; Chien 2003; Franz et al. 2001). Recent clinical research suggests that cardiotropic viruses are important environmental pathogenic factors in human DCM, which may thus be considered as a *chronic* viral cardiomyopathy (Bowles et al. 1986; Kühl et al. 2003a, 2005; Pauschinger et al. 1999a,b; Why et al. 1994; Frustaci et al. 2003; Matsumori et al. 1995, 2000). Since DCM is a far more frequent disorder than acute viral cardiomyopathy (i.e., classical myocarditis), the new data on viruses in DCM hearts greatly expand the number of patients classified as having a virus-associated cardiac disease. All these patients therefore come into the focus of virological research and may be considered for antiviral strategies. In the following review, we discuss several novel therapeutic approaches derived from this broadened perspective on virus-associated cardiac disorders. Clinical endpoints for any new therapy of this type are improvements of symptoms or survival (Kühl et al. 2003a). Cardiac functional improvement may be expected:

1. If the virus-infected cardiac cells were eliminated by an antiviral immune response without over-shooting inflammation (Kühl et al. 2003a).
2. If intracardiac viruses were functionally "silenced" in the sense that no new virus particles or viral genomes are synthesized, and furthermore, that any transcriptional activity of the viral genomes is suppressed. If a major fraction of the cardiac cells were infected, such "silencing" may be preferable over immune elimination, i.e., destruction of infected cells.
3. If de novo infection of cardiac cells by virus spreading within the heart or from extracardiac virus pools would occur, blockade of this process at the receptor level may be useful. Possible new molecular targets to achieve the goals are located (1) in the cellular virus uptake machinery, in particular in the heart (Bergelson et al. 1997; Gaggar et al. 2003; Noutsias et al. 2001; Fechner et al. 2003; Dörner et al. 2004), (2) in virus-induced cellular signal pathways of the innate immune system (Ulevitch 2004; Hertzog et al. 2003), or (3) in interac-

tions between virus-encoded proteins with structurally or functionally important target cell proteins.

The relative therapeutic potential of these targets is discussed while considering differences between acute vs chronic disease phases of a viral disorder, and between a human disease (Why et al. 1994; Baboonian et al. 1997) vs more or less analogous animal models of that disease (Yanagawa et al. 2004; Yanagawa et al. 2003). In contrast to some highly specialized animal models, very little is known so far in humans on the kinetics of the cardiac virus load, on possible migration within the heart or between cardiac and other, e.g., lymphatic cells, or on cellular signal pathway activation by the different cardiotropic viruses detected in human hearts. Certain therapeutic strategies suggested by cell biological (Dörner et al. 2004) or animal studies (Yanagawa et al. 2003, 2004) of coxsackievirus cardiomyopathy may possibly also work in humans, but for most cardiotropic viruses recently identified in human hearts (e.g., PVB19, HHV6, AdV) (Kühl et al. 2005), we do not have adequate animal models at the present time.

16.2 Virus–Host Interactions and Antiviral Therapy

1. Figure 1 provides a general overview on key interactions sites between viruses and their target cells. Most known cardiotropic viruses are taken up into the cell by receptor-mediated endocytosis (Bergelson et al. 1997; Gaggar et al. 2003; Noutsias et al. 2001; Fechner et al. 2003) and then further transported into the nucleus via the nuclear pore complex. The receptor level thus constitutes a first level of therapeutic targets (Fig. 2). If significant de novo infection of cardiac cells by virus spreading within the heart or from extracardiac virus pools occurs, therapeutic intervention at this level may be effective. *Acute* cardiac viral infections may not only be caused by CVB3 or AdV, but also by PVB19 (Kühl et al. 2003b; Rohayem et al. 2001; Nigro et al. 2000; Murry et al. 2001). In these often fulminant and ultimately fatal diseases, intervention at the receptor level may have particular therapeutic potential.
2. During the process of endocytosis, viral genomic DNA or RNA may activate signaling cascades of innate antiviral immunity via Toll-like

receptors (TLR). On one hand this contributes to immunological elimination of virus-infected cells, on the other it may cause damage by overshooting immune reactions. Innate immune signaling thus constitutes a second level of therapeutic targets (Fig. 3) (Ulevitch 2004; Hertzog et al. 2003). An optimal antiviral immune response would pinpoint infected cells without collateral damage by inadequate inflammation. Virus elimination solicited by interferon (INF)-β treatment has already been shown to result in improved cardiac function and New York Heart Association (NYHA) classification in humans (Kühl et al. 2003a). Even if immune elimination of infected cells is not achieved, the suppression of inflammatory damage by

▶

Fig. 1. Virus–host interactions and antiviral therapy. The frequencies at which cardiotropic viruses were detected in a series of 245 DCM patients are parvovirus B19 (*PVB19*) (51.4%), human herpes virus (*HHV*)6 (21.6%), enteroviruses (*EV*) (9.4%), Epstein-Barr virus (*EBV*) (2.0%), adenovirus (*AdV*) (1.6%), and hepatitis C virus (*HCV*) (0.8%). Also screened but not detected were HSV1/2, influenza (Inf)A/B, cytomegalovirus (CMV). Cellular entry of these viruses occurs via cell surface receptors and coreceptors, followed by receptor-mediated endocytosis of the virus particles, release of the viral RNA (CVB3) or DNA genomes (other viruses), and their nuclear important via the nuclear pore complex (*NPC*). Virus receptors form a first level for antiviral therapeutic interventions (see Fig. 2). Coxsackievirus–adenovirus receptor (CAR) shows high interindividual variability in human hearts (Noutsias et al. 2001; Fechner et al. 2003), but little is known on the cardiac expression of the other known receptors for cardiotropic viruses. Toll-like receptors (TLRs) 3,7,8,9 are located in the membrane of the endosomes and may trigger signaling cascades of innate antiviral immunity (INF-β dependent and independent). TLRs and TLR-dependent signal pathways form a second level for therapeutic interventions (see Fig. 3). Knowledge on innate immune signal molecules in human hearts is sparse. When the viral genomes are transcribed in the nucleus and translated in the cytosol, their products may act upon and damage structural and functional proteins of the cardiomyocyte. Whereas for most human cardiotropic virus-encoded proteins no data on interactions with structural and functional proteins are available, dystrophin has been shown to be markedly vulnerable to cleavage by enteroviral protease 2A (Badorff et al. 1999). Suppression of potentially damaging virus-encoded proteins may have therapeutic potential (Fig. 4)

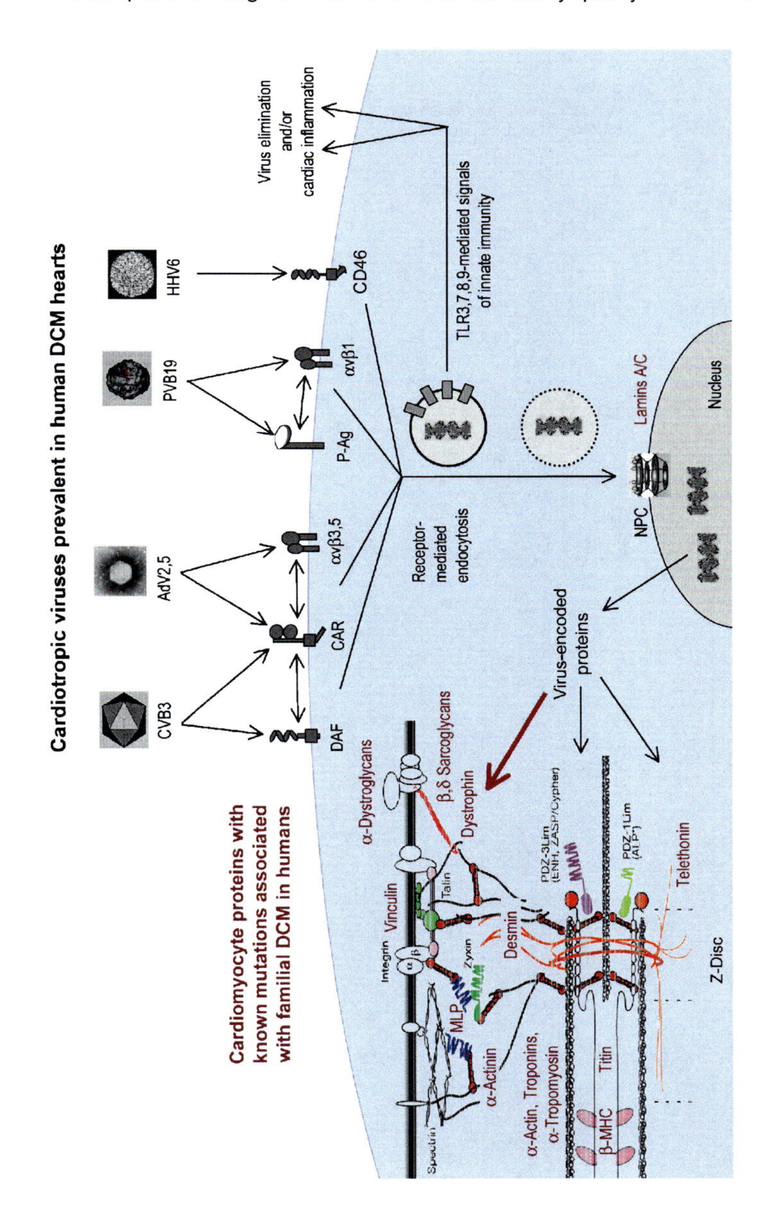
Cardiotropic viruses prevalent in human DCM hearts
CVB3
AdV2,5
PVB19
HHV6
DAF
CAR
αvβ3,5
P-Ag
αvβ1
CD46
Virus elimination
and/or
cardiac inflammation
TLR3,7,8,9-mediated signals
of innate immunity
Receptor-
mediated
endocytosis
NPC
Lamins A/C
Nucleus
Virus-encoded
proteins
Cardiomyocyte proteins with
known mutations associated
with familial DCM in humans
α-Dystroglycans
β,δ Sarcoglycans
Dystrophin
Vinculin
Talin
Integrin
Zyxin
MLP
Desmin
α-Actinin
Spectrin
α-Actin, Troponins,
α-Tropomyosin
β-MHC
Titin
PDZ-3Lim
(ENH, ZASP/Cypher)
PDZ-1Lim
(ALP)
Telethonin
Z-Disc

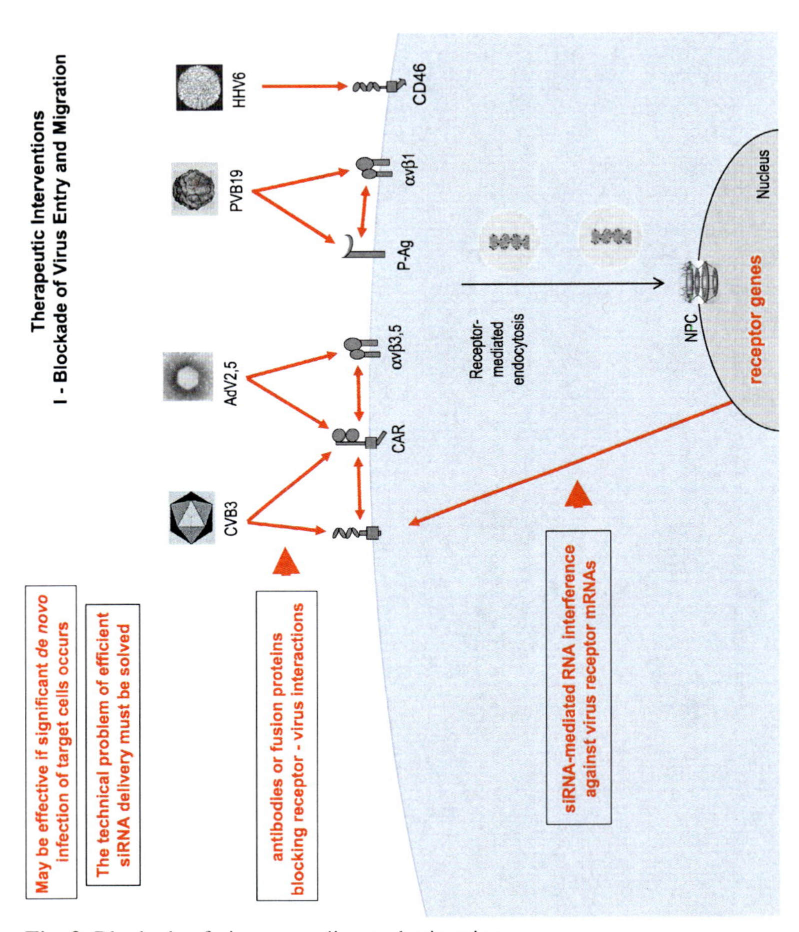

Fig. 2. Blockade of virus spreading and migration

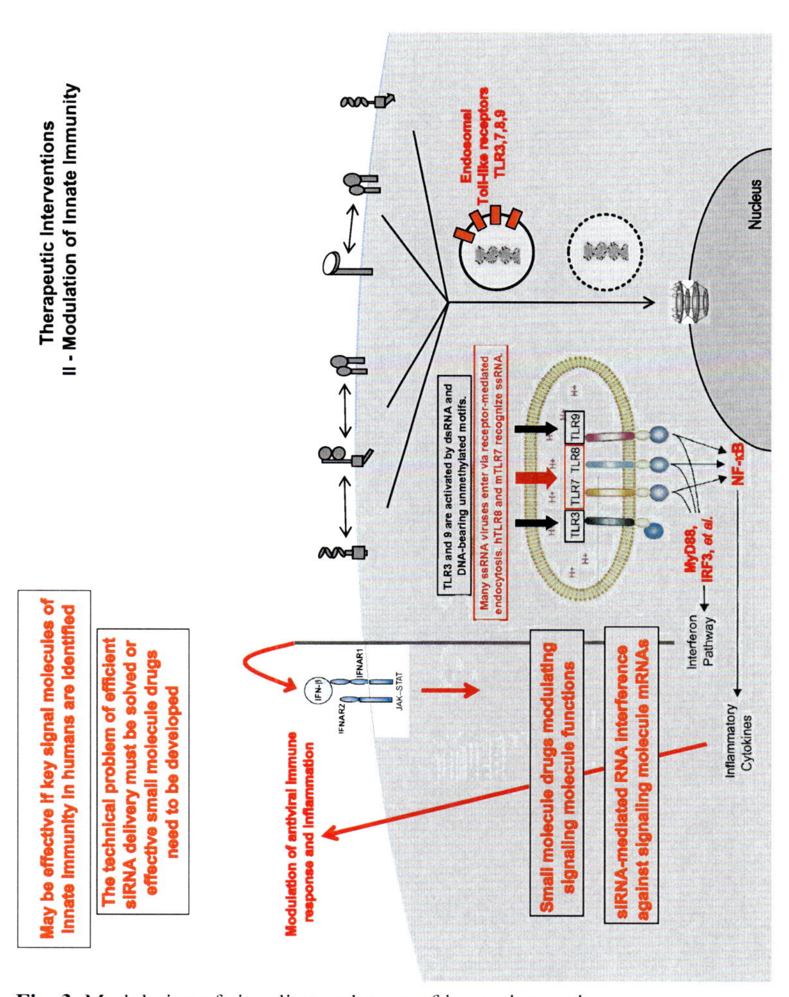

Fig. 3. Modulation of signaling pathways of innate immunity

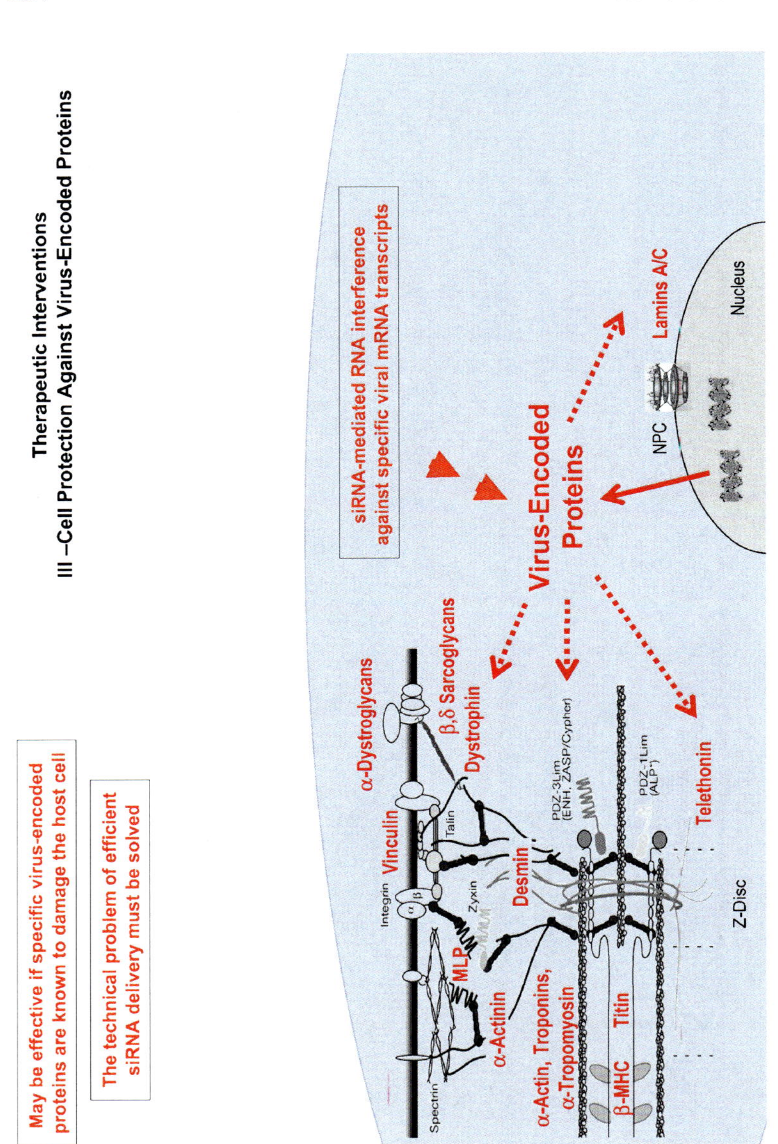

Fig. 4. Protection of host cells against virus-encoded proteins

modulation of innate immune signaling may be useful in chronic infections, where persistent; overshooting inflammation is maintained by chronic signal pathway activation.

3. In the chronic phase of a viral cardiomyopathy, even when no new virus particles or viral genomes are synthesized and therefore no cell-to-cell virus spreading can possibly occur via receptor-mediated processes, there may still be chronic viral transcriptional activity damaging the host cells in a slow, insidious process. This may go unnoticed by the clinician unless major, irreversible cardiac damage occurred. One example illustrating this pathogenic principle is the cleavage of cellular dystrophin by the enteroviral protease 2A encoded by CVB3 (Badorff et al. 1999; Xiong et al. 2002). Further proteins encoded by other cardiotropic viruses may also damage the infected heart even when synthesized at a low level only, but over long periods of time from cardiac-persistent viruses (Fig. 4). This may lead to an insidious decay of cardiac function proceeding unnoticed unless the patient is regularly and carefully monitored. "Silencing" of the respective viral genes, e.g., by RNA interference, constitutes a third target level for therapeutic interventions. Interventions of this type ("silencing strategies") may protect the chronically infected heart even after attempts at immune elimination of the virus-infected cells have remained unsuccessful.

16.2.1 Therapeutic Targets (I): Virus Uptake Mechanisms

Virus Uptake Mechanisms

Cellular virus entry, cell-to-cell spreading within the heart, and organ-to-organ migration within the host via the circulation are determined by virus receptor expression on target cells. The viral entry pathways into the human heart have not been fully delineated, but induction or inhibition of these pathways would decisively alter the individual susceptibility for cardiac viral infections. Induction of a common receptor (CAR) for coxsackieviruses and adenoviruses was observed in human DCM but not in further cardiomyopathies (Noutsias et al. 2001; Fechner et al. 2003). Receptor induction may open a path for infections with receptor-dependent viruses (EV/AdV) and contribute to dual infections by the respective viruses, which are further influenced by distinct coreceptors. Another common receptor (CD46), shared by otherwise unrelated viruses (HHV6 and CAR-independent group B AdV), has been identified (Gaggar et al. 2003). It is currently unknown what genetic or environmental factors cause the dysregulation of CAR in DCM, but its induction may critically influence pathogenesis. Since virus genomes are prevalent in DCM hearts, genetic variants of virus receptors or their regulatory pathways could provide a molecular basis for important genome–environment interactions influencing the disease process.

Therapeutic targets at the receptor level as outlined in Fig. 2 may be expected to work (1) if there is significant de novo entry of virus into new target cells by cell-to-cell spreading within the heart, or by organ-to-organ migration within the host via the circulation; or (2) if it is possible block key entry mechanisms with high efficiency. Whereas there is significant virus migration in CVB3 myocarditis models in mice, little is known about the kinetics of cardiotropic viruses [PVB19, HHV6, CVB3, AdV, cytomegalovirus (CMV), EBV, HCV] detected in human chronic viral cardiomyopathy (Kühl et al. 2005). It is necessary to better characterize these kinetics to decide if anti-receptor approaches may have applications in the human disease too. The tools currently available to block or downregulate virus receptors are:

1. Antibodies or fusion proteins directed against the receptor protein and/or virus epitopes involved in virus binding and/or internalization. With respect to CVB3, the best-characterized human cardiotropic virus, there are interesting recent studies on the use of a soluble recombinant decay accelerating factor (DAF) in the form of an IgG1-Fc-fusion protein (Yanagawa et al. 2003) or an analogous soluble CAR-Fc (Yanagawa et al. 2004). Importantly, however, these approaches work only in first few days after CVB3 infection. This a priori limits their possible use in humans to few cases of *acute* fulminant CVB3 myocarditis, but only if significant cell-to-cell migration of the CVB3 occurs in human too, similar to the situation in the animal model.
2. Antisense or decoy oligonucleotides (Yokoseki et al. 2001), which have been successfully employed in a myocarditis model. Short interfering RNAs (siRNAs) directed against an mRNA encoding a virus

receptor have been used, e.g., for human immunodeficiency (HIV) inhibition (Zhou et al. 2004).

16.2.2 Therapeutic Targets (II): Signaling Pathways of Innate Immunity

Signaling Pathways of Innate Immunity

During their transport through the endosomal pathway, the RNA or DNA viral genome might engage certain TLRs (3, 7, 8, 9) linked to key signaling cascades of innate immunity (Ulevitch 2004). Genetic alterations of immune pathways decisively alter the disease course in viral infections. Thus, the mortality after CVB3 infection is very low in mice lacking MyD88, a key component of TLR-dependent innate immune signaling, as compared to wild-type animals. Mice are protected from CVB3 myocarditis by gene-targeted knockout of p56Lck, the Src family kinase essential for T cell activation (Liu et al. 2000). In contrast, CVB3 infection of INF-β-deficient mice results in excessive mortality as compared to controls (Deonarain et al. 2004). Extracellular signal-regulated kinase 1 and 2 (ERK-1/2) is intense in the heart of myocarditis-susceptible A/J mice, in contrast to myocarditis-resistant C57BL/6 mice. The ERK-1/2 response to CVB3 may thus contribute to differential host susceptibility too (Opavsky et al. 2002). In rat experimental autoimmune myocarditis a decoy oligonucleotide against nuclear factor (NF)-κB was highly efficient against the development of the disease (Yokoseki et al. 2001), but to our knowledge so far no analogous study of a viral heart disease has been published.

Therapeutic targets in virus-activated signaling cascades of innate immunity as outlined in Fig. 3 may be derived from recent work in genetic knockout animals infected with CVB3. Analogous to the improved survival upon infection in MyD88 (see Chap. 8, this volume) or p56Lck knockout animals (Liu et al. 2000), suppression of the respective genes by siRNA-mediated RNA interference should be investigated with respect to its therapeutic potential. Since MyD88 is at a key point of innate immune signaling, its suppression by siRNAs or a small molecule drug inhibiting MyD88 function may be highly efficient if inappropriate overshooting inflammatory reactions in the heart accompany the antiviral response. An analysis of MyD88 and other important immune signaling molecules in virus-infected human hearts seems worthwhile, since recent animal data suggest a high therapeutic potential. It is currently unknown if, in chronic human viral cardiomyopathy, similar innate immune signaling pathways as targeted in the specific CVB3 animal model are relevant. Other viruses are encountered in chronic human viral cardiomyopathy far more frequently (e.g., PVB19, HHV6) and the kinetics of the chronic human disease are grossly different from the animal model. It is therefore of interest to investigate by cardiac expression profiling (based on human cardiac biopsies), if each virus induces specific alterations of innate immune signaling, or if there is a common pattern across many different virus species. In the latter case, approaches directed at innate immunity may become therapeutically relevant for a large number of patients.

16.2.3 Therapeutic Targets (III): Cellular Damage by Virus-Encoded Proteins

Cellular Damage by Virus-Encoded Proteins

Beyond the proteins involved in cellular virus entry and antiviral defense, several structural proteins of the cardiomyocyte building the sarcomere, sarcolemma cytoskeleton, Z-bands, and intermediate filaments are known to be mutated in cases of familial DCM (Poller et al. 2005; Franz et al. 2001). No population-wide comprehensive mutation scanning has so far been published for these genes, and thus the full extent of their genetic variability is unknown. However, one pioneering experimental study has already demonstrated the extent to which such mutations may aggravate the cardiac damage caused by a virus, when CoxB3-infection led to grossly aggravated disease in dystrophin-deficient mice (Badorff et al. 1999; Xiong et al. 2002). The CoxB3-encoded enteroviral protease 2A cleaves the cellular structural protein dystrophin not only in the dystrophin-deficient animals, but also in normal animals, although to a lesser extent. Most probably, a systematic search for further cellular sites vulnerable to attack by proteins which are encoded by human cardiotropic viruses will reveal further therapeutic targets. Protection of the respective vulnerable cellular proteins may, in principle, be achieved by siRNA-mediated suppression of the respective viral genes, or by specifically developed small-molecule drugs.

Specific Therapeutic Tools – RNA Interference

Whereas the very laborious development of small-molecule drugs blocking relevant virus-encoded proteins or their interaction site with a cellular protein must be done for every individual relevant protein, the application RNA interference by siRNAs is a more generalized approach. It has already been successfully employed against a number of different viruses both in vitro and in vivo (Zhou et al. 2004; Stevenson 2004; Ping et al. 2004; Dave and Pomerantz 2004; Gao et al. 2004; Zhang et al. 2004; Bhuyan et al. 2004; Takigawa et al. 2004; Kameoka et al. 2004; Wang et al. 2004; Lu et al. 2004; Li et al. 2004) including recent studies of CVB3 (Schubert et al. 2005), HHV6 (Yoon et al. 2004), and CMV inhibition in vitro (Wiebusch et al. 2004). One possible problem of siRNA therapy in vivo is the efficient transfer of the siRNA to all target cells. Recently, adeno-associated virus (AAV) vectors expressing siRNAs from specialized expression cassettes have been successfully used for various purposes (Pinkenburg et al. 2004; Grimm et al. 2005; Michel et al. 2005; Xia et al. 2004).

16.3 Summary and Perspective

As far as we currently understand chronic cardiac viral infections in humans, any therapeutic strategies inhibiting cellular virus uptake and migration (I) may only work if significant cell-to-cell traffic via receptor-mediated endocytosis of pathogenic virus still occurs in the chronic phase. At present we do not know if, e.g., cardiac reinfection from extracellular pools or intracardiac virus spreading occurs in the chronic disease in humans, and further clinical studies to solve this questions are warranted. Modulation of maladaptive innate immune signaling (II) was highly efficient in a CVB3 animal model. Further studies of the respective signaling pathways in humans should be performed, since the approach may possibly work across different virus species, thus providing a new generalized therapeutic principle. Finally, "silencing" (III) of viruses persisting in cardiac cells with very low transcriptional activity over long periods of time may inhibit the insidious damage to the heart as observed, e.g., in virus-positive DCM. If future molecular studies of cardiotropic viruses will identify virus-encoded proteins damaging key cellular proteins in cardiomyocytes or endothelial cells, this approach may gain major importance. In general, there are well-characterized CVB3 *animal* models, whereas there is a lack of animal models for other cardiotropic viruses and of data on their kinetics in chronic cardiac viral disease in *humans*. Clinical and animal studies addressing the latter issues are warranted, since they shall provide the basis for the development of improved therapies.

Acknowledgements. This publication has been supported by the Deutsche Forschungsgemeinschaft through Sonderforschungsbereich/Transregio 19.

References

Baboonian C, Davies M, Booth J, McKenna W (1997) Coxsackie B viruses and human heart disease. Curr Top Microbiol Immunol 223:31–52

Badorff C, Lee G, lamphear B, et al. (1999) Enteroviral protease 2A cleaves dystrophin: evidence of cytoskeletal disruption in an acquired cardiomyopathy. Nat Med 5:320–326

Bergelson J, Cunningham J, Droguett G, et al (1997) Isolation of a common receptor for coxsackie B viruses and adenoviruses 2 and 5. Science 275:1320–1323

Bhuyan PK, Kariko K, Capodici J, et al (2004) Short interfering RNA-mediated inhibition of herpes simplex virus type 1 gene expression and function during infection of human keratinocytes. J Virol 78:10276–10281

Bowles N, Richardson P, Olsen E, Archard L (1986) Detection of coxsackie-B virus-specific RNA sequences in myocardial biopsy samples from patients with myocarditis and dilated cardiomyopathy. Lancet 1:1120–1123

Chien K (2003) Genotype, phenotype: upstairs, downstairs in the family of cardiomyopathies. J Clin Invest 111:175–178

Dave RS, Pomerantz RJ (2004) Antiviral effects of human immunodeficiency virus type 1-specific small interfering RNAs against targets conserved in select neurotropic viral strains. J Virol 78:13687–13696

Deonarain R, Cerullo D, Fuse K, Liu PP, Fish EN (2004) Protective role for interferon-beta in coxsackievirus B3 infection. Circulation 110:3540–3543

Dörner A, Xiong D, Couch K, Yajima T, Knowlton K (2004) Alternatively spliced soluble Coxsackie-adenovirus receptors inhibit Coxsackievirus infection. J Biol Chem 279:18497–18503

Fechner H, Noutsias M, Tschoepe C, et al (2003) Induction of coxsackievirus-adenovirus-receptor expression during myocardial tissue formation and remodeling – identification of a cell–cell contact dependent regulatory mechanism. Circulation 107:876–882

Franz W, Muller O, Katus H (2001) Cardiomyopathies: from genetics to the prospect of treatment. Lancet 358:1627–1637

Frustaci A, Chimenti C, Calabrese F, Pieroni M, Thiene G, Maseri A (2003) Immunosuppressive therapy for active lymphocytic myocarditis: virologic and immunologic profile of responders versus non-responders. Circulation 107:857–863

Gaggar A, Shayakhmetov D, Lieber A (2003) CD46 is a cellular receptor for group B adenoviruses. Nat Med 9:1408–1412

Gao X, Wang H, Sairenji T (2004) Inhibition of Epstein-Barr virus (EBV) reactivation by short interfering RNAs targeting p38 mitogen-activated protein kinase or c-myc in EBV-positive epithelial cells. J Virol 78:11798–11806

Grimm D, Pandey K, Kay MA (2005) Adeno-associated virus vectors for short hairpin RNA expression. Methods Enzymol 392:381–405

Hertzog P, O'Neill L, Hamilton J (2003) The interferon in TLR signaling: more than just antiviral. Trends Immunol 24:534–539

Kameoka M, Nukuzuma S, Itaya A, et al. (2004) RNA interference directed against Poly(ADP-Ribose) polymerase 1 efficiently suppresses human immunodeficiency virus type 1 replication in human cells. J Virol 78:8931–8934

Kühl U, Pauschinger M, Schwimmbeck P, et al (2003a) Interferon-β treatment eliminates cardiotropic viruses and improves left ventricular function in patients with myocardial persistence of viral genomes and left ventricular dysfunction. Circulation 107:2793–2798

Kühl U, Pauschinger M, Bock T, et al. (2003b) Parvovirus B19 infection mimicking acute myocardial infarction. Circulation 108:945–950

Kühl U, Pauschinger M, Noutsias M, Seeberg B, Bock T, Lassner D, Poller W, Kandolf R, Schultheiss HP (2005) High prevalence of viral genomes and multiple viral infections in the myocardium of adults with "idiopathic" left ventricular dysfunction. Circulation 111:887–893

Li G, Li XP, Peng Y, Liu X, Li XH (2004) Effect of inhibition of EBV-encoded latent membrane protein-1 by small interfering RNA on EBV-positive nasopharyngeal carcinoma cell growth. Di Yi Jun Yi Da Xue Xue Bao 24: 241–246

Liu P, Aitken K, Kong YY, Opavsky MA, Martino T, Dawood F, Wen WH, Kozieradzki I, Bachmaier K, Straus D, Mak TW, Penninger JM (2000) The tyrosine kinase p56lck is essential in coxsackievirus B3-mediated heart disease. Nat Med 6:429–434

Lu A, Zhang H, Zhang X, et al. (2004) Attenuation of SARS coronavirus by a short hairpin RNA expression plasmid targeting RNA-dependent RNA polymerase. Virology 324:84–89

Matsumori A, Matoba Y, Sasayama S (1995) Dilated cardiomyopathy associated with hepatitis C virus infection. Circulation 92:2519–2525

Matsumori A, Yutani C, Ikeda Y, Kawai S, Sasayama S (2000) Hepatitis C virus from the hearts of patients with myocarditis and cardiomyopathy. Lab Invest 80:1137–1142

Michel U, Malik I, Ebert S, Bahr M, Kugler S (2005) Long-term in vivo and in vitro AAV-2-mediated RNA interference in rat retinal ganglion cells and cultured primary neurons. Biochem Biophys Res Commun 326:307–312

Murry C, Jerome K, Reichenbach D (2001) Fatal parvovirus myocarditis in a 5-year-old girl. Hum Pathol 32:342–345

Nigro G, Bastianon V, Colloridi V, et al (2000) Human parvovirus B19 infection in infancy associated with acute and chronic lymphocytic myocarditis and high cytokine levels: report of 3 cases and review. Clin Microbiol Infect 31:65–69

Noutsias M, Fechner H, Jonge Hd, et al (2001) Human coxsackie-adenovirus-receptor is co-localized with integrins avb3 and avb5 on the cardiomyocyte sarcolemma and upregulated in dilated cardiomyopathy – implications for cardiotropic viral infections. Circulation 104:275–280

Opavsky M, Martino T, Rabinovitch M, et al (2002) Enhanced ERK-1/2 activation in mice susceptible to coxsackievirus-induced myocarditis. J Clin Invest 109:1561–1569

Pauschinger M, Bowles N, Fuentes-Garcia J, et al. (1999a) Detection of adenoviral genome in the myocardium of adult patients with idiopathic left ventricular dysfunction. Circulation 99:1348–1354

Pauschinger M, Doerner A, Kuehl U, et al (1999b) Enteroviral RNA replication in the myocardium of patients with left ventricular dysfunction and clinically suspected myocarditis. Circulation 99:889–895

Ping YH, Chu CY, Cao H, Jacque JM, Stevenson M, Rana TM (2004) Modulating HIV-1 replication by RNA interference directed against human transcription elongation factor SPT5. J Acquir Immune Defic Syndr Hum Retrovirol 1:46

Pinkenburg O, Platz J, Beisswenger C, Vogelmeier C, Bals R (2004) Inhibition of NF-kappaB mediated inflammation by siRNA expressed by recombinant adeno-associated virus. J Virol Methods 120:119–122

Poller W, Kuhl U, Tschoepe C, Pauschinger M, Fechner H, Schultheiss HP (2005) Genome–environment interactions in the molecular pathogenesis of dilated cardiomyopathy. J Molec Med [Epub head of print 2 June 2005]

Rohayem J, Dinger J, Fischer R, Klingel K, Kandolf R, Rethwilm A (2001) Fatal myocarditis associated with acute parvovirus B19 and herpesvirus 6 coinfection. J Clin Microbiol 39:4585–4587

Schubert S, Grunert H-P, Zeichhardt H, Werk D, Erdmann V, Kurreck J (2005) Maintaining inhibition: siRNA double expression vectors against coxsackieviral RNAs. J Mol Biol 346:457–465
Stevenson M (2004) Therapeutic potential of RNA interference. N Engl J Med 351:1772–1777
Takigawa Y, Nagano-Fujii M, Deng L, et al (2004) Suppression of hepatitis C virus replicon by RNA interference directed against the NS3 and NS5B regions of the viral genome. Microbiol Immunol 48:591–598
Ulevitch R (2004) Therapeutics targeting the innate immune system. Nat Rev Immunol 4:512–520
Wang Z, Ren L, Zhao X, et al. (2004) Inhibition of severe acute respiratory syndrome virus replication by small interfering RNAs in mammalian cells. J Virol 78:7523–7527
Why HJ, Meany BT, Richardson PJ, et al (1994) Clinical and prognostic significance of detection of enteroviral RNA in the myocardium of patients with myocarditis or dilated cardiomyopathy. Circulation 89:2582–2589
Wiebusch L, Truss M, Hagemeier C (2004) Inhibition of human cytomegalovirus replication by small interfering RNAs. J Gen Virol 85:179–184
Xia H, Mao Q, Eliason SL, et al. (2004) RNAi suppresses polyglutamine-induced neurodegeneration in a model of spinocerebellar ataxia. Nat Med 10:816–820
Xiong D, Lee G-H, Badorff C, et al. (2002) Dystrophin deficiency markedly increases enterovirus-induced cardiomyopathy: A genetic predisposition to viral heart disease. Nat Med 8:782–877
Yanagawa B, Spiller OB, Choy J, et al. (2003) Coxsackievirus B3-associated myocardial pathology and viral load reduced by recombinant soluble human decay-accelerating factor in mice. Lab Invest 83:75–85
Yanagawa B, Spiller OB, Proctor DG, et al (2004) Soluble recombinant coxsackievirus and adenovirus receptor abrogates coxsackievirus b3-mediated pancreatitis and myocarditis in mice. J Infect Dis 189:1431–1439
Yokoseki O, Suzuki J, Kitabayashi H, et al (2001) cis Element decoy against nuclear factor-kappaB attenuates development of experimental autoimmune myocarditis in rats. Circ Res 89:899–906
Yoon JS, Kim SH, Shin MC, et al. (2004) Inhibition of herpesvirus-6B RNA replication by short interference RNAs. J Biochem Mol Biol 37:383–385
Zhang XN, Xiong W, Wang JD, Hu YW, Xiang L, Yuan ZH (2004) siRNA-mediated inhibition of HBV replication and expression. World J Gastroenterol 10:2967–2971
Zhou N, Fang J, Mukhtar M, Acheampong E, Pomerantz RJ (2004) Inhibition of HIV-1 fusion with small interfering RNAs targeting the chemokine coreceptor CXCR4. Gene Ther 11:1703–1712

17 Myocarditis and Inflammatory Cardiomyopathy: Histomorphological Diagnosis

F. Calabrese, A. Angelini, E. Carturan, G. Thiene

Abstract. Myocarditis is a non-ischemic inflammatory disease of the myocardium associated with cardiac dysfunction. It most often results from infectious agents, hypersensitivity responses, or immune-related injury. In spite of the development of various diagnostic modalities, early and definite diagnosis of myocarditis still depends on the detection of inflammatory infiltrates in endomyocardial biopsy specimens according to Dallas criteria. Routine application of immunohistochemistry (for characterization of inflammatory cell infiltration) and Polymerase Chain Reaction PCR analysis (for identification of infective agents) has become an essential part of the diagnostic armamentarium for a more precise biopsy report. A new morphological classification is advanced to overcome the limits of Dallas criteria. A semiquantitative assessment of myocyte damage/inflammation (grading) as well as of fibrosis (staging) is indicated, thus providing histopathological diagnosis useful to the clinician for more appropriate patient risk stratification and for the application of new therapies. Consequently, the final diagnosis of myocarditis should be mainly based on three features: etiology, grade, and stage of the disease.

17.1 Introduction

According to the 1995 World Health Organization/International Society and Federation of Cardiology Task Force on the Definition and Classification of Cardiomyopathy (Richardson et al. 1996), myocarditis is a specific cardiomyopathy defined as an inflammatory disease of the myocardium associated with cardiac dysfunction and diagnosed by histological, immunological, and immunohistochemical criteria.

Infectious, autoimmune, and idiopathic forms of inflammatory cardiomyopathies have been recognized.

The clinical diagnosis far exceeds the histopathological confirmation of the presence of myocarditis in endomyocardial biopsies (EMB). It has been estimated that myocardial biopsies are positive in only 30% of clinically suspected myocarditis (Billingham 1989). Thus, histology represents a fundamental step in the diagnosis of myocarditis. Although the disease is a diffuse process, it may be microscopically focal, causing a low diagnostic yield by biopsy sampling error (Edwards et al. 1982; Edwards 1984, 1985). The number of tissue pieces procured as well as the size and the processing of specimens has been demonstrated to influence the sensitivity of EMB in the detection of myocarditis. A greater sensitivity of the Stanford-Caves and Cordis bioptomes over other smaller bioptomes has been proved (Edwards et al. 1982; Edwards 1984). Particularly, the number of biopsy samples increases the likelihood of detecting foci of myocarditis. Chow et al. (1989) and Hauck et al. (1989) independently reported a sensitivity for detection of myocarditis of approximately 50% using four to five biopsy samples. When 17 biopsy specimens per case were considered, the sensitivity reached 79% (Chow et al. 1989).

It has also been shown that serial sectioning and multiple level examination of specimens increase the sensitivity of EMB in the evaluation of myocarditis (Burke et al. 1991; Thiene et al. 1996), thus facilitating the diagnosis of focal myocarditis with a rate of sensitivity similar to diffuse myocarditis.

Although some authors have demonstrated that biventricular EMB (obtaining several samples from multiple areas of both the left and right ventricles) may improve the sensitivity in the detection of myocarditis (Frustaci et al. 1994), this aggressive approach increases the risk of complications.

However the majority of experts in the field agree that an actual increase of sensitivity of EMB has now been reached by using immunohistochemistry together with routine histology. A large panel of monoclonal and polyclonal antibodies is now mandatory to identify and characterize the inflammatory cell population as well as the activated immunological processes (Herskowitz et al. 1990; Kühl et al. 1994, 1995; Angelini et al. 2002).

Based only on histopathological criteria, several distinct types of myocarditis have been identified: lymphocytic, eosinophilic, polymorphous, giant cell, and granulomatous myocarditis.

Lymphocytic myocarditis is the most common form of myocarditis in Western countries, and most of the cases are documented or presumed to be viral in origin. In order to develop uniform, reproducible morphological criteria for the pathological diagnosis of myocarditis, a panel of cardiac pathologists developed a classification of the disease based on histological features of EMB specimens (Aretz 1987).

Following this classification, known as the Dallas criteria, myocarditis is defined as an "inflammatory infiltrate of the myocardium with necrosis and/or degeneration of adjacent myocytes, not typical of ischemic damage associated with coronary artery disease".

Two separate classifications are used for the first and subsequent biopsies (Table 1). The Dallas criteria do not permit the confident diagnosis of resolved or resolving myocarditis on the first biopsy. The adjectives "resolved" or "resolving" therefore can only be used in instances in which unequivocal myocarditis has been previously diagnosed. Although the Dallas criteria include the advantage of the use of a simple, universally accepted and standardized terminology, they have some important limitations.

Table 1. Dallas criteria: morphological diagnosis of myocarditis

I Biopsy	II Biopsy
Active myocarditis	Ongoing myocarditis
Borderline myocarditis	Resolving myocarditis
No myocarditis	Resolved myocarditis

According to the Dallas criteria, fibrosis may be observed in any type of myocarditis, both at first and second biopsy (with or without fibrosis)

In the original Dallas classification, other histological types of inflammatory infiltrate (eosinophilic, neutrophilic, and giant cell) are just mentioned in view of possible differentiation between a primary (or idiopathic) form of myocarditis from a process secondary to a known cause (Aretz 1987). It is not specified if the terminology (such as active and borderline (with or without fibrosis) currently used in the more common form (lymphocytic type) can be equally applied to the other histotypes of inflammatory cardiomyopathy.

Myocyte injury, including frank necrosis and degeneration, has been described in the original report from Aretz (1987); however, specification of the type and extension of myocyte damage was not included in the classification.

Although it is quite difficult to know the natural history of the disease and the specific timing of its evolution, adding temporal information about the lesion when the tissue was obtained for analysis (particularly when the clinical history is known) could provide some advantages, such as an indication about the course of the disease and suggestions for therapy. Moreover, this approach may avoid the inclusion of borderline myocarditis, which sometimes comprises chronic forms, in favor of the more general term "inflammatory cardiomyopathy" (Table 2) Angelini et al. 2002).

The intensity and distribution of the inflammatory infiltrate are highly variable, ranging from a solitary small focus to multifocal aggregates up to diffuse myocardial involvement. Many interstitial cells may be difficult to characterize on routine hematoxylin-eosin stained sections, and normal components such as mast cells, fibroblast nuclei cut in cross section, pericytes, histiocytes, and endothelial cells may resemble lym-

Table 2. Active vs borderline myocarditis: clinical data

	Active	Borderline	p
No. of patients	26 (62%)	16 (38%)	N.S.
Age	37 ± 15	40 ± 13	N.S.
NYHA at onset	2.4 (1.3)	2.2 ± 1.2	N.S.
LVEDV (ml/m^2)	90 ± 42	128 ± 50	0.002
LVEF (%)	34 (12)	34 (15)	N.S.
Follow-up death/CT	4 (15%)	2 (12.5%)	N.S.
Follow-up NYHA	1.3 ± 0.6	1.1 ± 0.4	N.S.

CT, cardiac transplantation; LVEDV, left ventricular end diastolic volume; LVEF, left ventricular ejection fraction; N.S., not significant; NYHA, New York Heart Association
From Angelini et al. (2002)

phocytes (Hauck and Edwards 1992). Moreover, a small number of inflammatory cells, including lymphocytes, may be found in the normal myocardium. In this regard, different ranges have been reported in the literature. On examination of 20 high-power (×400) microscope fields on sections of normal myocardium, Edwards and colleagues found the mean number of inflammatory interstitial cells to be less than 5 per "high-power field" (Edwards et al. 1982). Linder et al. (1985) and Schnitt et al. (1987), using immunohistochemistry for leukocyte common antigen, detected an average of 13 immunoreactive cells/mm^2 in "uninflamed" biopsy specimens. Tazelaar and Billingham (1987) found foci of inflammatory cells (>5 inflammatory cells/"focus") in 9.3% of 86 cardiac transplant donor hearts (Table 3). Recently, a cut-off of less than 14 leucocytes/mm^2 with the presence of T lymphocytes less than 7 cells/mm^2 has been considered a more realistic value (Kühl et al. 1996; Maisch et al. 1998). Moreover, neither histotype of inflammatory cell infiltration nor fibrosis extension has been considered in the Dallas classification.

Another important limit of the Dallas classification is that it lacks any reference to etiological factors.

Even though sampling and processing techniques have significantly increased the sensitivity of EMB in the evaluation of myocarditis, classical morphological analysis has substantial limits in the detection of

Table 3. Quantification of inflammatory cell infiltration in the normal myocardium

Reference	Disease	No. of cases	mean age (years)	No. of inflammatory cells	Magnification
Edwards et al. 1982	Normal hearts	170	29	< 5/high power field	×400 (20 fields)
Linder et al. 1985	Normal hearts	20	61	< 13/mm^2	×500 (20 fields)
Tazelaar and Billingham 1987	Donor hearts	86	23	9.3% with focus of > 5 cells	×80
Maisch et al. 1998	Normal hearts	56	–	< 14/mm^2	×400

infective pathogens, particularly viral agents, the most common cause of inflammatory cardiomyopathy. Viral myocarditides usually lack specific cytopathic effects, especially those sustained by RNA viruses. With the exception of some rare forms of cytomegaloviral myocarditis, these effects, when observed, neither necessarily imply the presence of nor are useful in detecting the type of virus, since they may represent degenerative changes or myocyte nuclear hyperplasia.

The development of molecular biological techniques, particularly amplification methods like polymerase chain reaction (PCR) or nested-PCR, allows the detection of low-copy viral genomes even from an extremely small amount of tissue such as EMBs.

Numerous studies of patients with myocarditis have demonstrated the usefulness of PCR analysis for etiological diagnosis (Jin et al. 1990; Chapman et al. 1990; Weiss et al. 1992; Grasso et al. 1992; Petitjean et al. 1992; Hilton et al. 1993; Martin et al. 1994; Pauschinger et al. 1999; Calabrese et al. 1999, 2002; Chimenti et al. 2001; Calabrese and Thiene 2003; Kühl et al. 2005).

A new morphological classification should be advanced to overcome the limits of the Dallas criteria based upon a semi-quantitative histological diagnosis, including a system of grading and staging, which is currently applied in other inflammatory diseases and graft rejection (Desmet et al. 1994; Vitali et al. 1993; Billingham 1990). It should provide a histopathological diagnosis useful to the clinician for more appropriate patient risk stratification and for the application of new therapeutic trials.

17.2 Classification of Inflammatory Cardiomyopathy: A New Proposal

Once the etiology is established, which for infective myocarditis, particularly viral forms, can be easily investigated in EMB tissue samples through PCR analysis, the histological data should be described in detail and incorporated into the histological report. In the new classification, three fundamental morphological information should be indicated: (1) inflammatory cell *type*, (2) semi-quantitative assessment of myocyte damage/inflammation, that is, *grading*, and (3) semi-quantitative assessment of fibrosis, or *staging* (Table 4).

The concepts of grading and staging have been applied in neoplastic pathology for years. Grading describes the degree of differentiation of a neoplasm, while staging denotes the extent of its spread. The same concepts have been applied with some modification to non-neoplastic conditions such as chronic hepatitis (Desmet et al. 1994). As in hepatitis, grading might be used in inflammatory cardiomyopathy to describe the intensity of necro-inflammatory activity. Staging, on the other hand, may be a measure of fibrosis and architectural modifications, features currently believed to be the consequence of tissue injury and repair. Numerical score may be attributed to both staging and grading, thus providing a semiquantitative assessment of the analyzed features. EMBs, adequate in number, size, and quality are mandatory to perform a more accurate diagnosis.

Table 4. Inflammatory cardiomyopathy classification: new proposal

Histotype
Type of inflammatory infiltrate
Grading
Burden of myocyte damage and inflammation
Staging
Fibrosis

As in chronic hepatitis, the new classification of inflammatory cardiomyopathy should: (1) drive the therapy in a more appropriate way, as in chronic hepatitis, (2) give a more correct clinical stratification of patients, and (3) provide valuable research data.

17.3 Histological Features of Grading

In inflammatory cardiomyopathy, grading should take into consideration the known and readily evaluable forms of myocyte damage (necrosis and apoptosis) together with the extent and distribution of the inflammatory cell infiltration. In our view, grading should include assessment of myocyte damage (Table 5), whether focal or plurifocal (Fig. 1a, b). Inflammatory cell infiltration should be mainly focused on the count of T cells/mm^2: 7 T cells/mm^2 is the most appropriate cut-off to be considered for the diagnosis of the disease (Fig. 2a, b, c). To this purpose, the use of immunohistochemistry is mandatory. Endocardial involvement (inflammation and/or thrombosis) should also be noted.

Interstitial edema should not be considered a criterion to be applied for diagnosing inflammatory cardiomyopathy in EMB specimens, since artifactual separation of myocytes may simulate edema.

Table 5. Inflammatory cardiomyopathy classification: grading

A Myocyte damage	Absent	0
	Focal	1
	Plurifocal	2
B Interstitial inflammation	< 7 T cells/mm^2	0
	$> 7 < 14$ T cells/mm^2	1
	> 14 T cells/mm^2	2
C Endocardial involvement	Absent	0
(Inflammation, thrombosis)	Present	1
	Maximum possible score	5

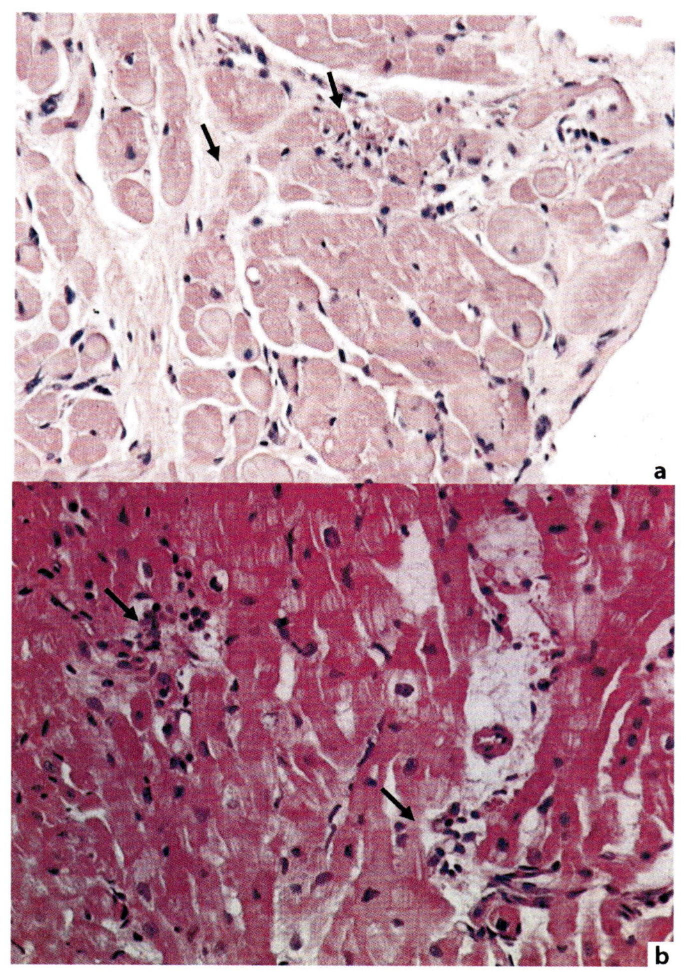

Fig. 1a,b. Explicative examples of grading for myocyte damage. **a** A small focus of myocyte necrosis – score 1 – (*arrow*). **b** Two foci of myocyte necrosis (*arrows*) – score 2. H&E, original magnification ×40

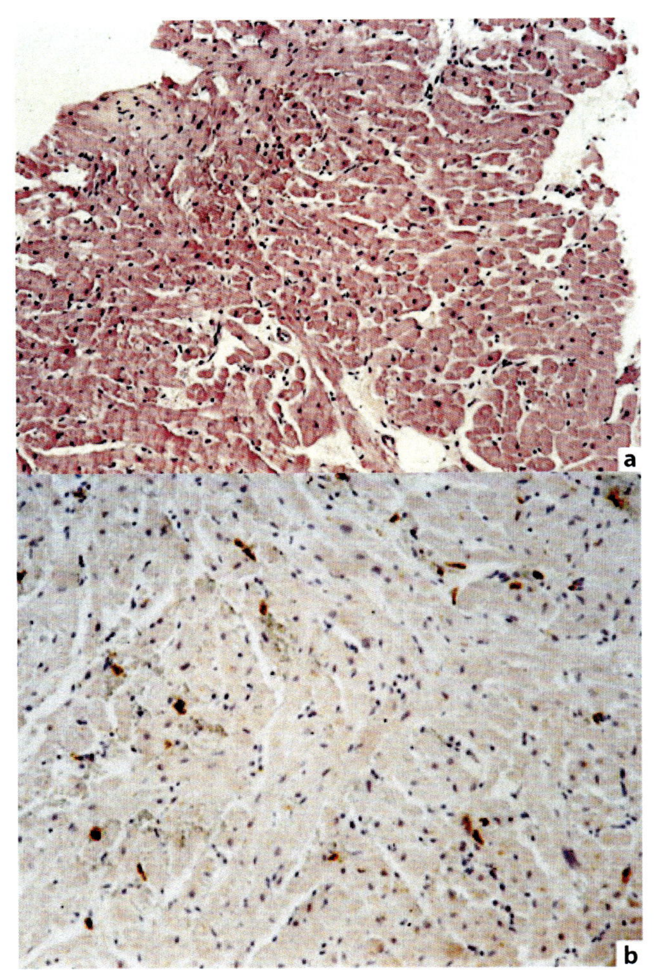

Fig. 2a–c. Explicative examples of grading for interstitial inflammation. **a**, **b** Inflammatory cell infiltration $> 7 < 14$ T (CD3-positive) cells/mm^2 – score 1. H&E (**a**), immunohistochemistry (**b**), original magnification ×20

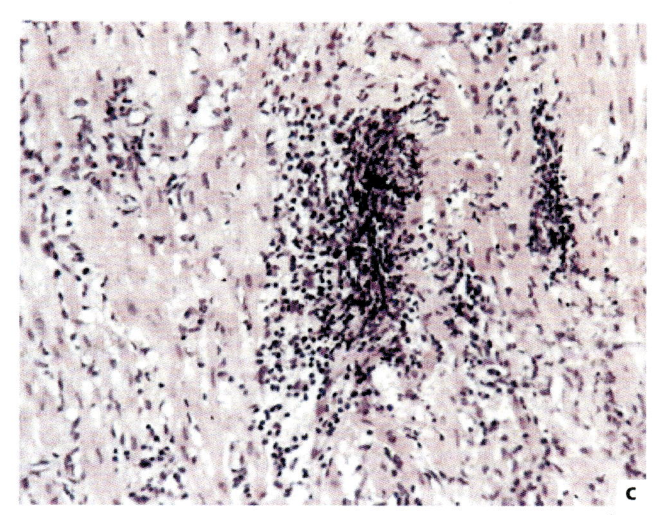

Fig. 2c. Inflammatory cell infiltration >14 T (CD3-positive) cells/mm^2 – score 2. H&E, original magnification ×25

The overall maximum score for grading should be 5. In the less-common forms of inflammatory cardiomyopathy with neutrophilic, eosinophilic, or giant cell infiltration, a higher score for inflammation might be used, since a more diffuse infiltrate is usually present in these histotypes.

Table 6. Inflammatory cardiomyopathy classification: staging

A Interstitial/replacement fibrosis	Absent	0
	> 10% < 20%	1
	> 20% < 40%	2
	> 40%	3
B Subendocardial fibrosis	Absent	0
	Present	1
C Endocardial fibroelastosis	Absent	0
	Present	1
	Maximum possible score	5

17.4 Histological Features of Staging

The features taken into consideration should include: interstitial fibrosis, subendocardial fibrosis, and endocardial fibroelastosis (Table 6; Fig. 3a, b, c). Replacement fibrosis should be considered abnormal when exceeding 10% of the area.

Fibrosis should always be specified as interstitial or replacement. Again, the overall maximum possible score should be 5.

17.4.1 Additional Pathological Features

Additional features which should be noted but not scored are:

- Myocyte hypertrophy.
- Nuclear abnormalities.
- Intracellular inclusions.
- Vessel abnormalities.
- Presence of adipocytes. (If fatty or fibrofatty replacement is present, it should be expressed as a percentage of the area.)

According to the new classification, grading and staging should be added to the conventional diagnostic report (Table 7).

Table 7. Example of conventional and new semi-quantitative biopsy reports

Conventional report

This endomyocardial biopsy shows diffuse and severe interstitial inflammatory infiltrates, mainly represented by T lymphocytes. Multifocal myocyte damage featured by frank myocytolysis is clearly seen. Mild interstitial and subendocardial fibrosis is also evident. *Summary:* Active myocarditis with fibrosis

New report:

Lymphocytic Inflammatory cardiomyopathy

The score appropriate for the diagnostic report according to Tables 3 and 4 would be

Grading: (A)2 + (B)2 + (C)0; total = 4
Staging: (A)1 + (B)1 + (C)0; total = 2

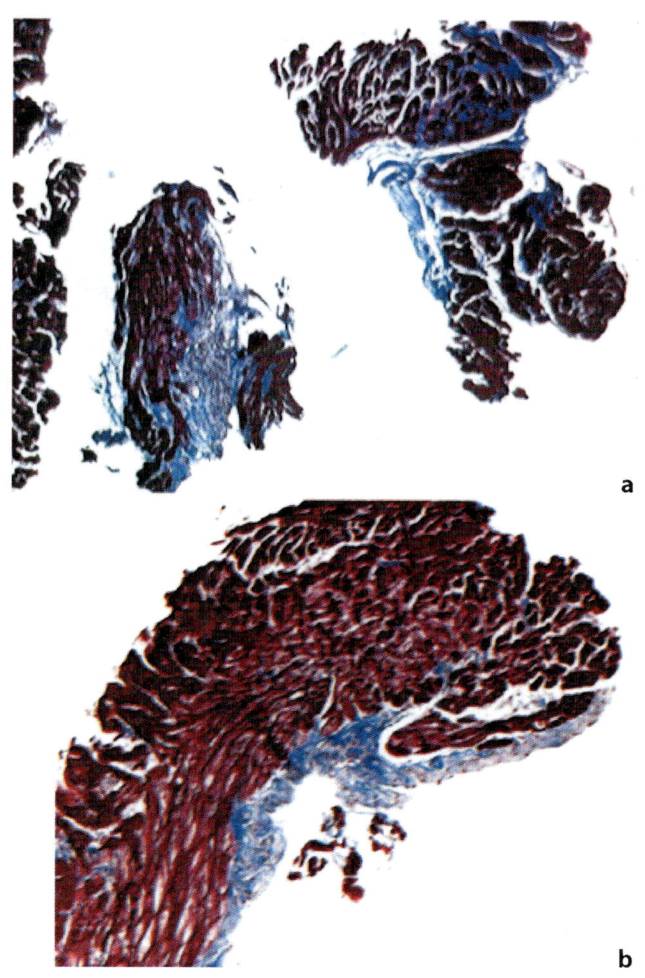

Fig. 3a,b. Explicative examples of staging for fibrosis. **a** Interstitial fibrosis > 20%. Azan-Mallory, original magnification ×2.5. **b** Subendocardial fibrosis. Azan-Mallory, original magnification ×2.5

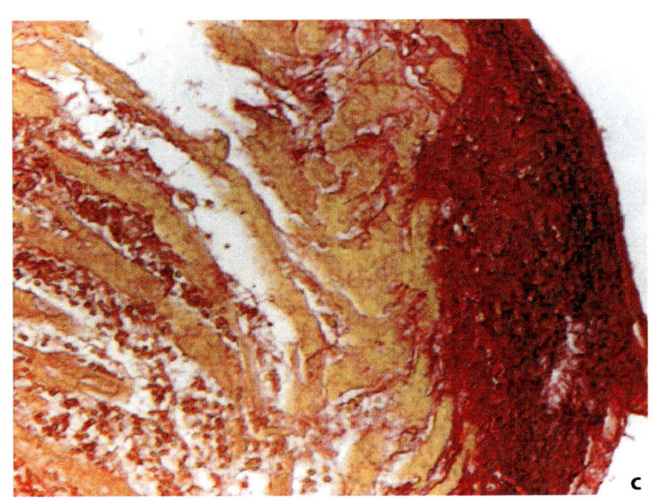

Fig. 3c. Endocardial fibroelastosis. Elastic van Gieson, original magnification ×5

17.5 Validation of Grading and Staging Systems

An acceptable system of grading and staging must have several features. First, it should include all aspects known to be of value in the evaluation of the severity and extent of the disease, as well as those believed to have prognostic implications. Second, the system to be utilized needs to be practical and not cumbersome, and should be reasonably reproducible, both by the individual pathologist at different times (low intra-observer variability) and by different pathologists (low inter-observer variability). Third, the system has to be useful to the clinician or evaluator of a clinical trial; reproducibility in itself is not of value if the reported data cannot be used by the clinician. A compromise between complexity and reproducibility is often unavoidable.

Supported by a grant of Ministry of Education, University and Research, Rome, Italy.

References

Angelini A, Crosato M, Boffa GM, et al. (2002) Active versus borderline myocarditis: clinicopathological correlates and prognostic implications. Heart 87:210–215

Aretz HT (1987) Myocarditis: the Dallas Criteria. Hum Pathol 18:619–624

Billingham ME (1989) Acute myocarditis: is sampling error a contraindication for diagnostic biopsies? J Am Coll Cardiol 14:921–922

Billingham ME (1990) Dilemma of variety of histopathologic grading systems for acute cardiac allograft rejection by endomyocardial biopsy. J Heart Transplant 9:272–276

Burke AP, Farb A, Robinowitz M, Virmani R (1991) Serial sectioning and multiple level examination of endomyocardial biopsies for the diagnosis of myocarditis. Mod Pathol 4:690–693

Calabrese F, Thiene G (2003) Myocarditis and inflammatory cardiomyopathy: microbiological and molecular biological aspects. Cardiovasc Res 60:11–25

Calabrese F, Valente M, Thiene G, et al. (1999) Enteroviral genome in native hearts may influence outcome of patients who undergo cardiac transplantation. Diagn Mol Pathol 8:39–46

Calabrese F, Rigo E, Milanesi O, et al. (2002) Molecular diagnosis of myocarditis and dilated cardiomyopathy in children. Clinico-pathologic features and prognostic implications. Diagn Mol Pathol 11:212–221

Chapman NM, Tracy S, Gauntt CJ, Fortmueller U (1990) Molecular detection and identification of enteroviruses using enzymatic amplification and nucleic acid hybridization. J Clin Microbiol 28:843–850

Chimenti C, Calabrese F, Thiene G, et al. (2001) Inflammatory left ventricular microaneurysms as a cause of apparently idiopathic ventricular tachyarrhythmias. Circulation 104:168–173

Chow LH, Radio SJ, Sears TD, McManus BM (1989) Insensitivity of right ventricular endomyocardial biopsy in the diagnosis of myocarditis. J Am Coll Cardiol 14:915–920

Desmet VJ, Gerber M, Hoofnagle JH, Manns M, Scheuer PJ (1994) Classification of chronic hepatitis: diagnosis, grading and staging. Hepatology 19:1513–1520

Edwards WD (1984) Myocarditis and endomyocardial biopsy. Cardiol Clin 2:647–656

Edwards WD (1985) Current problems in establishing quantitative histopathologic criteria for the diagnosis of lymphocytic myocarditis by endomyocardial biopsy. Heart Vessels 1:138–142

Edwards WD, Holmes DR, Reeders GS (1982) Diagnosis of active lymphocytic myocarditis by endomyocardial biopsy: quantitative criteria for light microscopy. Mayo Clin Proc 57:419–425

Frustaci A, Bellocci F, Olsen EG (1994) Results of biventricular endomyocardial biopsy in survivors of cardiac arrest with apparently normal hearts. Am J Cardiol 74:890–895

Grasso M, Arbustini E, Silini E, et al. (1992) Search for coxsackievirus B3 RNA in idiopathic dilated cardiomyopathy using gene amplification by polymerase chain reaction. Am J Cardiol 69:658–664

Hauck AJ, Edwards ED (1992) Histopathologic examination of tissues obtained by endomyocardial biopsy. In: Fowles RE (ed) Cardiac biopsy. Futura, Mount Kisco, pp 95–153

Hauck AJ, Kearney DL, Edwards WD (1989) Evaluation of postmortem endomyocardial biopsy specimens from 38 patients with lymphocytic myocarditis: implications for role of sampling error. Mayo Clin Proc 64:1235–1245

Herskowitz A, Ahmed-Ansari A, Neumann DA, et al. (1990) Induction of major histocompatibility complex antigens within the myocardium of patients with active myocarditis: a nonhistologic marker of myocarditis. J Am Coll Cardiol 15:624–632

Hilton DA, Variend S, Pringle JH (1993) Demonstration of coxsackie virus RNA in formalin-fixed tissue sections from childhood myocarditis cases by in situ hybridization and the polymerase chain reaction. J Pathol 170:45–51

Jin O, Sole MJ, Butany JW, et al. (1990) Detection of enterovirus RNA in myocardial biopsies from patients with myocarditis and cardiomyopathy using gene amplification by polymerase chain reaction. Circulation 82:8–16

Kuhl U, Seeberg B, Schultheiss HP, Strauer BE (1994) Immunohistological characterization of infiltrating lymphocytes in biopsies of patients with clinically suspected dilated cardiomyopathy. Eur Heart J 15:62–67

Kuhl U, Noutsias M, Schultheiss HP (1995) Immunohistochemistry in dilated cardiomyopathy. Eur Heart J 16:100–106

Kuhl U, Noutsias M, Seeber B, Shultheiss HP (1996) Immunohistological evidence for a chronic intramyocardial inflammatory process in dilated cardiomyopathy. Heart 75:295–300

Kuhl U, Pauschinger M, Noutsias M, et al. (2005) High prevalence of viral genomes and multiple viral infections in the myocardium of adults with "idiopathic" left ventricular dysfunction. Circulation 111:887–893

Linder J, Cassling RS, Rogler WC, et al. (1985) Immunohistochemical characterization of lymphocytes in uninflamed ventricular myocardium. Implications for myocarditis. Arch Pathol Lab Med 109:917–920

Maisch B, Bultmann B, Factor S, et al. (1998) Dilated cardiomyopathy with inflammation or chronic myocarditis: variability and consensus in the diagnosis. Eur Heart J 19:647 (abs)

Martin AB, Webber S, Fricker FJ, et al. (1994) Acute myocarditis. Rapid diagnosis by PCR in children. Circulation 90:330–339

Pauschinger M, Bowles NE, Fuentes-Garcia FJ, et al. (1999) Detection of adenoviral genome in the myocardium of adult patients with idiopathic left ventricular dysfunction. Circulation 99:1348–1354

Petitjean J, Kopecka H, Freymuth F, et al. (1992) Detection of enterovirus in endomyocardial biopsy by molecular approach. J Med Virol 37:76–82

Richardson P, McKenna WJ, Bristow M, et al. (1996) Report of the 1995 WHO/ISFC task force on the definition of cardiomyopathies. Circulation 93:841–842

Schnitt SJ, Ciano PS, Schoen FJ (1987) Quantitation of lymphocytes in endomyocardial biopsies: use and limitations of antibodies to leukocyte common antigen. Hum Pathol 18:796–800

Tazelaar HD, Billingham ME (1987) Myocardial lymphocytes. Fact, fancy or myocarditis? Am J Cardiovasc Pathol 1:47–50

Thiene G, Bartoloni G, Poletti A, Boffa GM (1996) Tecniche, indicazioni ed utilità della biopsia endomiocardica. In: Baroldi G, Thiene G (eds) Biopsia endomiocardica – Testo atlante. Edizioni Piccin, Padova, pp 5–45

Vitali C, Bombardieri S, Moutsopoulos HM, et al. (1993) Preliminary criteria for the classification of Sjogren's syndrome. Results of a prospective concerted action supported by the European Community. Arthritis Rheum 36:340–347

Weiss LM, Liu XF, Chang KL, Billingham ME (1992) Detection of enteroviral RNA in idiopathic dilated cardiomyopathy and other human cardiac tissues. J Clin Invest 90:156–159

18 Anti-viral Treatment in Patients with Virus-Induced Cardiomyopathy

U. Kühl, M. Pauschinger, W. Poller, H.-P. Schultheiss

Abstract. Ongoing viral persistence in the myocardium is associated with an adverse prognosis of cardiomyopathy eventually resulting in a reduced capacity for work and thus it is associated with enormous social costs. Experimental and clinical data highlight that an imbalance of the cytokine network and a defect in the cytokine-induced immune response may constitute major causes leading to the development of virus persistence and progression of myocardial dysfunction. Reversibility of cardiac impairment during the early stages of the disease and the arising chance of specific treatment options demand early diagnosis and treatment of the disease. Our pilot data on anti-viral treatment using INF-β showed beneficial clinical effects and suggest that some of the ventricular dysfunction and wall motion abnormalities resolved after elimination of the responsible agents. The data also suggest that elimination of cardiotropic viruses and associated clinical effects may occur even in DCM patients presenting with a long history.

18.1 Introduction

Idiopathic dilated cardiomyopathy (DCM) is a serious disorder and the most common cause of heart failure in young patients. A genetic origin of DCM has been reported in up to 25% of cases, but the majority are sporadic, and viral or immune pathogenesis is suspected in as much as 60% of idiopathic DCM (Michels et al. 1992; Richardson et al. 1996; Kühl et al. 1996, 2003a, 2005; Feldmann and McNamara 2000). The proof of a viral affliction of the myocardium warrants analysis of endomyocardial biopsies with sensitive molecular biological methods such as in situ hybridization or polymerase chain reaction (PCR) (Bowles et al. 1986; Why et al. 1994). These methods enable the detection of viral RNA and DNA in tissues with even low numbers of viral copies. The enteroviral genome was confirmed by PCR amplification in ca. 20% of EMBs from myocarditis and DCM patients (Baboonian and Treasure 1997). Adenovirus has been elucidated as another virus infecting cardiomyocytes (Pauschinger et al. 1999a), and there is evolving evidence for further cardiotropic viruses, especially parvovirus B19, human herpesvirus 6, hepatitis C and Epstein-Barr virus (Kühl et al. 2003a; Frustaci et al. 2003; Matsumori 2001).

Investigations on the tissue distribution of these viruses have revealed profound differences: Whereas the enteroviruses infect cardiomyocytes, endothelial cells are the primary targets of parvovirus B19 infection (Kandolf et al. 1987; Li et al. 2000a; Bultmann et al. 2003). In this context, the pattern of viral receptors might be a decisively important issue (Noutsias et al. 2001; Poller et al. 2002. At present, we are lacking detailed data on the prognostic impact, pathways of viral entry and persistence of these recently identified cardiotropic viruses.

18.2 Natural Course

During the acute stage of the disease, complete or partial recovery of myocardial function may occur even in severe cases of cardiac depression. The retrospective study by McCarthy et al. for the first time demonstrated that a fulminant inflammatory process is associated with a better long-term outcome compared with the nonfulminant presentation of myocarditis patients (93% vs 45%) after 5.6 years of follow-up (McCarthy et al. 2000). From that it may be deduced that intramyocardial inflammation is not detrimental per se and, moreover, that inflammation may merely reflect the attempt of the immune system to eliminate cardiotropic viruses. The reversibility of ventricular function suggests that initial impairment of myocardial contractility is not caused by an irreversible loss of cardiomyocytes or irreversible destruction of the cell-matrix integrity. The initial deterioration of ventricular function, to some extent, might be caused by the production of negative inotropic mediators of the immune response to viral infection, such as cytokines (TNF-α, IL-1β, IL-6, IFN-γ), or the induction of inducible nitric oxide synthase (iNOS), causing increased amounts of NO, which reversibly interferes with myocardial cell function and matrix integrity (Freeman et al. 1998; Lowenstein et al. 1996; Bachmaier et al. 1997; Li et al. 2001). Therefore, myocardial function may recover if these harmful substances decrease during spontaneous virus clearance and resolution of the inflammatory process (Finkel et al. 1992; Torre Amione et al. 1996; Testa et al. 1996; Bozkurt et al. 1998).

In a number of patients, however, immunologic processes fail to clear viruses completely. A considerable body of evidence, therefore, implicates viruses – classically regarded as agents of self-limiting infections – in the aetiology of chronic myocarditis and DCM (Baboonian and Treasure 1997; Pauschinger et al. 1999a; Kawai 1999). Chronic viral infection of the myocardium may be associated with progression of the disease. The adverse prognostic impact of enteroviral persistence in DCM was identified early (Dec and Fuster 1994), and recent results indicate a special importance of the replicative infection mode in a significant proportion of DCM patients (Pauschinger et al. 1999b; Fujioka et al. 2000). The restricted viral replication in even a relatively small number of infected myocardial cells has been shown to be sufficient for the maintenance of chronic inflammation, morphological alterations of the structural integrity of the myocardium and interference with myocyte function in animal models of coxsackievirus B3-infected mice, and thus may be responsible for progressive myocardial impairment (Wessely et al. 1998; Klingel and Kandolf 1993; Pauschinger et al. 1999c). Active enterovirus replication in the myocardium of DCM patients implies that enteroviruses may exert an active metabolism and consequently an ongoing myocytopathic effect even in late persistent infection (Pauschinger et al. 1999a,b; Wessely et al. 1998; Liu et al. 2000b; Kawai 1999; Klingel et al. 1998). One can assume that both the direct myocytopathic effect of cardiotropic viruses and the activation of the immune system against viral antigens and/or cryptic myocardial antigens released by viral infection perpetuate cardiac dysfunction in DCM patients. In addition, virus- and immune-mediated disturbance of myocardial cell metabolism, energy production or energy transfer may also contribute to myocardial dysfunction (Badorff et al. 1997; Schulze et al. 1999; Doerner et al. 2000). Recent data indicate that interference of virus- and immune-mediated processes with cell matrix integrity might be of considerable importance during the acute and chronic stage of the disease (Li et al. 2000c, 2001; Badorff et al. 1999; Mann and Taegtmeyer 2001; Libby and Lee 2000). In this regard, the extracellular matrix turnover, regulated by matrix metalloproteinases (MMPs) and a family of tissue inhibitors of metalloproteinases (TIMPs) may contribute to myocardial remodelling (Thomas et al. 1998).

18.3 Antiviral Immunity

Data from myocarditis animal models and other virus-induced human diseases suggest that, in addition to various different virus-dependent mechanisms, the nature of the local and systemic immune response controls viral infections of different organs including the heart (Biron et al. 1999; Hunter and Reiner 2000; Talon et al. 2000). The molecular pathomechanisms causing virus persistence in some patients while promoting virus clearance in others are currently unknown. Increased levels of various cytokines have been reported in patients with myocarditis and dilated cardiomyopathy, but the possible influence of antiviral cytokines on the natural course of cardiac viral disease has not been analysed (Matsumori et al. 1994).

18.4 Interferons: Experimental Data and Lessons from Animal Models

The important role played by the interferons (IFNs) as a natural defence against viruses is documented by three types of experimental and clinical observations: In many viral infections, a strong correlation has been established between IFN production and natural recovery; inhibition of IFN production or action enhances the severity of infection, and treatment with IFN protects against viral infection. The antiviral effect is independent of virus type and results in an intracellular block of the viral replication cycle. IFNs increase the resistance towards viral replication even in cells that neighbour infected cells but have not been infected yet. Immunomodulatory effects include activation of macrophages and natural killer cells as well as enhancement of major histocompatibility complex (MHC) antigen expression (Baron 1992).

An antiviral effect of IFNs has been reported in an in vitro model mimicking persistent enteroviral infection (Heim et al. 1992). The efficacy of IFNs in preventing viral myocarditis in susceptible rodents has also been demonstrated (Matsumori et al. 1988; Horwitz et al. 2000a). The analysis of cellular immunity and cytokine profiles in resistant versus susceptible mice has revealed dramatic differences in the levels of pro-inflammatory type I cytokine production in response to virus (Ramshaw et al. 1997). High systemic levels of antiviral IFN-γ, interleukin (IL)-2, IL-12 and tumour necrosis factor (TNF)-α are produced in virus-resistant C57BL/6 mice (Shioi et al. 1997; Yamada et al. 1994). Susceptible A/J or SWR mice, on the other hand, which develop a progressive virus-induced disease, show a predominantly Th_2-like cell response to infection. Pro-inflammatory, antiviral cytokines are absent or produced at low levels in these strains. IL-4 caused down-regulation of the antiviral immune response, reducing CTL activity, CTL precursor frequency, and expression of IL-2, IL-12 and IFN-γ mRNA (Ramshaw et al. 1997). In other infectious diseases, dysregulation of the immune response with an imbalance between pro- and anti-inflammatory cytokines and insufficient amounts of antiviral cytokines prevails over effective viral clearance (Katze et al. 2002; Yamamura et al. 1991; Sher and Coffman 1992; Clerici and Shearer 1993).

In most cases, myocarditis resistance or susceptibility is a relative rather than absolute condition, in that, e.g. immunomodulation of resistant mice completely restores disease susceptibility (see comments in Horwitz et al. 2000a; see also Shioi et al. 1996; Yamamoto et al. 1998). In addition to its direct antiviral effect, IFN-γ-induced activation of the immune system may suppress virus spreading and facilitate virus clearance (Ramshaw et al. 1997), with antigen-specific cytotoxic lymphocytes, natural killer cells, memory cells and cytokines being involved in this process (Seko et al. 1993; Huang et al. 1993).

18.5 Treatment of Patients with Virus-Induced Cardiomyopathy

In spite of the progress made in heart-failure therapy since the introduction of angiotensin converting enzyme (ACE) inhibitors, β-blockers, spironolactone and amiodarone/ICD (implantable cardioverter-defibrillator) treatment, mortality of both myocarditis and dilated cardiomyopathy is still 10% per year (Dec 1985; Grogan et al. 1995; McCarthy et al. 2000). Current heart-failure therapy is symptomatic. It may interfere with secondary pathogenetic processes, e.g. virus- or inflammation-triggered myocardial remodelling, but does not influence specific pathogenic mechanisms such as viral infections. Progression of left ventricular dysfunction therefore may be delayed but cannot be halted by a therapy with digitalis, diuretics, ACE inhibitors and β-blocking agents. A successful treatment of myocarditis and its sequela inflammatory cardiomyopathy warrants virus-specific treatment strategies.

18.6 Immunomodulatory Treatment Strategies

A retrospective analysis by Frustaci et al. confirmed that patients with persistence of cardiotropic viruses (except for hepatitis C) do not improve, or even deteriorate, under immunosuppression, which interferes with effective clearance of these viruses (Frustaci et al. 2003). Natural antiviral mediators of the immune response, among others, are IFNs, which may be a better choice for treatment in these virus-positive patients. Three types of IFNs have been identified which differ both in structure and in antigenic properties: IFN-α, derived from leucocytes, IFN-β from fibroblasts, and IFN-γ, which is derived from lymphocytes. IFNs react with specific receptors on cell surfaces to activate cytoplasmic signal proteins. These proteins enter the nucleus to stimulate cellular genes encoding a number of other proteins responsible for the defensive activity, including antiviral, antiproliferative, antitumour, and immunomodulatory effects. Since effective concentrations of IFN-β can be attained in vivo, IFN-β may become useful in the treatment of patients with viral cardiomyopathy.

18.7 Anti-viral Treatment Strategies in Man

IFN-α: Randomised studies demonstrating a benefit of a specific antiviral therapy in patients with virus-induced heart muscle disease do not exist so far. In one study, a 6-month IFN-α2a therapy in four patients with biopsy-proven enteroviral infection led to haemodynamic improvement in all patients, but virus was still detected in two of the four patients (Stille Siegener et al. 1995). One patient with biopsy-proven enteroviral infection underwent IFN-α2a therapy over 6 months. Left ventricular ejection fraction remained unchanged, but the patient improved from New York Heart Association (NYHA) class III to NYHA II, and pulmonary artery pressures were reduced. Control biopsies 1 month after end of therapy were negative for enterovirus. A 1-year follow-up demonstrated stable clinical and unchanged haemodynamic conditions (Heim et al. 1994). In an open-label study, 13 patients with clinical evidence of myocarditis or dilated cardiomyopathy were treated with IFN-α or thymomodulin (Miric et al. 1996). Left ventricular contractility improved after 6 months of treatment, but virus infection of the myocardium before or after treatment has not been documented in this study.

IFN-β: The antiviral potential of IFN-β against coxsackieviruses has been demonstrated in vitro (Heim et al. 1992) and in animal models

(Matsumori et al. 1988), while the effect of any of the IFNs against adenoviral infection has not yet been elucidated. Because IFN-β shows high antiviral potency, 22 patients with biopsy-proven persisting viral cardiomyopathy (adenovirus: 7 patients; enterovirus: 15 patients) were treated with recombinant IFN-β1a in an open pilot study (Kühl et al. 2003b).

Table 1. Demographic and clinical characteristics of 22 virus-positive patients with persistent left ventricular dysfunction included in the IFN-β treatment study

Demographic	
Number of patients	22
Mean (SD) age (years)	51.8 ± 13.6[a]
Sex (male/female)	13/9
Clinical characteristics	
NYHA functional class	
I	1 (5)[b]
II	10 (45)[b]
III	10 (45)[b]
IV	1 (5)[b]
Fatigue	18 (82)[b]
Angina	7 (32)[b]
Dyspnea	18 (82)[b]
Arrhythmias	11 (50)[b]
History of cardiac symptoms (months)	44 ± 27
Endomyocardial biopsy	
Enterovirus/adenovirus	15/7
Active/borderline myocarditis	0/0
Immunohistology (CD3-T lymphocytes/HPF)	5.7 ± 5.5
Medication	
ACE inhibitors	73%
β-blockers	64%
Glycosides	64%
Diuretics	59%
Warfarin	50%

[a] Mean ± standard deviation
[b] *n* (percentage)

Mean history of cardiac complaints and regional of global wall movement abnormalities of patients was 44 ± 27 months (Table 1). According to histological analysis (Dallas criteria), neither active nor borderline myocarditis was detected in any of the tissues. An inflammatory process could be detected in 7 patients presenting with increased numbers of CD3-positive lymphocytes in the immunohistological analysis (Fig. 1a).

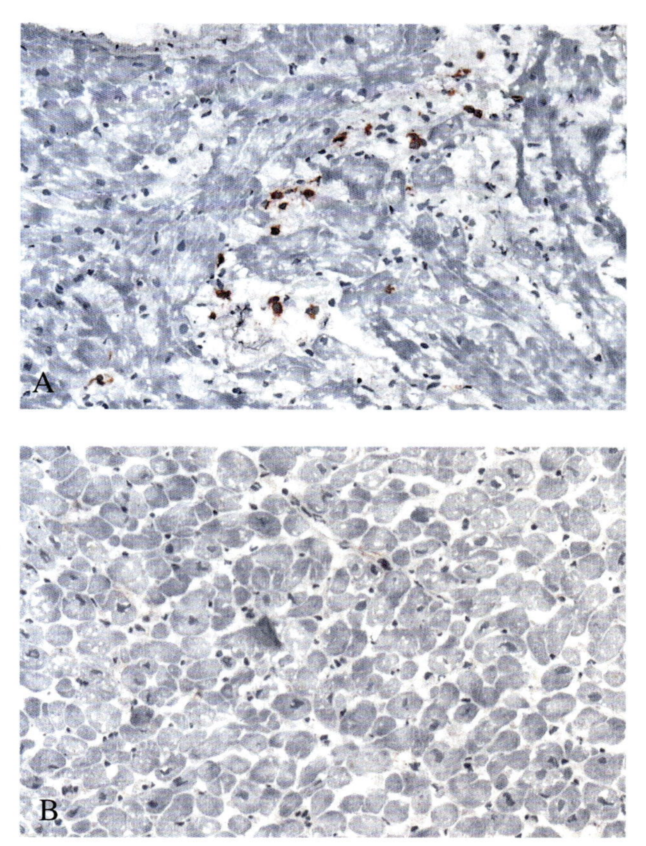

Fig. 1a,b. Immunohistological staining for $CD3^+$-T lymphocytes before (**a**) and after (**b**) interferon-β treatment

During the screening period preceding treatment, left ventricular end-diastolic and end-systolic diameters increased significantly and ejection fraction (EF) did not improve within 22 months before treatment in spite of extensive conventional heart failure medication. At baseline, most patients were in NYHA classes II and III.

18.8 Treatment Protocol

The IFN-β therapy (Beneferon, Renschler, Laupheim) followed a stepped regimen in order to reduce the flu-like side-effects typical of the initial phase of an IFN therapy. The subcutaneous administration was initiated at a dose of 2 million units IFN-β 3 times a week on alternate days, and was increased to 12 million units during the second and 18 million units during the third week. By the end of week 24, the IFN-β treatment was discontinued. The protocol was approved by the Human Research Committee of the Charité, Campus Benjamin Franklin, Berlin, and all patients gave written informed consent before treatment.

18.9 Results and Outcome

Treatment of these first 22 patients did not show any severe adverse side-effects of the IFN-β therapy necessitating cessation of treatment. Main complaints reported during the study were injection site reactions, fatigue, arthralgia, headache, myalgia, fever and other influenza-like symptoms (Fig. 2). Some disease-related symptoms such as dyspnea, angina pectoris and arrhythmia were also reported as adverse events during the first weeks of treatment. Most symptoms were of mild intensity and transient in nature.

Upon IFN-β treatment, both adenoviral and enteroviral genomes disappeared. Control biopsies within a period of 6 weeks after the end of therapy demonstrated that both enterovirus and adenovirus genomes were no longer detectable after IFN-β1a treatment (Table 2). In parallel

▶

Fig. 2. Patients with adverse events during the IFN-β pilot study (events which occurred in > 10% of patients)

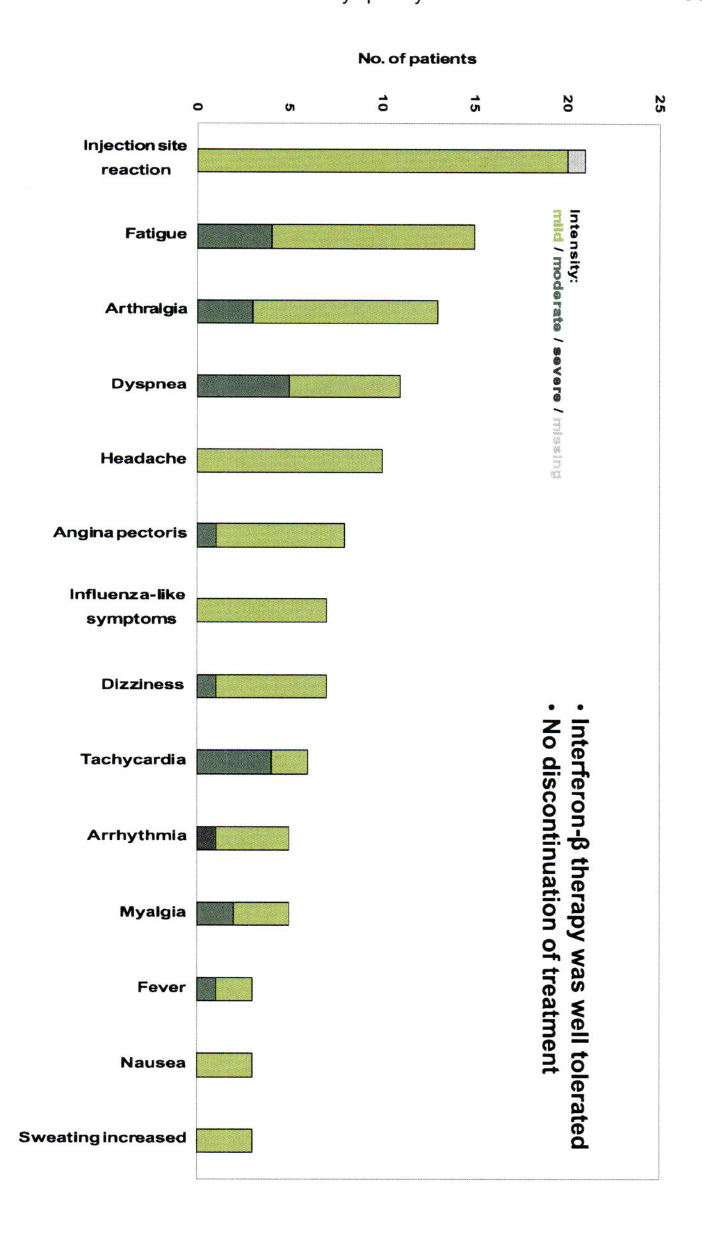
No. of patients
0
5
10
15
20
25
Injection site reaction
Fatigue
Arthralgia
Dyspnea
Headache
Angina pectoris
Influenza-like symptoms
Dizziness
Tachycardia
Arrhythmia
Myalgia
Fever
Nausea
Sweating increased
Intensity:
mild / moderate / severe / missing
• Interferon-β therapy was well tolerated
• No discontinuation of treatment

Table 2. Molecular biological assessment of viral genomes after 6 months of interferon-β treatment

	n	Virus elimination after treatment
Enterovirus	15	15/15 (100%)
Adenovirus	7	7/7 (100%)

Table 3. Hemodynamic changes after 6 months of interferon-β treatment

	Baseline	Follow-up	*p*-value
LVEF (%)	44.7 ± 15.5	53.1 ± 16.8	0.001
LVEDD (mm)	59.7 ± 11.1	56.5 ± 11.1	0.001
LVESD (mm)	43.4 ± 13.6	39.4 ± 12.1	0.001

EF, ejection fraction; LVEDD, left ventricular end-diastolic diameter; LVESD, left ventricular end-systolic diameter

with the virus clearance, myocardial inflammation resolved in all but one patient (Figs. 1a and 3). Of the patients, 67% showed an improvement of at least one NYHA class (for mean NYHA functional class improvement see Fig. 4). None of the patients showed a deterioration in NYHA. Decrease of left ventricular diameters and improvement of EF occurred in parallel with improvement of clinical symptoms (Table 3; Kühl et al. 2003b). Patients with regional as well as global contractile dysfunction improved, but improvement was more pronounced in patients with global left ventricular dysfunction. After a mean follow-up of 12 months, recovery of myocardial function remained preserved in all patients. No patients deteriorated during treatment or the follow-up period, but five patients further improved.

18.10 Conclusion

The findings upon INF-β treatment taken together with the lack of improvement before specific anti-viral treatment suggest that some of the ventricular dysfunction and wall motion abnormalities were caused by the persistent viral infection and subsequently resolved after elimination

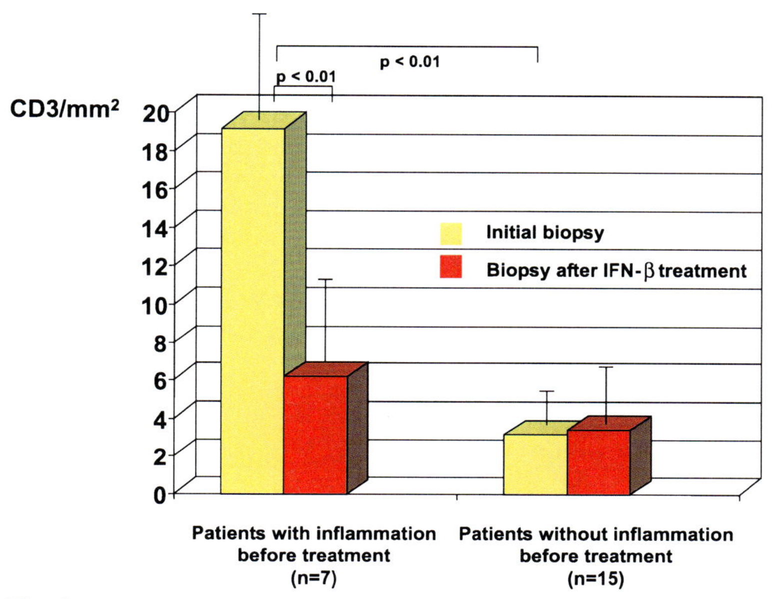

Fig. 3. Lymphocyte counts in endomyocardial biopsies before and after interferon-β treatment

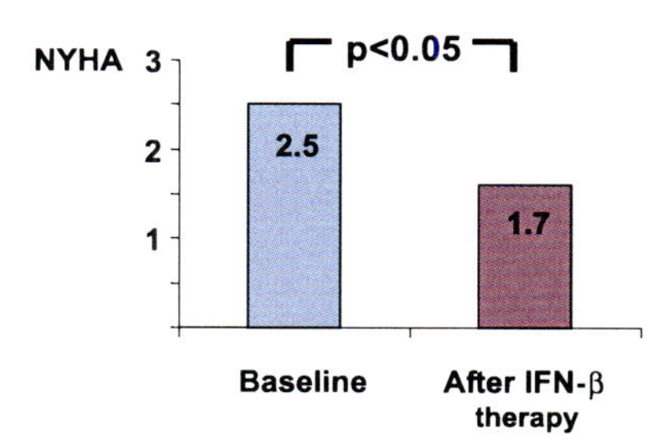

Fig. 4. Clinical improvement (NYHA) after 6 months of IFN-β therapy

of the responsible agents. The data furthermore suggest that the beneficial clinical effect of IFN-β, based upon the elimination of cardiotropic viruses, may occur even in DCM patients presenting with a long history.

Ongoing viral persistence is associated with an adverse prognosis such as early death or requirement for transplantation (Why et al. 1994). Experimental and clinical data highlight that an imbalance of the cytokine network and a defect in the cytokine-induced immune response may constitute major causes leading to the development of virus persistence and progression of myocardial dysfunction. The outcome of progressive viral-induced cardiomyopathy is worse and, even if not fatal, it reduces capacity for work and is associated with enormous social costs of conservative treatment and recurring hospitalizations. Spontaneous reversibility of cardiac impairment during early stages of the disease and the arising chance of specific treatment options demand early diagnosis and treatment of the disease. Based on the positive results of the above pilot study, a randomized, double-blind, placebo-controlled European-wide multicentre study [BICC: Betaferon (interferon-β) in Patients With Chronic Viral Cardiomyopathy] is currently underway.

Acknowledgements. This publication has been supported by the Deutsche Forschungsgemeinschaft through Sonderforschungsbereich/Transregio 19.

References

Baboonian C, Treasure T (1997) Meta-analysis of the association of enteroviruses with human heart disease. Heart 78:539–543

Bachmaier K, Neu N, Pummerer C, Duncan GS, Mak TW, Matsuyama T, Penninger JM (1997) iNOS expression and nitrotyrosine formation in the myocardium in response to inflammation is controlled by the interferon regulatory transcription factor 1. Circulation 96:585–591

Badorff C, Noutsias M, Kühl U, Schultheiss HP (1997) Cell-mediated cytotoxicity in hearts with dilated cardiomyopathy: correlation with interstitial fibrosis and foci of activated T lymphocytes. J Am Coll Cardiol 29:429–434

Badorff C, Lee GH, Lamphear BJ, Martone ME, Campbell KP, Rhoads RE, Knowlton KU (1999) Enteroviral protease 2A cleaves dystrophin: evidence of cytoskeletal disruption in an acquired cardiomyopathy [see comments]. Nat Med 5:320–326

Baron S (1992) Introduction to the interferon system. In: Baron S (ed) Interferon: principles and medical applications. The University of Texas Medical Branch at Galveston, Dept. of Microbiology, Galveston, pp 1–15

Biron CA, Nguyen KB, Pien GC, Cousens LP, Salazar-Mather TP (1999) Natural killer cells in antiviral defense: function and regulation by innate cytokines. Annu Rev Immunol 17:189–220

Bowles NE, Richardson PJ, Olsen EG, Archard LC (1986) Detection of coxsackie-B-virus-specific RNA sequences in myocardial biopsy samples from patients with myocarditis and dilated cardiomyopathy. Lancet 1:1120–1123

Bozkurt B, Kribbs SB, Clubb FJ Jr, Michael LH, Didenko VV, Hornsby PJ, Seta Y, Oral H, Spinale FG, Mann DL (1998) Pathophysiologically relevant concentrations of tumor necrosis factor-alpha promote progressive left ventricular dysfunction and remodeling in rats. Circulation 97:1382–1391

Bultmann BD, Klingel K, Sotlar K, Bock CT, Baba HA, Sauter M, Kandolf R (2003) Fatal parvovirus B19-associated myocarditis clinically mimicking ischemic heart disease: an endothelial cell-mediated disease. Hum Pathol 34:92–95

Clerici M, Shearer GM (1993) A Th1 to Th2 switch is a critical step in the etiology of HIV infection. Immunol Today 14:73–92

Dec GW, Fuster V (1994) Idiopathic dilated cardiomyopathy. N Engl J Med 331:1564–1575

Dec GW Jr, Palacios IF, Fallon JT, Aretz HT, Mills J, Lee DC, Johnson RA (1985) Active myocarditis in the spectrum of acute dilated cardiomyopathies. Clinical features, histologic correlates, and clinical outcome. N Engl J Med 312:885–890

Doerner A, Pauschinger M, Schwimmbeck PL, Kühl U, Schultheiss HP (2000) The shift in the myocardial adenine nucleotide translocator isoform expression pattern is associated with an enteroviral infection in the absence of an active T-cell dependent immune response in human inflammatory heart disease. J Am Coll Cardiol 35:1778–1784

Feldmann AM, McNamara D (2000) Myocarditis. N Engl J Med 343:1388–1398

Finkel MS, Oddis CV, Jacob TD, Watkins SC, Hattler BG, Simmons RL (1992) Negative inotropic effects of cytokines on the heart mediated by nitric oxide. Science 257:387–389

Freeman GL, Colston JT, Zabalgoitia M, Chandrasekar B (1998) Contractile depression and expression of proinflammatory cytokines and iNOS in viral myocarditis. Am J Physiol 274:H249–H258

Frustaci A, Chimenti C, Calabrese F, Pieroni M, Thiene G, Maseri A (2003) Immunosuppressive therapy for active lymphocytic myocarditis: virological and immunologic profile of responders versus nonresponders. Circulation 107:857–863

Fujioka S, Kitaura Y, Ukimura A, Deguchi H, Kawamura K, Isomura T, Suma H, Shimizu A (2000) Evaluation of viral infection in the myocardium of patients with idiopathic dilated cardiomyopathy. J Am Coll Cardiol 36:1920–1926

Grogan M, Redfield MM, Bailey KR, Reeder GS, Gersh BJ, Edwards WD, Rodeheffer RJ (1995) Long-term outcome of patients with biopsy-proved myocarditis: comparison with idiopathic dilated cardiomyopathy. J Am Coll Cardiol 26:80–84

Heim A, Canu A, Kirschner P, Simon T, Mall G, Hofschneider PH, Kandolf R (1992) Synergistic interaction of interferon-beta and interferon-gamma in coxsackievirus B3-infected carrier cultures of human myocardial fibroblasts. J Infect Dis 166:958–965

Heim A, Stille-Siegener M, Kandolf R, Kreuzer H, Figulla HR (1994) Enterovirus-induced myocarditis: hemodynamic deterioration with immunosuppressive therapy and successful application of interferon-alpha. Clin Cardiol 17:563–565

Horwitz MS, La Cava A, Fine C, Rodriguez E, Ilic A, Sarvetnik N (2000a) Pancreatic expression of interferon-gamma protects mice from lethal coxsackievirus B3 infection and subsequent myocarditis. Nat Med 6: 693–697

Huang S, Hendrix W, Althage A, Bluethmann H, Kamijo R, Vilcek J, Zinkernagel RM, Aguet M (1993) Immune response in mice that lack the interferon-gamma receptor. Science 259:1742–1745

Hunter CA, Reiner SL (2000) Cytokines and T cells in host defense. Curr Opin Immunol 12:413–418

Kandolf R, Ameis D, Kirschner P, Canu A, Hofschneider PH (1987) In situ detection of enteroviral genomes in myocardial cells by nucleic acid hybridization: an approach to the diagnosis of viral heart disease. Proc Natl Acad Sci U S A 84:6272–6276

Katze MG, He Y, Gale M (2002) Viruses and interferon: a fight for supremacy. Nat Rev Immunol 2:675–687

Kawai C (1999) From myocarditis to cardiomyopathy: mechanisms of inflammation and cell death: learning from the past for the future. Circulation 99:1091–1100

Klingel K, Kandolf R (1993) The role of enterovirus replication in the development of acute and chronic heart muscle disease in different immunocompetent mouse strains. Scand J Infect Dis Suppl 88:79–85

Klingel K, Rieger P, Mall G, Selinka HC, Huber M, Kandolf R (1998) Visualization of enteroviral replication in myocardial tissue by ultrastructural in situ hybridization: identification of target cells and cytopathic effects. Lab Invest 78:1227–1237

Kühl U, Noutsias M, Seeberg B, Schultheiss HP (1996) Immunohistological evidence for a chronic intramyocardial inflammatory process in dilated cardiomyopathy. Heart 75:295–300

Kühl U, Pauschinger M, Bock T, Klingel K, Schwimmbeck PL, Seeberg B, Krautwurm L, Schultheiß HP, Kandolf R (2003a) Parvovirus B19 infection mimicking acute myocardial infarction. Circulation 108:945–950

Kühl U, Pauschinger M, Schwimmbeck PL, Seeberg B, Lober C, Noutsias M, Poller W, Schultheiss HP (2003b) Interferon-beta treatment eliminates cardiotropic viruses and improves left ventricular function in patients with myocardial persistence of viral genomes and left ventricular dysfunction. Circulation 107:2793–2798

Kühl U, Pauschinger M, Michel Noutsias M, Seeberg B, Bock T, Lassner D, Kandolf R, Schultheiss HP (2005) High prevalence of viral genomes and multiple viral infections in the myocardium of adults with "idiopathic" left ventricular dysfunction. Circulation (in press)

Li Y, Bourlet T, Andreoletti L (2000a) Enteroviral capsid protein VP1 is present in myocardial tissues from some patients with myocarditis or dilated cardiomyopathy. Circulation 101:231–234

Li YY, McTiernan CF, Feldman AM (2000c) Interplay of matrix metalloproteinases, tissue inhibitors of metalloproteinases and their regulators in cardiac matrix remodeling. Cardiovasc Res 46:214–224

Li YY, Feng Y, McTiernan CF, Pei W, Moravec CS, Wang P, Rosenblum W, Kormos RL, Feldman AM (2001) Downregulation of matrix metalloproteinases and reduction in collagen damage in the failing human heart after support with left ventricular assist devices. Circulation 104:1147–1152

Libby P, Lee RT (2000) Matrix matters. Circulation 102:1874–1876

Liu P, Aitken K, Kong Y, Opavsky MA, Martino T, Dawood F, Wen W, Kozieradzki I, bachmaier K, Straus D, Mak TW, Penninger JM (2000b) The tyrosine kinase p56lck is essential in coxsackievirus B3-mediated heart disease. Nat Med 6:429–434

Lowenstein CJ, Hill SL, Lafond Walker A, Wu J, Allen G, Landavere M, Rose NR, Herskowitz A (1996) Nitric oxide inhibits viral replication in murine myocarditis. J Clin Invest 97:1837–1843

Mann DL, Taegtmeyer H (2001) Dynamic regulation of the extracellular matrix after mechanical unloading of the failing human heart: recovering the missing link in left ventricular remodeling. Circulation 104:1089–1091

Matsumori A (2001) Hepatitis C virus and cardiomyopathy. Intern Med 40:78–79

Matsumori A, Tomioka N, Kawai C (1988) Protective effect of recombinant alpha interferon on coxsackievirus B3 myocarditis in mice. Am Heart J 115:1229–1232

Matsumori A, Yamada T, Suzuki H, Matoba Y, Sasayama S (1994) Increased circulating cytokines in patients with myocarditis and cardiomyopathy. Br Heart J 72:561–566

McCarthy RE, Boehmer JP, hruban RH, Hutchins GM, Kasper EK, Hare JM, Baughman KL (2000) Long-term outcome of fulminant myocarditis as compared with acute (non-fulminant) myocarditis. N Engl J Med 342:734–735

Michels V, Moll P, Miller FA, Tajik AJ, Chu JS, Driscoll DJ, Burnett JC, Rodeheffer RJ, Chesebro JH, Tazelaar HD (1992) The frequency of familial dilated cardiomyopathy in a series of patients with idiopathic dilated cardiomyopathy. N Engl J Med 326:77–82

Miric M, Vasiljevic J, Bojic M, Popovic Z, Keserovic N, Pesic M (1996) Long-term follow up of patients with dilated heart muscle disease treated with human leucocytic interferon alpha or thymic hormones initial results. Heart 75:596–601

Noutsias M, Fechner H, de Jonge H, Wang X, Dekkers D, Houtsmuller AB, Pauschinger M, Bergelson J, Warraich R, Yacoub M, Hetzer R, Lamers J, Schultheiss HP, Poller W (2001) Human coxsackie-adenovirus receptor is colocalized with integrins alpha(v)beta(3) and alpha(v)beta(5) on the cardiomyocyte sarcolemma and upregulated in dilated cardiomyopathy: implications for cardiotropic viral infections. Circulation 104:275–280

Pauschinger M, Bowles NE, Fuentes-Garcia FJ, Pham V, Kühl U, Schwimmbeck PL, Schultheiss HP, Towbin JA (1999a) Detection of adenoviral genome in the myocardium of adult patients with idiopathic left ventricular dysfunction. Circulation 99:1348–1354

Pauschinger M, Doerner A, Kuehl U, Schwimmbeck PL, Poller W, Kandolf R, Schultheiss HP (1999b) Enteroviral RNA replication in the myocardium of patients with left ventricular dysfunction and clinically suspected myocarditis. Circulation 99:889–895

Pauschinger M, Knopf D, Petschauer S, Doerner A, Poller W, Schwimmbeck PL, Kühl U, Schultheiss HP (1999c) Dilated cardiomyopathy is associated with significant changes in collagen type I/III ratio. Circulation 99:2750–2756

Poller W, Fechner H, Noutsias M, Tschoepe C, Schultheiss HP (2002) Highly variable expression of virus receptors in the human cardiovascular system. Implications for cardiotropic viral infections and gene therapy. Z Kardiol 91:978–991

Ramshaw IA, Ramsay A, Karupiah G, Rolph MS, Mahalingam S, Ruby JC (1997) Cytokines and immunity to viral infection. Immunol Rev 159:119–135

Richardson P, McKenna W, Bristow M, Maisch B, Mautner B, J OC, Olsen E, Thiene G, Goodwin J, Gyarfas I, Martin I, Nordet P (1996) Report of the 1995 World Health Organization/International Society and Federation of

Cardiology Task Force on the Definition and Classification of cardiomyopathies [news]. Circulation 93:841–842

Schulze K, Witzenbichler B, Christmann C, Schultheiss HP (1999) Disturbance of myocardial energy metabolism in experimental virus myocarditis by antibodies against the adenine nucleotide translocator. Cardiovasc Res 44:91–100

Seko Y, Shinkai Y, Kawasaki A, Yagita H, Okumura K, Yazaki Y (1993) Evidence of perforin-mediated cardiac myocyte injury in acute murine myocarditis caused by coxsackie virus B3. J Pathol 170:53–58

Sher A, Coffman RL (1992) Regulation of immunity to parasites by T cells and T-cell derived cytokines. Annu Rev Immunol 10:385–409

Shioi T, Matsumori A, Sasayama S (1996) Persistent expression of cytokine in the chronic stage of viral myocarditis in mice. Circulation 94:2930–2937

Shioi T, Matsumori A, Nishio R, Ono K, Kakio T, Sasayama S (1997) Protective role of interleukin-12 in viral myocarditis. J Mol Cell Cardiol 29:2327–2334

Stille Siegener M, Heim A, Figulla HR (1995) Subclassification of dilated cardiomyopathy and interferon treatment. Eur Heart J 16 Suppl O:147–149

Talon J, Horvath CM, Polley R, Basler CF, Muster T, Palese P, Garcia-Sastre A (2000) Activation of interferon regulatory factor 3 is inhibited by the influenza A virus NS1 protein. J Virol 74:7989–7996

Testa M, Yeh M, Lee P, Fanelli R, Loperfido F, Berman JW, LeJemtel TH (1996) Circulating levels of cytokines and their endogenous modulators in patients with mild to severe congestive heart failure due to coronary artery disease or hypertension. J Am Coll Cardiol 28:964–971

Thomas CV, Coker ML, Zellner JL, Handy JR, Crumbley AJ 3rd, Spinale FG (1998) Increased matrix metalloproteinase activity and selective upregulation in LV myocardium from patients with end-stage dilated cardiomyopathy. Circulation 97:1708–1715

Torre Amione G, Kapadia S, Lee J, Durand JB, Bies RD, Young JB, Mann DL (1996) Tumor necrosis factor-alpha and tumor necrosis factor receptors in the failing human heart. Circulation 93:704–711

Wessely R, Henke A, Zell R, Kandolf R, Knowlton KU (1998) Low-level expression of a mutant coxsackieviral cDNA induces a myocytopathic effect in culture: an approach to the study of enteroviral persistence in cardiac myocytes. Circulation 98:450–457

Why HJ, Meany BT, Richardson PJ, Olsen EG, Bowles NE, Cunningham L, Freeke CA, Archard LC (1994) Clinical and prognostic significance of detection of enteroviral RNA in the myocardium of patients with myocarditis or dilated cardiomyopathy. Circulation 89:2582–2589

Yamada T, Matsumori A, Sasayama S (1994) Therapeutic effect of anti-tumor necrosis factor-alpha antibody on the murine model of viral myocarditis induced by encephalomyocarditis virus. Circulation 89:846–851
Yamamoto N, Shibamori M, Ogura M, Seko Y, Kikuchi M (1998) Effects of intranasal administration of recombinant murine interferon-gamma on murine acute myocarditis caused by encephalomyocarditis virus. Circulation 97:1017–1023
Yamamura M, Uyemura K, Deans RJ, Weinberg K, Rea TH, Bloom BR, Modlin RL (1991) Defining protective response to pathogens: cytokine profile in leprosy lesions. Science 254:277–279

19 Immunosuppressive Treatment of Chronic Non-viral Myocarditis

A. Frustaci, M. Pieroni, C. Chimenti

Abstract. Inflammatory cardiomyopathy defined as myocarditis associated with cardiac dysfunction, represents a main cause of heart failure. Despite the improvement of diagnostic techniques, a specific standardized treatment of myocarditis is not yet available. The immunohistochemical detection of myocardial HLA up-regulation has been demonstrated useful in the identification of a subgroup of autoimmune inflammatory dilated cardiomyopathy (DCM) in part susceptible to immunosuppression. Recently, in a retrospective study, we defined the virologic and immunologic profile of responders and non-responders to immunosuppressive therapy of active lymphocytic myocarditis and chronic heart failure in patients who had failed to benefit from conventional supportive treatment. Non-responders were characterized by high prevalence (85%) of viral genomes in the myocardium and no detectable cardiac autoantibodies in the serum. Conversely, 90% of responders were positive for autoantibodies, while only 3 (15%) of them presented viral particles at PCR analysis on frozen endomyocardial tissue. With regard to the type of virus involved in non-responders, enterovirus, adenovirus, or their combination was associated with the worst clinical outcome. Hepatitis C virus (HCV) was the only viral agent of our series associated with detectable cardiac autoantibodies, suggesting a relevant immunomediated mechanism of damage by HCV and explaining the relief of myocardial inflammation after immunosuppressive treatment. The assessment of virologic and immunologic features of patients with biopsy-proven inflammatory cardiomyopathy may allow us to identify a specific treatment leading to recovery of cardiac function.

19.1 Introduction

Inflammatory cardiomyopathy according to the WHO/ISFC definition is characterized by myocarditis associated with cardiac dysfunction (WHO/ISFC 1996). Clinical presentation includes acute heart failure and cardiogenic shock (Feldman and McNamara 2000), conduction disturbances, or ventricular tachyarrhythmias (Chimenti et al. 2001). In addition, myocardial inflammation may mimic an acute myocardial infarction (Narula et al. 1993; Angelini et al. 2000) and may even be responsible for global biventricular dysfunction in patients with severe coronary artery disease (Frustaci et al. 1999).

The mortality rate for patients with myocarditis with heart failure is 20% at 1 year and 56% at 4 years (Mason et al. 1995). A definite diagnosis of myocarditis is possible only with endomyocardial biopsy (Aretz et al. 1986) as clinical history and findings, as well as standard non-invasive techniques, usually do not allow investigators to identify the inflammatory origin of myocardial damage. In recent years, the histologic diagnosis of myocarditis has been significantly improved by the introduction of immunohistochemical and molecular biology techniques (Kühl et al. 1994; Chimenti et al. 2001). Immunohistochemistry on myocardial biopsies is actually considered mandatory for the quantification and characterization of inflammatory infiltrates that represent a main step in the diagnosis of myocarditis. In addition, the systematic application of molecular biology techniques (PCR or in situ hybridization) in myocardial tissue specimens allowed the recognition of the fundamental pathogenic role played by cardiotropic viruses and the complex multiple interactions with the immune system (Pauschinger et al. 1999a,b). Specific surface receptors allowing the viruses to enter the myocardial cell (Noutsias et al. 2001) and the production of viral proteolytic enzymes capable of cleaving structural myocardial proteins, such as dystrophin, have been described (Badorff et al. 1999).

With regard to the effective pathogenic role played by viruses in myocardial damage, important contributions may derive from the combination of molecular biology with recently developed techniques such as laser-capture microdissection. This technique allows us to dissect from the same paraffin section different cellular populations recognized by immunohistochemical staining; the isolated cluster of cells may be therefore separately used for cellular-specific molecular biology and proteomic studies (Gjerdrum et al. 2001).

In a recent study, we applied this new technique in patients with myocarditis and detectable Epstein-Barr virus (EBV) genome in the endomyocardial biopsies by conventional PCR (Chimenti et al. 2004).We studied 9 patients with EBV-related inflammatory cardiomyopathy: Lymphocytes and myocytes were microdissected from paraffin sections and analyzed separately by PCR on DNA extracted from the collected cells. Blood and myocardial samples from patients with positive and negative serology for EBV were used as controls. The EBV genome was detected in myocytes but not in infiltrating lymphocytes of patients, nor in myocardial samples of controls. At a mean follow-up of 31 ± 14 months, despite full conventional heart-failure therapy, a progressive cardiac dilation and dysfunction was observed in all patients. Intramyocyte detection of the EBV genome supports a direct cytopathic role for this virus and suggests the opportunity for an antiviral/immunomodulatory therapy. According to these studies, it is clear that detection of viral genome in myocardial tissue of patients with myocarditis represents the first step toward identifying the best treatment for each subgroup of patients.

19.2 Immunosuppressive Treatment in Inflammatory Cardiomyopathy

Despite the improvement of diagnostic techniques in defining the characteristics and the etiology of the inflammatory process and the more specific comprehension of the mechanisms leading to myocardial damage, a specific standardized treatment of myocarditis is not yet available.

In particular, immunosuppressive treatment of myocarditis has drawn much attention in the past but less enthusiasm in recent times because of controversial results obtained both in children (Chan et al. 1991; Matitiau et al. 1994; Lee et al. 1999; Kleinert et al. 1997) and in adults (Parrillo et al. 1989; Kühl and Schultheiss 1995). In the absence of specific markers of eligibility for immunosuppressive therapy, large trials produced misleading results (Mason et al. 1995), showing the absence of an evident improvement of survival in myocarditis patients treated with immunosuppressive drugs versus placebo. Actually, immunosuppression is recommended essentially for the treatment of eosinophilic (Frustaci et al. 1998), granulomatous (Badorff et al. 1997), giant-cell myocarditis (Cooper et al. 1997; Frustaci et al. 2000), and lymphocytic myocarditis associated with connective tissue diseases (Frustaci et al. 1996) or with the rejection of a transplanted heart (Billingham 1997) (Table 1).

With regard to idiopathic lymphocytic myocarditis, beyond the acute phase, where a spontaneous resolution has been reported in up to 40% of the cases (Dec et al. 1985), there is growing evidence that many patients with idiopathic myocarditis and chronic heart failure are likely to benefit from immunosuppression. Recently, Wojnicz et al., in a randomized placebo-controlled study, suggested that an up-regulation of HLA in the myocardial tissue of patients with lymphocytic myocarditis may identify a homogeneous subgroup of inflammatory dilated cardiomyopathy (DCM) sustained by an autoimmune mechanism of damage and represent a marker of susceptibility to beneficial effects of immunosuppression (Wojnicz et al. 2001). The evidence that only 70% of patients with myocarditis and HLA up-regulation really had an improvement

Table 1. Responsiveness of myocarditis to immunosuppressive therapy

Type of myocarditis	Response
Myocarditis associated with hypereosinophilic syndrome	+ + +
Myocarditis associated with connective tissue disorders	++
Rejection of transplanted heart	++
Giant-cell myocarditis	+/−
Viral/idiopathic myocarditis	+/−

of cardiac function from immunosuppression vs 30% of the group on placebo could be in part due to the limit of the immunohistochemical semiquantitative method used to evaluate HLA up-regulation.

Recently, we defined in a retrospective study the virologic and immunologic profile of responders and non-responders to immunosuppressive therapy of patients with active lymphocytic myocarditis and chronic heart failure who failed to benefit from conventional supportive treatment (Frustaci et al. 2003).

We studied 41 patients with a histologic diagnosis of active lymphocytic myocarditis and characterized by a progressive heart failure with an ejection fraction (EF) of less than 40%, lasting over 6 months in spite of conventional supportive therapy. All patients were similar in terms of duration and severity of cardiac disease, histologic findings, and poor responsiveness to full conventional therapy. They received immunosuppressive therapy including prednisone 1 mg $\times$ kg^{-1} $\times$ d^{-1} for 4 weeks followed by 0.33 mg $\times$ kg^{-1} $\times$ d^{-1} for 5 months and azathioprine 2 mg $\times$ kg^{-1} $\times$ d^{-1} for 6 months. The patients were classified as responders if they had a decrease of at least 1 NYHA class and an improvement in $EF \geq 10\%$ compared with baseline measures. The patients were classified as non-responders if New York Heart Association (NYHA) class and EF failed to improve or deteriorated and if they faced major events like cardiogenic shock, heart transplantation, or cardiac death. Of the patients, 21 responded with prompt improvement in left ventricular EF and showed evidence of healed myocarditis at control biopsy. Conversely, 20 patients failed to respond and remained stationary (12 patients), underwent cardiac transplantation (3 patients), or died (5 patients), showing a histologic evolution toward dilated cardiomyopathy. Retrospective PCR on frozen endomyocardial tissue and evaluation of circulating cardiac autoantibodies on patients sera showed that non-responders had high prevalence (85%) of viral genomes in the myocardium and no detectable cardiac autoantibodies in the serum, while 90% of responders were positive for autoantibodies but only 3 (15%) of them presented viral particles at PCR analysis on frozen endomyocardial tissue. Among non-responders, the myocardial persistence of enterovirus and adenovirus or their combination was associated with the worst clinical outcome. Pilot studies have demonstrated that these patients may benefit from the administration of interferon-β with complete myocardial clearance of viral genome and long-term improvement of clinical symptoms and cardiac function (Kühl et al. 2003). Interestingly, serology for cardiotropic viruses failed to predict the presence of viral genome in the myocardium, being positive in both patients with positive and negative PCR results, suggesting as in other studies (Martin et al. 1994) that this tool cannot be used as an alternative to endomyocardial tissue PCR for the diagnosis of viral myocarditis.

In order to confirm these encouraging results in a prospective manner, we decided to perform a further study enrolling patients with myocarditis and chronic heart failure and submitting to immunosuppressive treatment all patients showing no evidence of viral genome in myocardial tissue. At least two frozen myocardial samples from each patient were used for PCR for the most common cardiotropic viruses (adenovirus, enterovirus, parvovirus B19, EBV, cytomegalovirus, herpes simplex virus, hepatitis C virus, influenza virus). Among 50 consecutive patients 35 resulted virus-negative at PCR. Among these, two patients were excluded because of contraindications to steroidal treatment while 33 were treated with prednisone $1\ mg \times kg^{-1} \times d^{-1}$ for 4 weeks followed by $0.33\ mg \times kg^{-1} \times d^{-1}$ for 5 months and azathioprine $2\ mg \times kg^{-1} \times d^{-1}$ for 6 months. The patients were classified as responders and non-responders according to the clinical and instrumental criteria adopted in our previously mentioned retrospective study (Frustaci et al. 2003).

After 6 months of immunosuppressive treatment, 29 patients (87%) showed a significant improvement of cardiac function and dimensions [LVEF from 27.8 ± 4.6 to 49.1 ± 6.0%, LVEDD from 65.9 ± 6.0 to 47.5 ± 6.3 mm, FS (fractional shortening) from 24.6 ± 4.9 to 45.0 ± 6.3%] while 4 patients maintained stable clinical picture and cardiac function parameters. Comparing baseline features of responders and non-responders, we were unable to identify any clinical, echocardiographic, or histologic marker predicting inefficacy of treatment.

These data obtained prospectively, basing the decision to use immunosuppressive treatment on the basis of myocardial evidence of viral genome, confirm our previous findings. The lack of response in a minority (13%) of virus-negative patients raises questions on the possible presence of additional viruses not screened for or the persistence of mechanisms inducing myocyte damage and dysfunction not susceptible to immunosuppression.

19.3 Conclusions

In conclusion our results showed that immunosuppressive therapy is an effective option in a majority of patients with chronic non-viral myocarditis. Molecular studies for the detection of viral genome in myocardial tissue represent the first step to characterize myocarditis patients and to direct therapeutic strategies. Further studies improving our knowledge on viral and immuno-mediated mechanisms of cardiac damage will allow us to further personalize the treatment of these patients.

References

Angelini A, Calzolari V, Calabrese F, et al. (2000) Myocarditis mimicking acute myocardial infarction: role of endomyocardial biopsy in the differential diagnosis. Heart 84:245–250

Aretz H, Billingham ME, Edwards WD, Factor SM, Fallon JT, Fenoglio JJ, Olsen EGJ, Schoen F (1986) Myocarditis. A histopathologic definition and classification. Am J Cardiovasc Pathol 1:3–14

Badorff C, Schwimmbeck PL, Kühl U, et al. (1997) Cardiac sarcoidosis: diagnostic validation by endomyocardial biopsy and therapy with corticosteroids. Z Kardiol 86:9–14

Badorff C, Lee GH, Lamphear BJ, et al. (1999) Enteroviral protease 2A cleaves dystrophin: evidence of cytoskeletal disruption in an acquired cardiomyopathy. Nat Med 5:320–326

Billingham ME (1997) Can histopathology guide immunosuppression for cardiac allograft rejection in the light of new techniques? Transplant Proc 29:35S–36S

Chan KY, Iwahara A, Benson LN, et al. (1991) Immunosuppressive therapy in the management of acute myocarditis in children; a clinical trial. J Am Coll Cardiol 17:458–460

Chimenti C, Calabrese F, Thiene G, et al. (2001) Inflammatory left ventricular microaneurysms as a cause of apparently idiopathic ventricular tachyarrhythmias. Circulation 104:168–173

Chimenti C, Russo A, Pieroni M, Calabrese F, Verardo R, Russo MA, Maseri A, Frustaci A (2004) Intramyocyte detection of Epstein Barr virus genome by laser capture microdissection in patients with inflammatory cardiomyopathy. Circulation 110:3534–3539

Cooper LT Jr, Berry GJ, Shabetai R (1997) Idiopathic giant-cell myocarditis-natural history and treatment. Multicenter Giant Cell Myocarditis Study Group Investigators. N Engl J Med 336:1860–1866

Dec GW Jr, Palacios IF, Fallon JT, et al. (1985) Active myocarditis in the spectrum of acute dilated cardiomyopathies. Clinical features, histologic correlates, and clinical outcome. N Engl J Med 312:885–890

Feldman AM, McNamara D (2000) Myocarditis. N Engl J Med 343:1388–1398

Frustaci A, Gentiloni N, Caldarulo M (1996) Acute myocarditis and left ventricular aneurysm as presentations of systemic lupus erythematosus. Chest 109:282–284

Frustaci A, Gentiloni N, Chimenti C, et al. (1998) Necrotizing myocardial vasculitis in Churg-Strauss syndrome: clinicohistologic evaluation of steroids and immunosuppressive therapy. Chest 114:1484–1489

Frustaci A, Chimenti C, Maseri A (1999) Global bi-ventricular dysfunction in patients with symptomatic coronary artery disease may be caused by myocarditis. Circulation 99:1295–1299

Frustaci A, Chimenti C, Pieroni M, et al. (2000) Giant cell myocarditis responding to immunosuppressive therapy. Chest 117:905–907

Frustaci A, Chimenti C, Calabrese F, Pieroni M, et al. (2003) Active lymphocytic myocarditis: virologic and immunologic profile of responders vs non-responders to immunosuppressive therapy. Circulation 107:857–863

Gjerdrum LM, Lielpetere I, Rasmussen LM, et al. (2001) Laser-assisted microdissection of membrane-mounted paraffin sections for polymerase chain reaction analysis. Identification of cell populations using immunohistochemistry and in situ hybridization. J Mol Diagn 3:105–110

Kleinert S, Weintraub RG, Wilkinson JL, et al. (1997) Myocarditis in children with dilated cardiomyopathy: incidence and outcome after dual therapy immunosuppression. J Heart Lung Transplant 16:1248–1254

Kühl U, Schultheiss HP (1995) Treatment of chronic myocarditis with corticosteroids. Eur Heart J 16 Suppl O:168–172

Kühl U, Seeberg B, Schultheiss HP, et al. (1994) Immunohistological characterization of infiltrating lymphocytes in biopsies of patients with clinically suspected dilated cardiomyopathy. Eur Heart J 15:62–67

Kühl U, Pauschinger M, Schwimmbeck PL, et al. (2003) Interferon-β treatment of patients with enteroviral and adenoviral cardiomyopathy causes effective virus clearance and long-term clinical improvement. Circulation 107:2793–2798

Lee KJ, McCrindle BW, Bohn DJ, et al. (1999) Clinical outcomes of acute myocarditis in childhood. Heart 82:226–233

Martin AB, Webber S, Fricker FJ, et al. (1994) Acute myocarditis. Rapid diagnosis by PCR in children. Circulation 90:330–339

Mason JW, O' Connell JB, Herskowitz A, et al. (1995) A clinical trial of immunosuppressive therapy for myocarditis. N Engl J Med 333:269–275

Matitiau A, Perez-Atayde A, Sanders SP, et al. (1994) Infantile dilated cardiomyopathy. Relation of outcome to left ventricular mechanics, hemodynamics, and histology at the time of presentation. Circulation 90:1310–1318

Narula J, Khaw BA, Dec GW Jr, et al. (1993) Brief report: recognition of acute myocarditis masquerading as acute myocardial infarction. N Engl J Med 328:100–104

Noutsias M, Fechner H, de Jonge H, et al. (2001) Human coxsackie-adenovirus receptor is colocalized with integrins alpha(v)beta(3) and alpha(v)beta(5) on the cardiomyocyte sarcolemma and upregulated in dilated cardiomyopathy: implications for cardiotropic viral infections. Circulation 104:275–280

Parrillo JE, Cunnion RE, Epstein SE, et al. (1989) A prospective, randomized, controlled trial of prednisone for dilated cardiomyopathy. N Engl J Med 321:1061–1068

Pauschinger M, Bowles NE, Fuentes-Garcia FJ, Pham V, Kühl U, Schwimmbeck PL, Schultheiss HP, Towbin JA (1999a) Detection of adenoviral genome in the myocardium of adult patients with idiopathic left ventricular dysfunction. Circulation 99:1348–1354

Pauschinger M, Doerner A, Kuehl U, et al. (1999b) Enteroviral RNA replication in the myocardium of patients with left ventricular dysfunction and clinically suspected myocarditis. Circulation 99:889–895

WHO/ISFC (1996) Report of the 1995 WHO/ISFC task force on the definition and classification of cardiomyopathies. Circulation 93:841–842

Wojnicz R, Nowalany-Kozielska E, Wojciechowska C, et al. (2001) Randomized, placebo-controlled study for immunosuppressive treatment of inflammatory dilated cardiomyopathy: two-year follow-up results. Circulation 104:39–45

20 Immunoadsorption in Dilated Cardiomyopathy

S.B. Felix

Abstract. Dilated cardiomyopathy (DCM) is characterized by progressive reduction in contractile function and by dilatation of the right and left ventricles. Abnormalities of the cellular and humoral immune system are present in patients with myocarditis and DCM. Several antibodies against cardiac structures have been detected in DCM patients. The functional significance of cardiac autoantibodies is under debate. For certain antibodies, in vitro data indicate a negative effect on cardiac performance. Furthermore, recent data have provided evidence that cardiac antibodies themselves induce DCM. Cardiac antibodies belong to the IgG fraction and can be eliminated by immunoadsorption therapy. Recent clinical studies showed that removal of antibodies by immunoadsorption improves cardiac function of patients with DCM, indicating that activation of the humoral immune system plays a functional role in DCM.

20.1 Abnormalities of the Cellular Immune System in Dilated Cardiomyopathy

Dilated cardiomyopathy (DCM) is a myocardial disease that is characterized by progressive depression of myocardial contractile function, and by ventricular dilatation in the absence of abnormal loading conditions or ischemic heart disease sufficient to cause global myocardial contractile dysfunction (Richardson et al. 1996).

An association between viral myocarditis and DCM has been hypothesized for a subset of patients with DCM. Both experimental and clinical data indicate that viral infection and inflammatory processes may be involved in the pathogenesis of myocarditis and DCM, and may represent important factors that cause progression of ventricular dysfunction (Feldman and McNamara 2000; Liu and Mason 2001). In patients with DCM, immunohistological methods have been introduced for the diagnosis of myocardial inflammation (Kühl et al. 1995; Wojnicz et al. 1998; Noutsias et al. 1999). Infiltration with lymphocytes and mononuclear cells, as well as increased expression of cell adhesion molecules, is a frequent phenomenon in DCM (Kühl et al. 1995; Wojnicz et al. 1998; Noutsias et al. 1999). These findings support the hypothesis that the immune process is still active. In many patients with DCM, the term "inflammatory cardiomyopathy" may therefore be applicable in describing the pathogenesis of the disease process (Richardson et al. 1996).

20.2 Role of the Humoral Immune System in Dilated Cardiomyopathy

In addition to abnormalities of the cellular immune system, disturbances of humoral immunity have been described in DCM patients. A number of antibodies against various cardiac cell proteins have been identified in DCM; these include antibodies against mitochondrial proteins, contractile proteins, cardiac β1-receptors and muscarinergic receptors (Schultheiss et al. 1988; Caforio et al. 1992; Limas et al. 1989; Magnusson et al. 1994; Fu et al. 1993). The functional role of cardiac autoantibodies in DCM remains to be elucidated. Moreover, it is unclear whether these antibodies play an active role in the disease process, or whether they reflect an inflammatory response to myocyte necrosis, thereby representing an epiphenomenon. On the other hand, in vitro data indicate that certain cardiac autoantibodies may be directly pathogenic in DCM: Antibodies against the ADP/ATP carrier interact with the calcium channel and have cardiotoxic properties (Schultheiss et al. 1988). In the plasma of patients with DCM, negative inotropic antibodies are detectable that decrease calcium transients of isolated cardiomyocytes (Felix et al. 2002). Immunization of rodents against peptides derived from cardiovascular G protein receptors induces morphological changes of myocardial tissue resembling DCM (Matsui et al. 1997). A recent study investigated the pathogenic significance of autoantibodies targeting cardiac β1-adrenoceptors in DCM: Rats immunized against the second extracellular loop of cardiac β1-receptors develop a b1-adrenergic receptor-directed autoimmune attack as a cause of DCM (Jahns et al. 2004). Interestingly, a recent study showed that mice deficient in the programmed cell detah-1 (PD-1) immunoinhibitory coreceptor develop autoimmune DCM with production of high-titer circulating IgG autoantibodies reactive to a 33-kDa protein expressed specifically on the surface of cardiomyocytes (Nishimura et al. 2001). This 33-kDa protein was recently identified as cardiac troponin I (Okazaki et al. 2003).

20.3 Immunoadsorption as a New Therapeutic Principle for Treatment of Dilated Cardiomyopathy

If cardiac antibodies do in fact contribute to cardiac dysfunction in DCM, their removal would be expected to improve the hemodynamics of patients with DCM. Antibodies are extractable by immunoadsorption (IA). IA has been successfully used for treatment of a number of autoimmune diseases such as Goodpasture syndrome or lupus erythematodes (Bygren et al. 1985; Palmer et al. 1988). An initial uncontrolled study was conducted by us with the purpose of ascertaining the short-term hemodynamic effects of immunoadsorption in patients with severe heart failure due to DCM (Dörffel et al. 1997). Immunoglobulin extraction from the plasma of these patients induced a significant increase in cardiac index (CI), which was accompanied by a fall in systemic vascular resistance

(SVR). These data suggest that removal of antibodies may improve hemodynamics in DCM.

A prospective randomized, open, and controlled study was performed by us to investigate the prolonged hemodynamic effects of IA (Felix et al. 2000). In this study, Ig was substituted after IA, and subsequent IgG depletion, in order to reduce the risk of infection. IA and IgG substitution was repeated at 1-month intervals until month 3. Patients with DCM (NYHA III-IV, left ventricular ejection fraction $< 30\%$) were randomly assigned either to the treatment group with IA and subsequent IgG substitution (IA/IgG group, $n = 9$), or to the control group without IA/IgG ($n = 9$). In the IA/IgG group, the patients were initially treated in one IA session daily, on three consecutive days. After the final IA session, 0.5 g/kg of polyclonal IgG was substituted. IA was then repeated for three further courses, until the third month Hemodynamic measurements were performed using a Swan-Ganz thermodilution catheter. Left ventricular function was ascertained by echocardiography. Hemodynamics and left ventricular ejection fraction (LVEF) did not change throughout the 3 months in the control group. In contrast, in the IA/IgG group's CI and stroke volume index (SVI) increased significantly – by 30%. The improvement in CI and SVI persisted after 3 months. LVEF furthermore increased significantly in the IA/IgG group, which indicates that IA and subsequent IgG substitution improves cardiovascular function in DCM.

In a recent case-controlled study performed by others, IA was conducted in one course of five consecutive days without IA substitution after depletion (Müller et al. 2000). Furthermore, IA was not repeated during follow-up. In this study, 1 year after IA, left ventricular ejection fraction had increased from 22% to 40% 1 year after IA: which was significantly different from the control group without IA therapy.

The possibility exists that immunoadsorption not only improves hemodynamics but also modulates myocardial inflammation in DCM. Accordingly, we recently conducted a randomized study in order to investigate the immunohistological changes induced by immunoadsorption therapy and subsequent IgG substitution (administered in four courses at 1-month intervals) in patients with DCM, in comparison with control individuals without immunomodulatory therapy (Staudt et al. 2001). For immunohistological analysis, right ventricular biopsies were obtained from all patients at baseline and after 3 months. Among control patients, the number of lymphocytes (CD3, CD4, and CD8) and of leukocyte common antigen (LCA)-positive cells in the myocardium remained stable over 3 months. Furthermore, no changes in expression of HLA class II antigens were observed. In contrast, immunoadsorption therapy and subsequent IgG substitution induced a significant decrease in lymphocytes and LCA-positive cells in the myocardium during follow-up, which was paralleled by a significant decline in HLA class II antigen expression.

A further study of ours investigated potential mechanisms of beneficial acute hemodynamic effects induced by immunoadsorption, by testing the effects of the antibodies in isolated rat cardiomyocytes (Felix et al.). After passage of plasma, the immunoadsorption columns were regenerated and the column eluent was collected and dialyzed (molecular weight cut-off 100 kDa). Confocal laser scanning microscopy was used to analyze the effects of column eluent on cell contraction and on calcium-dependent fluorescence in isolated field-stimulated rat cardiomyocytes. Column eluent obtained from blood of healthy donors (controls) did not influence calcium transients and cell shortening of the cardiomyocytes. In contrast, the column eluent from DCM patients collected during the first regeneration cycle of the first immunoadsorption session brought about an immediate and concentration-dependent decrease in calcium transients and cell shortening. Acute hemodynamic improvement among the patients correlated with the cardiodepressant effect of column eluent on the isolated cardiomyocytes. These data indicate that removal of circulating negative inotropic antibodies from plasma may contribute to the early beneficial hemodynamic effects of immunoadsorption in patients with DCM.

IgG subclasses differ from one another immunologically and functionally. The antibodies that trigger effector functions and that are most likely to be involved in immunoregulatory activity are IgG-3 and IgG-1. IgG-3 is the most active complement-fixing IgG subclass (Bruggemann et al. 1987). Furthermore, IgG-3 antibodies are more efficient than IgG-1 as mediators of antibody-dependent cellular cytotoxicity. It was recently shown that DCM patients have elevated levels of IgG-3 antibodies against α- and β-myosin heavy chains. The level of these antibodies correlates with the degree of left ventricular dysfunction (Warraich et al. 2002).

A recent study investigated the role of antibodies belonging to different immunoglobulin G subclasses to cardiac dysfunction in DCM. According to the data of this study, the negative inotropic effect of antibodies obtained from DCM patients is mainly attributable to antibodies belonging to the IgG-3 subclass. Removal of antibodies of the IgG-3 subclass may represent the essential mechanism of IA in DCM (Staudt et al. 2002).

The detection of cardiodepressant antibodies may be of essential therapeutic relevance, because the contribution of humoral activity with production of cardiodepressant antibodies may differ among DCM patients. For treatment of patients with severe heart failure due to DCM, it is important to identify those patients who will receive hemodynamic benefit from IA. Hence, predictors for hemodynamic improvement are necessary. A recent study accordingly investigated systematically for the first time, and in a larger number of patients, whether detection of cardiodepressant effects of antibodies received from patients' plasma before IA may be predictive for short- and long-term hemodynamic improvement during IA (Staudt et al. 2004). Because, furthermore, IA effectively removes cardiac autoantibodies from plasma, such a procedure enables evaluation of the role played by the humoral immune system in cardiac dysfunction among DCM patients. Before IA, antibodies were purified from plasma of 45 DCM patients (LVEF $< 30\%$). The functional effects of antibodies (300 mg/l) on calcium transients and on systolic cell shortening were analyzed in rat cardiomyocytes. After this in vitro analysis, IA was performed in four courses at 1-month intervals until month 3. Antibodies of the majority of these patients induced a significant reduction of calcium transients and cell shortening of cardiomyocytes (cardiodepressant group, $n = 29$). Antibodies from 16 patients did not influence calcium transients and cell shortening (non- cardiodepressant group). Only the cardiodepressant group demonstrated a significant hemodynamic improvement after IA. In contrast, IA did not show any beneficial effects in the non-cardiodepressant group. These data indicate that evidence of cardiodepressant antibodies predicts hemodynamic benefits during IA.

In conclusion, immunoadsorption may represent a therapeutic option for a subset of patients with DCM. The data from the above-cited studies indicate that activation of the humoral immune system, with production of cardiac autoantibodies, may play a functional role in cardiac dysfunction of patients with DCM. Modulation of the humoral immune system by immunoadsorption may represent a promising therapeutic approach for intervention in this autoimmune process. However, large-scale studies are necessary to evaluate the effects of immunoadsorption on prognostic clinical endpoints such as mortality and morbidity of patients with DCM.

References

Bruggemann M, Williams GT, Bindon CI, Clark MR, Walker MR, Jefferis R, Waldmann H, Neuberger MS (1987) Comparison of the effector functions of human immunoglobulins using a matched set of chimeric antibodies. J Exp Med 166:1351–1361

Bygren P, Freiburghaus C, Lindholm T, Simonsen O, Thysell H, Weislander J (1985) Goodpasture's syndrome treated with staphylococcal protein A immunoadsorption. Lancet 2:1295–1296

Caforio AL, Grazzini M, Mann JM, Feeling PJ, Bottazzo GF, McKenna WJ, Schiaffino S (1992) Identification of alpha- and beta-cardiac myosin heavy chain isoforms as major autoantigens in dilated cardiomyopathy. Circulation 85:1734–1742

Dörffel WV, Felix SB, Wallukat G, Brehme S, Bestvater K, Hofmann T, Kleber FX, Baumann G (1997) Short-term hemodynamic effects of immunoadsorption in dilated cardiomyopathy. Circulation 95:1994–1997

Feldman AM, McNamara DM (2000) Myocarditis. N Engl J Med 343:1388–1398

Felix SB, Staudt A, Landsberger M, Grosse Y, Stangl V, Spielhagen T, Wallukat, G, Wernecke K-D, Baumann G, Stangl K (2002) Removal of cardiodepressant antibodies in dilated cardiomyopathy by immunoadsorption. J Am Coll Cardiol 39:646 –652

Felix SB, Staudt A, Dörffel WV, Stangl V, Merkel K, Pohl M, Döcke WD, Morgera S, Neumayer HM, Wernecke K-D, Wallukat G, Stangl K, Baumann G (2000) Hemodynamic effects of immunoadsorption and subsequent immunoglobulin substitution in dilated cardiomyopathy: three-month results from a randomized study. J Am Coll Cardiol 35:1590–1598

Fu LX, Magnusson Y, Bergh CH (1993) Localization of a functional autoimmune epitope on the muscarinic acetylcholine receptor-2 in patients with idiopathic dilated cardiomyopathy. J Clin Invest 91:1964–1968

Jahns R, Boivin V, Hein L, Triebel S, Angermann CE, Ertl G, Lohse MJ (2004) Direct evidence for a b1-adrenergic receptor-directed autoimmune attack as a cause of idiopathic dilated cardiomyopathy. J Clin Invest 113:1419–1429

Kühl U, Noutsias M, Schultheiss HP (1995) Immunohistochemistry in dilated cardiomyopathy. Eur Heart J 16 Suppl O:100–106

Limas CJ, Goldenberg IF, Limas C (1989) Autoantibodies against beta-adrenoceptors in human idiopathic dilated cardiomyopathy. Circ Res 64:97–103

Liu PP, Mason JW (2001) Advances in the understanding of myocarditis. Circulation 104:1076–1082

Magnusson Y, Wallukat G, Waagstein F, Hjalmarson A, Hoebeke J (1994) Autoimmunity in idiopathic dilated cardiomyopathy. Characterization of antibodies against the beta 1-adrenoceptor with positive chronotropic effect. Circulation 89:2760–2767

Matsui S, Fu MLX, Katsuda S, Teraoka K, Kurihara I, Takekoshi N, Murakami E, Hoebeke J, Hjalmarson A (1997) Peptides derived from cardiovascular G-protein-coupled receptors induce morphological cardiomyopathic changes in immunized rabbits. J Mol Cell Cardiol 29:641–655

Müller J, Wallukat G, Dandel M, Bieda H, Brandes K, Spiegelsberger S, Nissen E, Kunze R, Hetzer R (2000) Immunoglobulin adsorption in patients with idiopathic dilated cardiomyopathy. Circulation 101:385–391

Nishimura H, Okazaki T, Tanaka Y, Nakatani K, Hara M, Matsumori A, Sasayama S, Mizoguchi A, Hiai H, Minato N, Honjo1 T (2001) Autoimmune dilated cardiomyopathy in PD-1 receptor-deficient mice. Science 291:319–322

Noutsias M, Seeberg B, Schultheiss HP, Kühl U (1999) Expression of cell adhesion molecules in dilated cardiomyopathy: evidence for endothelial activation in inflammatory cardiomyopathy. Circulation 99:2124–2131

Okazaki T, Tanaka Y, Nishio R, Mitsuiye T, Mizoguchi A, Wang J, Ishida M, Hiai H, Matsumori A, Minato N, Honjo T (2003) Autoantibodies against cardiac troponin I are responsible for dilated cardiomyopathy in PD-1-deficient mice. Nat Med 9:1477–1483

Palmer A, Gjorstrup P, Severn A, Welsh K, Traube D (1988) Treatment of systemic lupus erythematodes by extracorporeal immunoadsorption. Lancet 2:272

Richardson P, McKenna W, Bristow M, Maisch B, Mautner B, O'Connel J, Olsen E, Thiene G, Goodwin J, Gyarfas I, Martin I, Nordet P (1996) Report of the 1995 World Health Organization/International Society and Federation of Cardiology Task Force on the Definition and Classification of Cardiomyopathies. Circulation 93:841–842

Schultheiss HP, Kuhl U, Janda I, Melzner B, Ulrich G, Morad M (1988) Antibody-mediated enhancement of calcium permeability in cardiac myocytes. J Exp Med 168:2102–2119

Staudt A, Schaper F, Stangl V, Plagemann A, Böhm M, Merkel K, Wallukat G, Wernecke K-D, Stangl K, Baumann G, Felix SB (2001) Immunohistological changes in dilated cardiomyopathy induced by immunoadsorption therapy and subsequent immunoglobulin substitution. Circulation 103:2681–2686

Staudt A, Bohm M, Knebel F, Grosse Y, Bischoff C, Hummel A, Dahm JB, Borges A, Jochmann N, Wernecke KD, Wallukat G, Baumann G, Felix SB (2002) Potential role of autoantibodies belonging to the immunoglobulin G-3 subclass in cardiac dysfunction among patients with dilated cardiomyopathy. Circulation 106:2448–2453

Staudt A, Staudt Y, Dörr M, Böhm M, Knebel F, Hummel A, Wunderle L, Tiburcy M, Wernecke K-D, Baumann G, Felix SB (2004) Potential role of humoral immunity in cardiac dysfunction of patient suffering from dilated cardiomyopathy. J Am Coll Cardiol 44:829–836

Warraich RS, Noutsias M, Kazak I, Seeberg B, Dunn MJ, Schultheiss HP, Yacoub MH, Kuhl U, Kasac I (2002) Immunoglobulin G3 cardiac myosin autoantibodies correlate with left ventricular dysfunction in patients with dilated cardiomyopathy: immunoglobulin G3 and clinical correlates. Am Heart J 143:1076–1084

Wojnicz R, Nowalany-Kozielska E, Wodniecki J, Szczurek-Katanski K, Nozynski J, Zembala M, Rozek MM (1998) Immunohistological diagnosis of myocarditis. Potential role of sarcolemmal induction of the MHC and ICAM-1 in the detection of autoimmune mediated myocyte injury. Eur Heart J 19:1564–1572

Ernst Schering Research Foundation Workshop

Editors: Günter Stock
Monika Lessl

Vol. 1 (1991): Bioscience ⇌ Societly Workshop Report
Editors: D.J. Roy, B.E. Wynne, R.W. Old

Vol. 2 (1991): Round Table Discussion on Bioscience ⇌ Society
Editor: J.J. Cherfas

Vol. 3 (1991): Excitatory Amino Acids and Second Messenger Systems
Editors: V.I. Teichberg, L. Turski

Vol. 4 (1992): Spermatogenesis – Fertilization – Contraception
Editors: E. Nieschlag, U.-F. Habenicht

Vol. 5 (1992): Sex Steroids and the Cardiovascular System
Editors: P. Ramwell, G. Rubanyi, E. Schillinger

Vol. 6 (1993): Transgenic Animals as Model Systems for Human Diseases
Editors: E.F. Wagner, F. Theuring

Vol. 7 (1993): Basic Mechanisms Controlling Term and Preterm Birth
Editors: K. Chwalisz, R.E. Garfield

Vol. 8 (1994): Health Care 2010
Editors: C. Bezold, K. Knabner

Vol. 9 (1994): Sex Steroids and Bone
Editors: R. Ziegler, J. Pfeilschifter, M. Bräutigam

Vol. 10 (1994): Nongenotoxic Carcinogenesis
Editors: A. Cockburn, L. Smith

Vol. 11 (1994): Cell Culture in Pharmaceutical Research
Editors: N.E. Fusenig, H. Graf

Vol. 12 (1994): Interactions Between Adjuvants, Agrochemical
and Target Organisms
Editors: P.J. Holloway, R.T. Rees, D. Stock

Vol. 13 (1994): Assessment of the Use of Single Cytochrome
P450 Enzymes in Drug Research
Editors: M.R. Waterman, M. Hildebrand

Vol. 14 (1995): Apoptosis in Hormone-Dependent Cancers
Editors: M. Tenniswood, H. Michna

Vol. 15 (1995): Computer Aided Drug Design in Industrial Research
Editors: E.C. Herrmann, R. Franke

Vol. 16 (1995): Organ-Selective Actions of Steroid Hormones
Editors: D.T. Baird, G. Schütz, R. Krattenmacher

Vol. 17 (1996): Alzheimer's Disease
Editors: J.D. Turner, K. Beyreuther, F. Theuring

Vol. 18 (1997): The Endometrium as a Target for Contraception
Editors: H.M. Beier, M.J.K. Harper, K. Chwalisz

Vol. 19 (1997): EGF Receptor in Tumor Growth and Progression
Editors: R.B. Lichtner, R.N. Harkins

Vol. 20 (1997): Cellular Therapy
Editors: H. Wekerle, H. Graf, J.D. Turner

Vol. 21 (1997): Nitric Oxide, Cytochromes P 450,
and Sexual Steroid Hormones
Editors: J.R. Lancaster, J.F. Parkinson

Vol. 22 (1997): Impact of Molecular Biology
and New Technical Developments in Diagnostic Imaging
Editors: W. Semmler, M. Schwaiger

Vol. 23 (1998): Excitatory Amino Acids
Editors: P.H. Seeburg, I. Bresink, L. Turski

Vol. 24 (1998): Molecular Basis of Sex Hormone Receptor Function
Editors: H. Gronemeyer, U. Fuhrmann, K. Parczyk

Vol. 25 (1998): Novel Approaches to Treatment of Osteoporosis
Editors: R.G.G. Russell, T.M. Skerry, U. Kollenkirchen

Vol. 26 (1998): Recent Trends in Molecular Recognition
Editors: F. Diederich, H. Künzer

Vol. 27 (1998): Gene Therapy
Editors: R.E. Sobol, K.J. Scanlon, E. Nestaas, T. Strohmeyer

Vol. 28 (1999): Therapeutic Angiogenesis
Editors: J.A. Dormandy, W.P. Dole, G.M. Rubanyi

Vol. 29 (2000): Of Fish, Fly, Worm and Man
Editors: C. Nüsslein-Volhard, J. Krätzschmar

Vol. 30 (2000): Therapeutic Vaccination Therapy
Editors: P. Walden, W. Sterry, H. Hennekes

Vol. 31 (2000): Advances in Eicosanoid Research
Editors: C.N. Serhan, H.D. Perez

Vol. 32 (2000): The Role of Natural Products in Drug Discovery
Editors: J. Mulzer, R. Bohlmann

Vol. 50 (2005): Animal Models of T Cell-Mediated Skin Diseases
Editors: T. Zollner, H. Renz, K. Asadullah

Vol. 51 (2005): Biocombinatorial Approaches for Drug Finding
Editors: W. Wohlleben, T. Spellig, B. Müller-Tiemann

Vol. 52 (2005): New Mechanisms for Tissue-Selective Estrogen-Free Contraception
Editors: H.B. Croxatto, R. Schürmann, U. Fuhrmann, I. Schellschmidt

Vol. 53 (2005): Opportunities and Challanges of the Therapies Targeting CNS Regeneration
Editors: D. Perez, B. Mitrovic, A. Baron Van Evercooren

Vol. 54 (2005): The Promises and Challenges of Regenerative Medicine
Editors: J. Morser, S.I. Nishikawa

Vol. 55 (2006): Chronic Viral and Inflammatory Cardiomyopathy
Editors: H.-P. Schultheiss, J.-F. Kapp, G. Grötzbach

Vol. 56 (2006): Cytokines as Potential Therapeutic Target for Inflammatory Skin Diseases
Editors: R. Numerof, C.A. Dinarello, K. Asadullah

Supplement 1 (1994): Molecular and Cellular Endocrinology of the Testis
Editors: G. Verhoeven, U.-F. Habenicht

Supplement 2 (1997): Signal Transduction in Testicular Cells
Editors: V. Hansson, F.O. Levy, K. Taskén

Supplement 3 (1998): Testicular Function:
From Gene Expression to Genetic Manipulation
Editors: M. Stefanini, C. Boitani, M. Galdieri, R. Geremia,
F. Palombi

Supplement 4 (2000): Hormone Replacement Therapy
and Osteoporosis
Editors: J. Kato, H. Minaguchi, Y. Nishino

Supplement 5 (1999): Interferon: The Dawn of Recombinant
Protein Drugs
Editors: J. Lindenmann, W.D. Schleuning

Supplement 6 (2000): Testis, Epididymis and Technologies
in the Year 2000
Editors: B. Jégou, C. Pineau, J. Saez

Supplement 7 (2001): New Concepts in Pathology and
Treatment of Autoimmune Disorders
Editors: P. Pozzilli, C. Pozzilli, J.-F. Kapp

Supplement 8 (2001): New Pharmacological Approaches
to Reproductive Health and Healthy Ageing
Editors: W.-K. Raff, M.F. Fathalla, F. Saad

Supplement 9 (2002): Testicular Tangrams
Editors: F.F.G. Rommerts, K.J. Teerds

Supplement 10 (2002): Die Architektur des Lebens
Editors: G. Stock, M. Lessl

Supplement 11 (2005): Regenerative and Cell Therapy
Editors: A. Keating, K. Dicke, N. Gorin, R. Weber, H. Graf

Supplement 12 (2005): Von der Wahrnehmung zur Erkenntnis –
From Perception to Understanding
Editors: M. Lessl, J. Mittelstraß

This series will be available on request from
Ernst Schering Research Foundation, 13342 Berlin, Germany